X. Rodó F. A. Comín

Global Climate

Springer
Berlin
Heidelberg
New York
Barcelona
Hong Kong
London
Milan
Paris
Tokyo

Xavier Rodó   Francisco A. Comín (Eds.)

Global Climate

Current Research and Uncertainties in the Climate System

With 86 Figures

Springer

DR. XAVIER RODÓ
University of Barcelona
Climate ResearchGroup
Baldiri Reixac, 4-6, Torre D
08032 Barcelona
Spain
e-mail: xrodo@porthos.bio.ub.es

DR. FRANCISCO A. COMÍN
Instituto Pirenaico de Ecologia-CSIC
Avda. Montanana 1005
50059 Zaragoza
Spain
e-mail: comin@ipe.csic.es

ISBN 3-540-43820-3 Springer-Verlag Berlin Heidelberg New York

Library of Congress Cataloging-in-Publication Data
Global climate / Xavier Rodó and Francisco A. Comín (eds.).
p.cm.
Includes bibliographical references and index.
ISBN 3540438203 (alk. paper)
1. Climatic changes. 2. Climatology. I. Rodó, Xavier, 1965- II. Comín, F.A. (Francisco A.)

Springer-Verlag Berlin Heidelberg New York
a member of Springer Science+Business Media
http://www.springer.de

Printed in Italy

Camera ready by authors
Cover design: E. Kirchner, Heidelberg
Printed on acid-free paper SPIN 10992203 30/3111/as 5 4 3 2 1

Foreword

Uncertainty for Everyone

The one thing that is certain about the world is that the world is uncertain.

I have here, the question that a part of the matter, living matter, has to resolve in each and every one of its moments of existance. The environment of a living being is a part of the living being where it turns out, the rest of the living beings live. This is the drama of life on earth. Every living individual debates with his environment, exchanging matter, energy and information in the hope of staying alive, the same as all living beings who share that same environment. The adventure of a living being (of all living beings) is to maintain reasonable independence in face of the fluctuations of uncertainty within the environment.

The range of restrictions and mutual relationships is colossal. How is the transcendental pretension of staying alive regulated? There is an equation imposed by the laws of thermodynamics and the mathematical theory of information about the interaction of a living being with his environment which we could state like this:

The complexity of a living individual plus his capacity for anticipation in respect to his environment is identical to the uncertainty of the environment plus the capacity of that living being to change the environment.

We could reconsider the question: what alternatives has a living being to continue living when the uncertainty of his environment grows? Let's speak about uncertainty first. There are many kinds of uncertainty. For example, even though it might sound strange, there is an uncertainty which we could say is predictable, that is, it acts with a certain amount of frequency, enough that natural selection acts over a long period of time "inventing" for example, the intelligence of a bacteria or a squid. However, there are more uncertain uncertainties that require the capacity of anticipation greater than that of an octopus, a dog or a primate....or that which begins with abstract knowledge and continues with science! Thus, with an increasing capacity to know the world, a part of the biosphere, the human biosphere has also managed to touch the other term of the equation, that of the capacity to change the world. Let us call this technology.

The spiral causes dizziness: the pressure of uncertainty pushes an individual to know what technology acquires in order to change the world. But two things happen. The changes affect the uncertainty of the particular environment of each individual, and all individuals do the same! Something very special occurs with climate in the second part of the equation.

While the complexity, anticipation and action vary a lot from one individual to the other (whether that individual be an organism, family, flock, society, or nation) with each day more, the increase of climatic uncertainty is the same for all. There is only one way out. If science and technology affect global uncertainty, only more science and more technology will allow us, perhaps, to control it.

J. Wagensberg
Barcelona Science Museum

Preface

Climate research has undergone dramatic changes in the last decades and a majority of the knowledge being produced tends to be lost among the vast amount of journals and dissemination papers or, at most has a short virtual life from time to time on the internet. It was in an attempt to summarize the state of the art in the climate sciences that a conference was first held in Barcelona in 1999 from which this book eventually saw light. The meeting was held with the aim of enriching the scientific debate on the climate sciences in our geographical area. The conference had in its initials, my own personal experience in several climate labs abroad and, a willingness to bring what was enlightening from that experience. The support and encouragement I received from some of the reputed colleagues who later attended the meeting and eventually contributed to this book was particularly important. This book was first devised as a recopilatory of the talks given in that symposium, but I sincerely think that the final product is even considerably better than what was originally thought of at first. In fact this improvement is clear as, to the original seven chapters, four new ones were added at several different stages of development. The focus that each of the authors was asked to give to their chapter, intended to highlight those areas where there were gaps in climate knowledge and where research was currently pushing strongly.

In other thematic areas of interest, unfortunately, the limited extent of the book fails to cover equally important topics that perhaps should have had a place in a book like this. However, from the very beginning, it was not the main purpose of the book to cover all the key issues that climate research currently faces. Opinions from experts in the different areas give a detailed view of what are, at present, the main 'hotspots' in the functioning of the climate system of our planet. The book is intended to give the reader a brief overview of the current status of climate research, trying to take him/her from the basics of climate science in every field to where research is currently being made in that particular area. The next years will, for sure, see rapid advances in many of these issues and quite probably, many of the questions raised here will eventually see a response.

In our planet both climate and biosphere evolved together, influencing each other for millennia. The physical limits that make possible the existence of an extraordinary variety of ecosystems and of life itself as we know it, arise from the ongoing interaction among phenomena as familiar as winds, clouds, light, air, land and water throughout the Earth's history. This same interaction produces large-scale phenomena that, like El Niño-Southern Oscillation (ENSO), exchange huge amounts of energy. It is also the way by which small changes in solar radiation are internalized and translated into dramatic climatic episodes such as the Ice Ages.

This interaction also largely determines the fate and the response to the increasing quantities of greenhouse gases that we are continuously throwing to the atmosphere as the result of our unlimited economic and industrial activity.

Our climate system has a great inertia and this fact implies that, in the near future, we will have to live with the consequences of our acts in the past and in the present in the form of unprecedented climate changes. And, as we cannot escape such changes, we will have to learn how to adapt to them. But to do so, we need to increase our knowledge on how the climate system works, how it worked in the past and on the basis of this, forecast how it may evolve in the future. In essence, we must begin to admit that our changing earth is now facing us with a new picture in which we are a fundamental part of the climate system.

The book has a set of introductory chapters: F.A. Comín and M.A. Rodríguez-Arias, in Chap. 1, give an overview of the current status and future prospects of climate research while S.G.H. Philander explains in Chap. 2, the basics of global warming and discusses why this subject is a controversial issue, beyond scientific results. In Chap. 3, the same author talks about the historical relations between El Niño and the Southern Oscillation, and the current status and prospects for its predictability. C.F. Ropelewki and B. Lyon talk in Chap. 4 about what should be an ideal Climate Information System for monitoring climate change and discuss the available forecasting and dissemination products in an overview of what are the main societal aspects of climate change. Later, S. M. Griffies in Chap. 5, analyzes what the main key areas of research and improvement in current ocean modeling are and describes what will possibly be the next generation of ocean models. In a second chapter (Chap. 6), S.M. Griffies discusses, on the basis of recent results, the role of ocean memory on climate predictability. P. Ciais analyzes the *new* and *past* global carbon cycles in Chap. 7, and discusses the sources, fluxes and sinks of carbon in the different compartments of the climate system. He also makes some remarks on the impacts of future climate scenarios on the global carbon cycle. T. Stocker relates changes in the global carbon cycle with the ocean circulation in Chap. 8 and discusses the possibility of abrupt climate changes taking place in certain future scenarios. In Chap. 9, K. Alverson and C. Kull talk about past climates and the detection of anthropogenic effects in historical records, reviewing the information given by paleoproxies of climate data. This information can be used to depict a baseline climate upon which to predict future climate changes. The authors eventually attempt to assess what the significance of the past is for the future. Krishnamurty and Kinter, in Chap. 10, describe what is currently known and what is not, of the Indian Monsoon System and relate the monsoonal activity with global climate variability. They further analyze the role of large-scale phenomena like ENSO and NAO in monsoon dynamics. Finally, X. Rodó in Chap. 11 discusses the relations between the Tropics and the Extratropics, and analyzes how new approaches to data analysis and modeling could also be used to increase predictability at midlatitudes.

Acknowledgements. The book we present here is the result of a series of lectures given within the scope of 'Global Climate', held at the Barcelona Science Museum, under the auspices of 'La Caixa' Foundation in March 1999. The conference was co-sponsored by the Departament d'Universitats i Recerca of the Generalitat de Catalunya and the University of Barcelona.

Among the main people to thank, I wish to express my most sincere appreciation to Miquel Àngel Rodríguez-Arias whose expertise, diligence and patience have underpinned the successful edition of this book, on time. I also wish to thank Josep-Anton Morguí for his useful comments which improved some of the chapters and to Paquita Ciller, whose scientific curiosity made it possible for this undertaking to see light. Thanks also to Anne Zanatta, for her careful correction of English style and grammar. I am also grateful to M. Prats, Anna Coll and in particular to the Director of the Science Museum, Jorge Wagensberg, for his support and contribution in organizing the symposium here in Barcelona. And of course, thanks to the authors of the different chapters for their prompt contribution to this book. I also wish to thank the following scientific journals for allowing us to reproduce some of the published material appearing in this book. And in particular, Science for figures 11.7, 11.15, 11.17; Nature, for figures 11.1, 11.11, 11.16, 11.18, 11.19; Journal of Climate for figures 11.5, 11.8, 11.9, 11.10; Climate Dynamics for figures 11.4, 11.6, 11.12, 11.13 and Table 11.1; Int. J. Clim. For figure 11.3 and Sedimentary Geology for figure 11.14.

As a last point, prior to finishing this presentation, I would like to recall Steinbeck (East of Eden, 1952) when he says:

'*Maybe the knowledge is too great and maybe men are growing too small... Maybe, kneeling down to atoms, they're becoming atomized in their souls. Maybe a specialist is only a coward, afraid to look out of his little cage. And think what any specialist misses: the whole world over his fence.*'

X. Rodó
Barcelona, August 2002

Som d'aquells que creuen que una curiositat envers el passat es troba justificada, per damunt de tot, per la vida del present.

(We are of those who think that curiosity towards the past is justified, above all things, by life in the present)

R. Lafont and Ch. Anatole,
History of Occitan literature, 1973

To Lluïsa, Pau, Aina and Lluc, my climate system.

X. R.

Table of Contents

Part I:

Part II:

List of contributors

Alverson, K.
PAGES International Project Office. Bärenplatz 2, Bern, CH-3011, Switzerland
alverson@pages.unibe.ch

Ciais, P.
Laboratoire des Sciences du Climate et l'Environement, LSCE (CNRS). Orme des Merisiers, Bât. 709, 91191 Gif-sur-Yvette, Cedex, France
ciais@lsce.saclay.cea.fr

Comín, F.A
Instituto Pirenaico de Ecología (CSIC). Avda. Montañana 1005, 50059 Zaragoza, Spain
comin@porthos.bio.ub.es

Griffies, S.M.
Geophysical Fluid Dynamics Laboratory, NOAA. Princeton Forrestal Campus Rte. 1, P.O. Box 308, Princeton, NJ 08542-0308, USA
smg@gfdl.noaa.gov

Kinter III, J.L.
Center for Ocean-Land-Atmosphere Studies, Institute of Global Environment and Society, Inc. 4041 Powder Mill Road, Suite 302, Carlverton, MD 20705-3106, USA
kinter@cola.iges.org

Krishnamurthy, V.
Center for Ocean-Land-Atmosphere Studies, Institute of Global Environment and Society, Inc. 4041 Powder Mill Road, Suite 302, Carlverton, MD 20705-3106, USA
krishna@cola.iges.org

Kull, C.
PAGES International Project Office. Bärenplatz 2, Bern, CH-3011, Switzerland
kull@pages.unibe.ch

Lyon, B
International Research Institute for Climate Prediction, Columbia University. 61 Rt. 9W, Monell Building, P.O. Box 1000, Palisades, NY 10964-8000, USA
blyon@iri.columbia.edu

Philander, S.G.H.
Department of Geosciences, Princeton University. Guyot Hall, Washington Rd., Princeton, NJ 08544-1003, USA
gphlder@Princeton.edu

Rodó, X.
GRC – Grup de Recerca del Clima, Centre de Climatologia i Meteorologia, Parc Científic de Barcelona, Universitat de Barcelona. Baldiri i Reixach 4-6, Torre D, 08028 Barcelona, Catalonia
xrodo@porthos.bio.ub.es

Rodríguez-Arias, M.A.
GRC – Grup de Recerca del Clima, Centre de Climatologia i Meteorologia, Parc Científic de Barcelona, Universitat de Barcelona. Baldiri i Reixach 4-6, Torre D, 08028 Barcelona, Catalonia
mar@porthos.bio.ub.es

Ropelewski, C.F.
International Research Institute for Climate Prediction, Columbia University. 61 Rt. 9W, Monell Building, P.O. Box 1000, Palisades, NY 10964-8000, USA
chet@iri.columbia.edu

Stocker, T.F.
Climate and Environmental Physics, Physics Institute, University of Bern. Sidlerstrasse 5, CH-3012, Bern, Switzerland
stocker@climate.unibe.ch

List of abbreviations

ACC	Antarctic Circumpolar Current
acf	Autocorrelation function
acvf	Autocovariance function
ACW	Antarctic Circumpolar Wave
AGCM	Atmospheric General Circulation Model
AMIP	Atmospheric Model Intercomparison Project
AO	Artic Oscillation
AOGCM	Atmospheric Ocean General Circulation Model
BGC	Biogeochemical Cycles
CLIVAR	Climate Variability and Predictability
CNRS	Centre Nationale de la Recherche Scientifique
COLA	Center for Ocean-Land-Atmosphere Studies
COADS	Comprehensive Ocean Atmosphere Data Set
CPTEC	Centro de Previsäo de Tempo e Estudos Climáticos
CT	Equatorial Cold Tongue Region
CZ	Cane-Zebiak Model
DARE	Data Rescue Project
DGF	Deep Geostrophic Flow
DIC	Dissolved Inorganic Carbon
DJF	December-January-February Season
DSP	Dynamical Seasonal Prediction
DWBC	Deep Western Boundary Current
EA	East Atlantic Pattern
EA-JET	East Atlantic Jet Pattern
EATL/WRU	East Atlantic / Western Russia Pattern
EBM	Energy Balance Model
ECMWF	European Center for Medium Range Weather Forecasts
EIMR	Extended Indian Monsoon Rainfall Index
ENSIP	El Niño Simulation Intercomparison Project
ENSO	El Niño Southern Oscillation
EOF	Empirical Orthogonal Function
EP	East Pacific Pattern
GCM	General Circulation Model
GCOS	Global Climate Observation System
GFD	Geophysical Fluids Dynamics
GFDL	Geophysics Fluid Dynamics Laboratory
GM90	Gent and McWillians 1990

GMS	**G**eostationary **M**eteorological **S**atellite
GOES	**G**eostationary **O**perational **E**nvironmental **S**atellite
GOGA	**G**lobal **O**cean **G**lobal **A**tmosphere
GOOS	**G**lobal **O**cean **O**bserving **S**ystem
GPP	**G**ross **P**rimary **P**roduction
GR	**G**lobal Domain **R**esiduals
HR	**H**eterotrophic **R**espiration
IDB	**I**nteramerican **D**evelopment **B**ank
IMD	**I**ndia **M**eteorological **D**epartment
IMR	**I**ndian **M**onsoon **R**ainfall Index
INSAT	**In**dian **Sat**ellite
IPCC	**I**nternational **P**anel for **C**limate **C**hange
IRI	**I**nternational **R**esearch **I**nstitute for Climate Prediction
ITCZ	**I**nter**T**ropical **C**onvergence **Z**one
JJA	**J**une-**J**uly-**A**ugust Season
JJAS	**J**une-**J**uly-**A**ugust-**S**eptember Season
JPL	**J**et **P**ropulsion **L**aboratory
LAI	**L**eaf **A**rea **I**ndex
LRF	**L**ong **R**ange **F**orecasts
LSCE	**L**aboratoire des **S**ciences du **C**limat et l'**E**nvironment
LUCC	**L**and **U**se and **C**over **C**hange
MAM	**M**arch-**A**pril-**M**ay Season
METEOSAT	**Meteo**rological **Sat**ellite
MH	**M**onsoon **H**adley Index
MJO	**M**adden-**J**ulian **O**scillation
MODIS	**Mod**erate Resolution **I**maging **S**pectroradiometer
MOGA	**M**id-latitude **O**cean **G**lobal **A**tmosphere
MONEG	**Mon**soon **N**umerical **E**xperimentation **G**roup
MOM	**M**odular **O**cean **M**odel
NAE	**N**orth **A**tlantic **E**uropean Sector
NAS	**N**orthern **A**frica – **S**outhern Europe
NASA	**N**ational **A**eronautics and **S**pace **A**dministration
NCAR	**N**ational **C**enter for **A**tmospheric **R**esearch
NCEP	**N**ational **C**enter for **E**nvironmental **P**rediction
NGO	**N**on **G**overnment **O**rganizations
NH	**N**orthern **H**emisphere
NOA	**N**orth **A**tlantic **O**scillation
NOAA	**N**ational **O**ceans and **A**tmosphere **A**dministration
NP	**N**orth **P**acific Pattern
NPP	**N**et **P**rimary **P**roduction
NTSG	**N**umerical **T**erradynamic **S**imulation **G**roup
OLR	**O**utgoing **L**ongwave **R**adiation
PAGES	**Pa**st Global Chan**ges**
PAR	**P**hotosynthetic **A**ctive **R**adiation
PC	**P**rincipal **C**omponents
PDF	**P**robability **D**istribution **F**unction

PDO	**P**acific **D**ecadal **O**scillation
PIRATA	**Pi**lot **R**esearch Moored **A**rray in the **T**ropical **A**tlantic
PNA	**P**acific/**N**orth **A**merican Pattern
POES	**P**olar **O**rbiting **E**nvironmental **S**atellite
PR	**P**lant **R**espiration
PSA	**P**acific/**S**outh **A**merican Pattern
PV	**P**otential **V**orticity
QB	ENSO **Q**uasi-**b**iennial band
QG	**Q**uasi-**g**eostrophic
QQ	ENSO **Q**uasi-**q**uadrennial band
RAS	**R**elaxed **A**rakawa-**S**chubert Scheme
RCM	**R**adiative-**C**onvective **M**odel
SAR	**S**econd **A**ssessment **R**eport (IPCC)
SCAND	**Scand**inavian Pattern
SDC	**S**cale **D**ependent **C**orrelation Analysis
SGS	**S**ub-**g**rid **S**cales
SH	**S**outhern **H**emisphere
SLP	**S**ea **L**evel **P**ressure
SMIP	**S**easonal Prediction **M**odel **I**ntercomparison **P**roject
SO	**S**outhern **O**scillation
SON	**S**eptember-**O**ctober-**N**ovember Season
SPCZ	**S**outh **P**acific **C**onvergence **Z**one
SRES	IPCC **S**pecial **R**eport on **E**missions **S**cenarios
SSH	**S**ea **S**urface **H**eight
SST	**S**ea **S**urface **T**emperature
SSTA	**S**ea **S**urface **T**emperature **A**nomaly
SVD	**S**ingular **V**alue **D**ecomposition
TAO	**T**ropical **A**tmosphere **O**cean
TAR	**T**hird **A**ssessment **R**eport (IPCC)
TBO	**T**ropospheric **B**iennial **O**scillation
TCR	**T**ransient **C**limate **R**esponse
TCZ	**T**ropical **C**onvergence **Z**one
THC	**T**hermo**h**aline **C**irculation
TOGA	**T**ropical **O**cean **G**lobal **A**tmosphere
TP	**T**ropical **P**acific
WBGU	**W**issenschaftlicher **B**eirat des Bundesregierung **G**lobale **U**nwelt-veränderungen
WETS	**W**orkshop on **E**xtra**t**ropical **S**STAnomalies
WGC	**W**ind-driven **G**eostrophic **C**urrent
WMO	**W**orld **M**eteorological **O**ffice
WP	**W**estern **P**acific pattern

Part I

I després de creure tant, t'arribaren tots els dubtes. Tenia raó el filòsof: presuposa la certesa el mateix fet de dubtar.

(And after believing so much, all the doubts arrive. The philosopher was right: certainty presupposes the very act of doubting).

Raimon,
I després de creure tant, 1981

1 What we know about the Climate System? A brief review of current research

Francisco A. Comín[1], Miguel Angel Rodríguez-Arias[2]

[1]Instituto Pirenaico de Ecología (CSIC). Avda. Montañana 1005, 50059 Zaragoza, Spain
comin@porthos.bio.ub.es
[2]GRC – Grup de Recerca del Clima, Centre de Climatologia i Meteorologia, Parc Científic de Barcelona, Universitat de Barcelona. Baldiri i Reixach 4-6, Torre D, 08028 Barcelona, Catalonia
mar@porthos.bio.ub.es

1.1 Introduction

The Earth is an oasis of life in the Universe because of its unique climatic conditions. However, the Earth's climate has experienced many changes during its history and, along with the biosphere, has continuously evolved throughout the past and in recent times. Rapidly increasing human impact is what is new to the history of the Earth and one of its consequences, the change of climatic conditions, is what causes enough worry to question the capacity of the Earth's system to buffer and assimilate it (Chap. 2). This leads to the discussions in this book on how to handle the state of the art climatology and the major current uncertainties about the functioning of the Earth's climate system. This knowledge is essential in order to provide answers about future change and correct management decisions for human activities related to climate.

Most of the general interest in the climate system lies in the benefits that can be derived from improved climate predictions. Although advancement in meteorological regional modeling is extending weather forecasts in some places to several days, acceptable predictions are restricted to a few days. Beyond this term, the rapid non linear interactions characteristic of atmospheric fluid dynamics cause unpredictable meteorological variations. In addition, a major interest exists in knowing the effects of human activities on climate. Changes in the atmosphere composition related to anthropogenic gas emissions have been recognized for a long time (Barnola et al. 1995), but human impact on biogeochemical cycles also affect other components of the climate system (Turco 1995).

A relevant advancement in weather forecast based on large scale atmospheric circulation modeling was promoted by international organizations and national governments during the 60s and the 70s. It focused on observations of the upper atmosphere and provided five-day forecasts for the structure of global atmospheric circulation (WMO 1975). The next step to provide a further lengthening and improvement of weather forecasts involved the observations in the lower atmosphere boundary layer to improve the estimation of the parameters feeding the model equations (Houghton and Morel 1984). This approach led naturally to the study of ocean-atmosphere interactions. In fact, earlier observations on the relationships between meteorological variables all around the world and sea surface temperature as a tracer of ocean circulation were the seeds for this growing interest (Bjerknes 1969, Glantz et al. 1991). Since then, an increasing number of significant interactions between ocean and climate have been recognized all around the world especially in the Tropics. In this area, both the investigation of atmospheric teleconnections between ocean basins and the role of tropical oceans in general atmospheric circulation, which account for most of the total energy released to the atmosphere, have become very important topics of research (Chaps. 3 and 10). Extratropical teleconnections, linked to the seasonal and interannual climate variability at mid-latitudes, has also been defined (Chap. 11). However, at longer timescales –from decades to centuries- the climate variability is strongly related to the thermal inertia of the oceans and to the heat transport from the Tropics to higher latitudes through ocean currents. So, variations in oceans-atmosphere coupling and dynamics are the basis for major global climate variability at several time scales. Finally, the oceans also play a major role accumulating CO_2 and thus contributing to counter-balance the warming trend induced by the increase of CO_2 atmospheric concentrations during the last century (Chap. 7).

Climate research has also incorporated two large and important subjects during the last decades. The first, paleoclimatology, studies past global climate and past climate change. Interpretation of past climate dynamics and its comparison with present conditions improves our forecasts of future climate impact as a consequence of the present climate change (Chap. 9). The other subject, global change, studies the modifications in the global biogeochemical cycles due to changes in the extent of major biomes, land-use, and human activities affecting the components of the Earth system.

1.2 The Earth Climate System

The present conditions of the Earth are remarkably different from those of its neighboring planets (Tables 1.1 and 1.2). The estimation from the Stefan-Boltzmann law of the blackbody radiation temperature of the Earth-atmosphere system is -18 °C. But we know that the average surface temperature of the Earth is about +15 °C (Chap. 8). The resulting 33 °C increase is due to the greenhouse effect of clouds and some atmospheric active gases which are semitransparent to the incoming solar radiation but nearly opaque to the outgoing infrared radiation. The Earth surface temperature and atmospheric composition are unique in the Solar System –and, up to now, in the Universe. These conditions appear extremely favorable and even necessary for life persistence, but are indeed the result of millions of years of contingent biological evolution shaping the biogeochemical interactions governing the different components of the Earth's climate system (Lovelock 1989a,b). The energetic driver of the Earth climate is the planetary heat exchange but it depends on the way the incoming energy of solar radiation distributes and dissipates in the different components of the climate system. Basically, the Earth heats up in the Tropics and cools down in the Polar Regions. The movement of the Earth fluids through convective cells (atmosphere) and currents (ocean), transports the heat from the Equator to the Poles.

Any alteration of the Earth's climate system (Fig. 1.1) will translate into adjustments among its components that are tied reciprocally by complex interactions resulting from a multitude of feedback mechanisms.

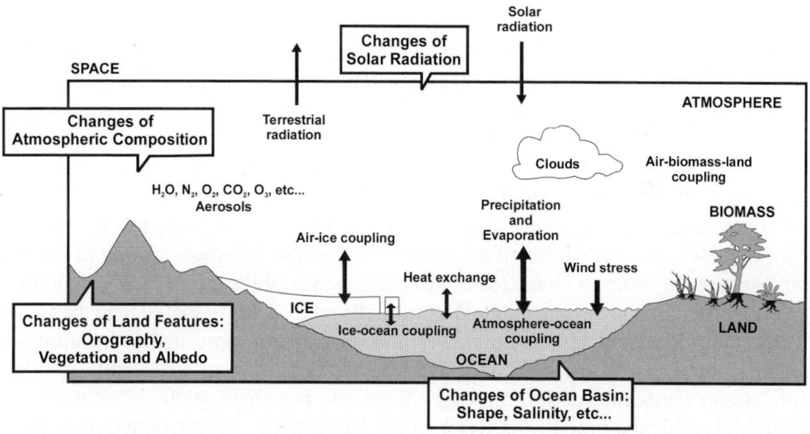

Fig. 1.1. Schematic illustration of the Earth's climate system, with some examples (boxes) of physical and biological processes related with climate and climate change.

Table 1.1. Physical characteristics of the planets Venus, Earth and Mars (adapted from Graedel and Crutzen 1993).

Physical characteristic	Venus	Earth	Mars
Total mass ($x10^{27}$g)	5	6	0.6
Radius (km)	6049	6371	3390
Mass of atmosphere[a]	100	1	0.06
Distance to Sun ($x10^6$ km)	108	150	228
Solar Constant (W m^{-2})	2613	1367	589
Albedo (%)	75	30	15
Average cloud cover (%)	100	50	Variable
Blackbody radiation temperature (°C)	-39	-18	-56
Surface temperature (°C)	427	15	-53
Greenhouse effect (°C)[b]	466	33	3

[a]With respect to the mass of Earth atmosphere.
[b]Increase of temperature due to greenhouse effect.

Table 1.2. Atmospheric concentration of active gases in the planets Venus, Earth and Mars (adapted from Graedel and Crutzen 1993).

Atmospheric concentration	Venus	Earth	Mars
N_2 (%)	< 2	78	< 2.5
O_2 (%)	< 1	21	< 0.25
CO_2 (%)	> 98	0.035	> 96
H_2O (%)	$1x10^{-4} – 0.3$	$3x10^{-4} – 4$	< 0.001
SO_2	150 ppmv	< 1 ppbv	nil

1.2.1. Climate variability and predictability

During the history of the Earth, climate fluctuations took place at different timescales (Fig. 1.2), because both the changes in incoming solar radiation and interactions among components of the system, take place at many temporal scales (Fig. 1.3). For example, changes of the orbital parameters of the Earth with their equivalence in terms of incoming solar radiation are responsible for long-term climatic cycles (Kutzbach 1995).

The atmospheric variability ranges at timescales from weeks and to a year, because the dynamic properties of air facilitate a fast and complete mixing of the full atmosphere. In order to characterize atmospheric variability, it is thus very important to measure the meteorological conditions with a high spatial and temporal resolution. Ocean processes are slower and have much more inertia because sea water is more viscous than air and, in consequence, it moves differently. As a result, ocean temporal variability ranges from one month to 4000 years, the latter being the residence time of the water in the larger ocean basins. The dynamics of the cryosphere are even calmer due to the long period of the glacial-interglacial fluctuations.

Many of the atmosphere-ocean interactions vary from decades to centuries, that is a time span of major interest for climate predictions. Knowledge of the mechanisms controlling these interactions has become the main objective of climate studies, requiring both observational data and climate modeling, the latter being the critical approach in climate studies to understand the past and to predict the future.

1.3 Atmospheric Models

The use of empirical statistical methods extrapolating historical data and processes to the near future has a restricted value as a forecasting tool. Fortunately, climate modeling infers climate dynamics from past and present observations applying the physical, chemical and biological equations representing the climate system at the scales of interest. As a result, climate models are the most promising approach to climate prediction. Once the models have proved to be accurate in representing present or past conditions, they can be used to forecast future ones.

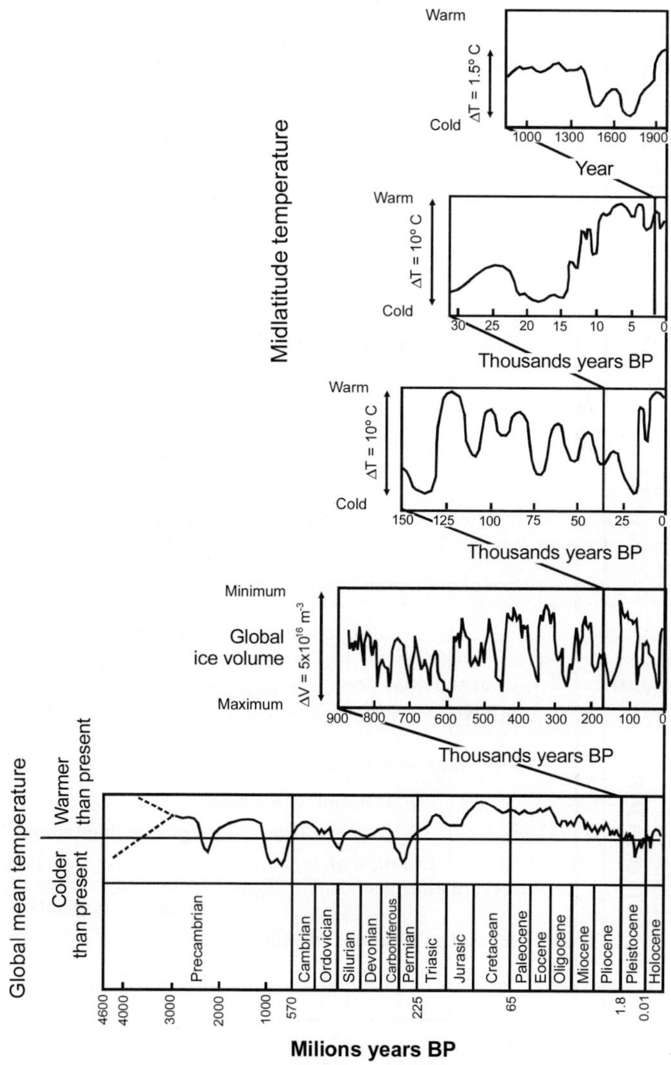

Fig. 1.2. Schematic representation of past climate change at several time-scales. Since the Pliocene the Earth has experienced an alternation of ice-ages and interglacials. Human society flourished during the last intergla-cial, the Holocene. Paleotemperatures have been estimated from oxygen isotopes ratios measured in bubbles from ice cores (adapted from Frakes 1980, Gribbin 1986 and, Schneider and Londer 1984).

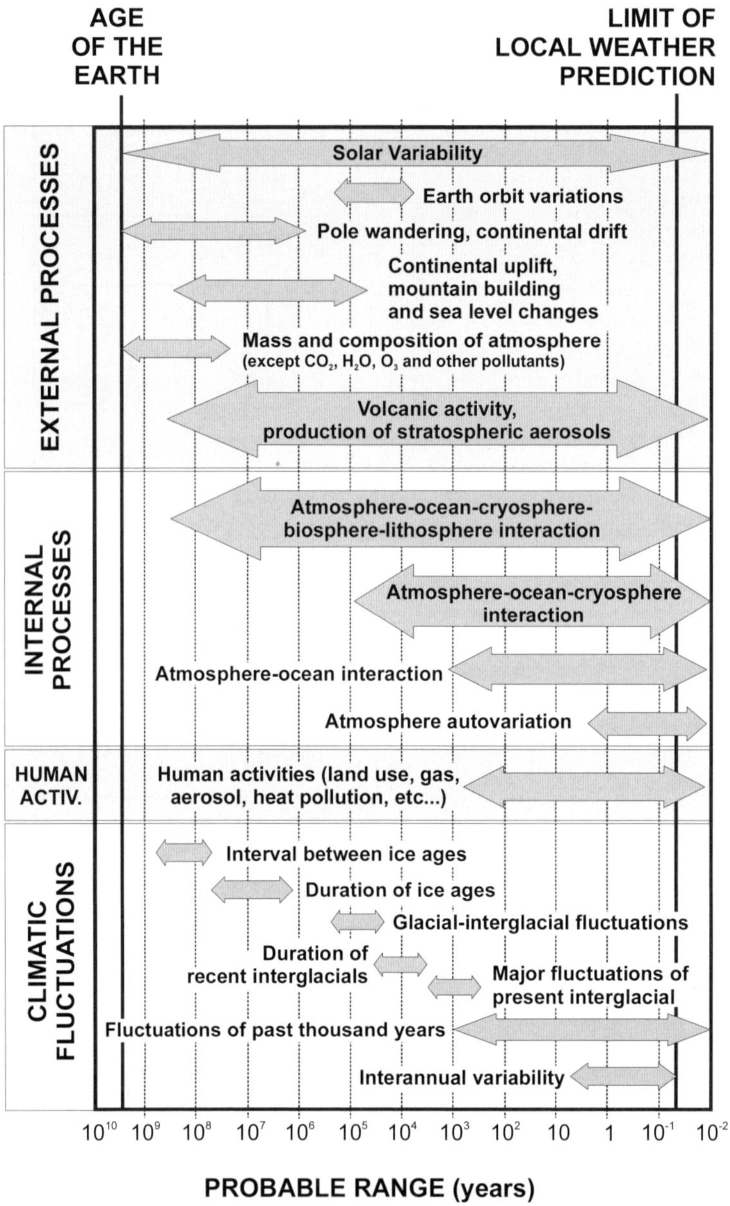

Fig. 1.3. Characteristic timescales for the climatic fluctuations and forcing factors (external, internal, human) of the climate system (adapted from Peixoto and Oort 1992)

A very simple model typifies the energy balance for the Earth:

$$S - F = 0 \tag{1.1}$$

where S is the net solar radiation absorbed by the Earth-atmosphere system and F is the long-wave radiation emitted to space. The outgoing radiation F, takes the form $F = Q\,(1\text{-}\alpha) = \sigma\,T^4_p$, (where Q is the incoming solar radiation, α is the albedo of the planet, $\sigma = 5.67\text{x}10^8\ Wm^{-2}\,{}^oK^{-4}$ is the Stefan-Boltzmann constant, and Tp is the effective blackbody radiation temperature of the Earth-atmosphere system, in ºK). This energy balance is affected by several feed-back mechanisms (see Chap. 7) and has a large sensitivity to changes in the albedo or in the solar radiation. For example, a larger incoming solar radiation increases surface temperature and enhances evaporation. The resulting increment of water vapor in the atmosphere magnifies greenhouse effect that raise even more the surface temperature. This positive feedback is, however, counterbalanced by a larger reflectivity (albedo) related with the increase of cloud cover as a consequence of the enhanced evaporation. Although many other positive and negative feedbacks can be introduced in Eq. 1.1, it cannot provide details of the spatial and temporal changes of the temperature because is a zero-order (non dimensional) model. Basically, this simple model derives from the flux form of the thermodynamic equation of the atmosphere (Trenberth 1992):

$$c_p\,(\partial T/\partial t) = -\ c_p\,\nabla.\,(\mathbf{v}T) - (c_p\,\partial(\omega T))/\partial p + c_p\,(\kappa\omega T/p) + Q_{rad} + Q_{con} \tag{1.2}$$

where T is the temperature, p is the altitude measured as geopotential height (in fact is the height in pressure coordinates), c_p is the specific heat of dry air at constant pressure, $\kappa \approx 0.286$ for dry air is a dimensionless constant, **v** is the horizontal wind vector, ω is the vertical velocity in p-coordinates, Q_{rad} the net radiative heating, and Q_{con} the heating due to condensational processes. In summary the first and second terms on the right of Eq. 1.2 are the variation of temperature due to horizontal and vertical movements, respectively; the third term represents the conversion of kinetic to potential energy, and the last two are heat fluxes. Averaging Eq. 1.2. over horizontal and vertical dimensions all the terms cancel and yields:

$$c_p\,(\partial T/\partial t) = S - F \tag{1.3}$$

If a stable climate prevails, a long term mean of Eq. 1.3 gives the simple Earth balance model of Eq. 1.1.

To have a deeper insight into the Earth energy balance, we need one-dimensional models. If averaging is performed only over horizontal directions (latitudinal and longitudinal) in Eq. 1.2, then the thermodynamic equation takes the form:

$$c_p\,(\partial T/\partial t) = Q_{rad} + Q_{con} \tag{1.4}$$

where all the quantities are a function of of the geopotential height, p. Given expressions for Q_{rad} and Q_{con} this radiative-convective model (RCM) gives a globally averaged vertical profile of T. Eq. 1.4 has been used to ascertain global effects of greenhouse gases and climate change. This one-dimensional model permits also to estimate the response of surface temperature to increases in high cloud cover (where longwave warming effect is more important than albedo), and in low and mid-level clouds cover (with albedo larger than longwave warming).

On the other way, if averaging of Eq. 1.2 is performed in the vertical and in the horizontal longitudinal direction:

$$c_p\,(\partial T/\partial t) = -\,(c_p\,\partial([vT]\cos\phi))/(R_E\cos\phi\,\partial\phi) + S - F \qquad (1.5)$$

where v is the meridional (northward) wind velocity, R_E the Earth radius and ϕ is the latitude. The first term on the right-hand side of the equation is the poleward heat transport. Eq. 1.5 gives an expression of T as a function of latitude (ϕ). This energy balance one-dimensional model (EBM) has been used to study climate feedbacks and stability, ice-age climates, dependency on poleward heat transport and coupled ocean-atmosphere problems (Trenberth 1992).

A two-dimensional model can be obtained averaging the thermodynamic equation (Eq. 1.2) only in longitude:

$$c_p\,(\partial T/\partial t) = -\,(c_p\,\partial([vT]\cos\phi))/(R_E\cos\phi\,\partial\phi) - (c_p\,\partial(\omega T))/\partial p + c_p\,(\kappa\omega T/p) + Q_{rad} + Q_{con} \qquad (1.6)$$

or only in height (averaging over p):

$$c_p\,(\partial T/\partial t) = -\,c_p\,\nabla.\,(\mathbf{v}T) + S - F \qquad (1.7)$$

In the first two-dimensional model (Eq. 1.6), T is a function of height and latitude and the prognostic variables are simply expressed as zonal means and deviations (e.g., for all the longitudes in a given latitude). In spite of its limitations, this model has been very popular to solve chemical and dynamical questions regarding the middle atmosphere. The second (Eq. 1.7) is a two-dimensional energy balance model, and has been used to study past climate (North et al. 1983).

Three-dimensional models do not average Eq. 1.2 over any spatial dimension, and yield temperatures as a function of altitude, latitude and longitude. In order to fully describe the atmospheric circulation, a part from the thermodynamic equation (Eq. 1.2), we need all the other equations describing the atmospheric motion: the momentum equation, the equation of state, the continuity equation for dry air, and the continuity equation for moist air. This set of five primitive equations characterizes any atmospheric general circulation model (AGCM).

The implementation of an AGCM requires the definition of the spatial resolution and integration length (number of iterations of a model run), with respect to the temporal and spatial variability of the problem to be solved. Then a parameterization of the internal primitive equations and of the boundary processes affecting the circulation should be performed at the selected spatial resolution. Finally, an AGCM run, should use a given atmospheric structure with its corresponding solutions of the primitive equations as initial conditions. The maximum resolution available in the first AGCMs was that corresponding to a coarse grid with cells of 2x2° in latitude/longitude and of 2 km in height. Fortunately, in the most recent models, the spatial resolution has increased enormously and the grid cells are quite small, allowing the resolution of a broader range of problems.

The parameterization of the boundary processes relevant to atmospheric circulation is one of the keys for the predictive success of the AGCMs. The parameterization of atmospheric radiation -the net result of incoming solar radiation and longwave radiation from the Earth, is one of the most challenging aspects of the present improvements of AGCMs. Recent developments coping with this problem incorporate into the models both the scattering of radiant energy by aerosols, which is very important for transferring radiant energy through the atmosphere and, the effect of longwave absorption by greenhouse gases and clouds. They are also challenging the parameterization of the coupling between the free atmosphere and the atmospheric boundary layer (the lower 2 km of turbulent atmosphere in contact with the surface of land and sea). The new research in these subjects will enhance the prediction capability of present AGCMs. Fortunately, these new developments can be easily validated thanks to the growing data set of direct measurements and satellite observations.

1.4 Interactions of the atmosphere with other components of the climate system

The development of successful climate models should not forget the interactions between the atmosphere and the other components of the climate system: oceans, sea-ice, snow cover, land processes and biosphere (Fig. 1.4).

Chap. 5 is a general explanation the role of the oceans on the Earth's climate and about ocean-climate modeling. At present, experts in ocean modeling repeatedly mention two necessary improvements to provide more realistic results. The first improvement of the models should be the implementation of the temporal overlapping of different water mixing processes (from 10 years in the upper water column to 1 000 years in the Pacific Ocean deep waters). The other improvement should be the incorporation of sea bottom topography and coastlines –particularly in the straits communicating ocean basins, with finer detail and better spatial resolution. A similar hierarchy to that explained for atmospheric models (Sect. 1.3) also exits for coupled ocean-atmosphere general circulation models (AOGCMs). While the most simple models compute sea surface temperatures (SST) using a straightforward surface energy balance, the most complex ones compute SST also taking into account heat storage, ocean currents and upwelling processes.

In the coupled models, it is very important to develop a good interface between the atmosphere and the ocean, with a clear parameterization of the fluxes between both compartments (Fig. 1.5). Obviously, the difference in time responses of both fluids should be adjusted, and the same happens with the systematic errors arising during the running of the models, before and after the coupling. To this purpose, many countries and international research programs are involved in the procurement of long-term data sets from a more compact network of atmospheric and oceanic stations.

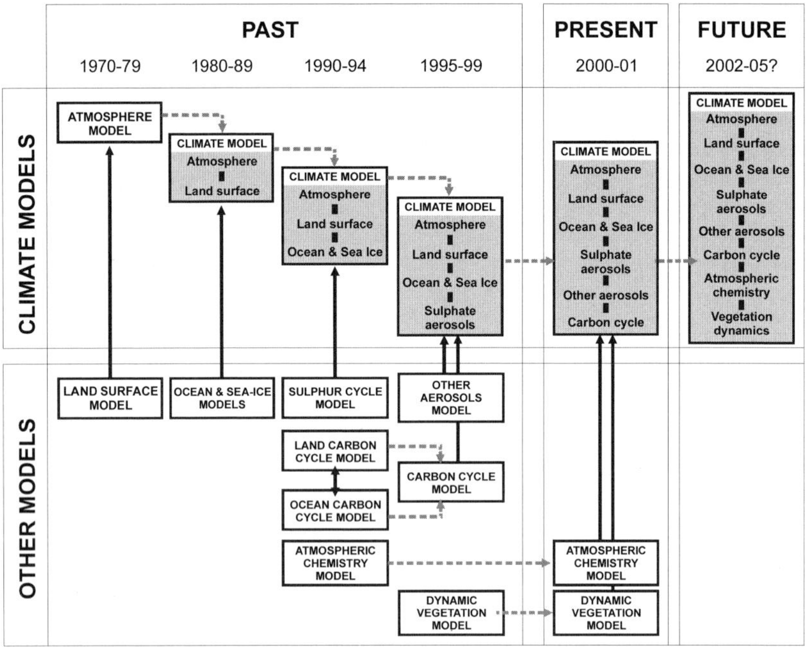

Fig. 1.4. The development of climate models over the last 25 years showing how the different components are first developed separately and later coupled into comprehensive climate models. Black arrows denote integration of a compartment model into climate models, while gray broken arrows indicate model evolution. Adapted from IPCC 2001.

Despite these simplifications, coupled models had provided a much better understanding of large scale climatic events such as those arising from ENSO (Chap. 3), monsoons (Chap. 10), carbon cycle related processes (Chap. 7) and thermohaline circulation (Chap. 8).

The interactions between the cryosphere and the other components of the climate system should also be taken into account. Ice-related processes play a very important role on a global scale through the ice-albedo feedback, but also control the heat and moisture fluxes on a regional scale, and may influence the atmospheric circulation and precipitation patterns in huge areas. Sea ice also influences the thermohaline circulation through its role in deep water formation and as a source of cold freshwater in polar seas (Oerlemans 1989, Van der Veen 1992, Untersteiner 1986). At present, however, we still have limited knowledge of the physics and thermodynamics of land and sea ice.

Land represents one third of the Earth surface and also exchanges moisture, momentum and heat with the atmosphere -although at lower amounts than the oceans. Biospheric processes happen both in land and oceans, but they are faster in the continents. A biospheric process exclusive to the oceans is the flux to the atmosphere of the dimethyl-sulfide, $(CH_3)_2S$. The phytoplankton and other marine primary producers release this gas that, once in the atmosphere, contributes to aerosol formation. As a result, the dimethyl-sulfide affects the radiative energy balance of the atmosphere. Furthermore, the ocean plays a major role in long term buffering of the atmospheric CO_2 increase, due to its high solubility and peculiar reactivity in seawater and to the large carbon storage capacity of the oceans related to the long residence of deep waters (Chap. 8). In contrast, terrestrial ecosystems seem to play a major role in the short-term uptake and sink of anthropogenic CO_2 emissions (Chap. 7, Tans et al. 1990, Taylor and Lloyd 1992).

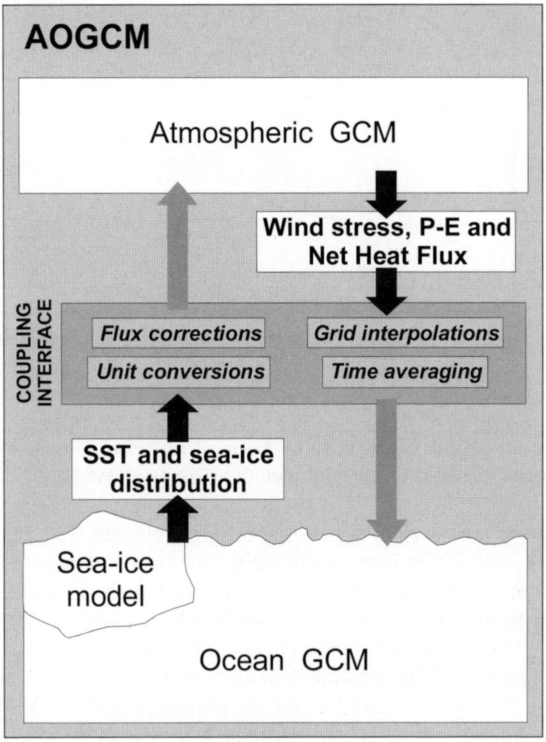

Fig. 1.5. Schematic of the ocean-atmosphere coupling interface of a AOGCM (adapted from Trenberth 1992)

1.4.1. Land related global climate processes

Many biospheric processes in the continents take place at a wide range of temporal and spatial scales (Fig. 1.6). This behavior should be considered when coupling land and atmosphere. Other relevant processes also occur through the oceans as a result of the direct transport of materials from the continents.

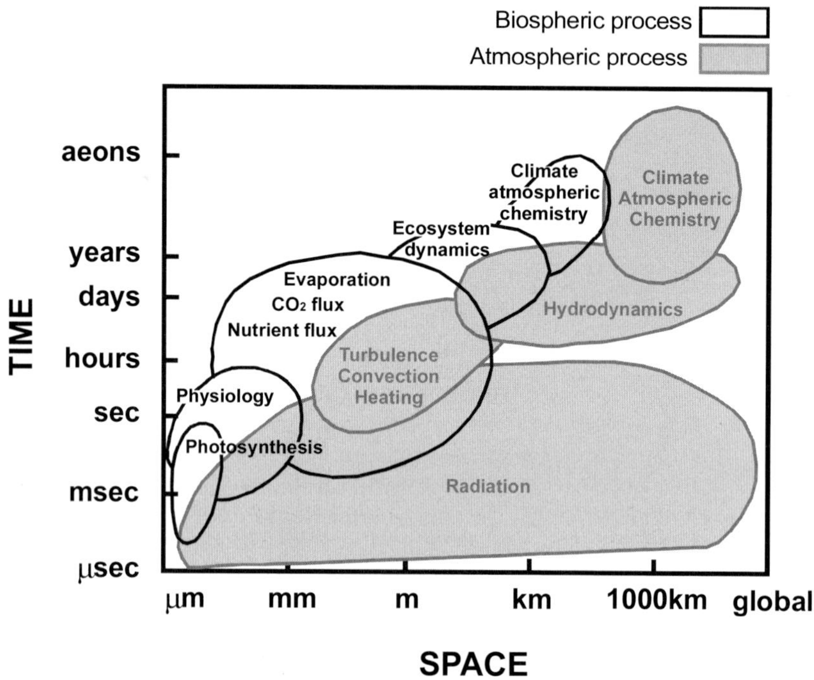

Fig. 1.6. Temporal and spatial scale of the most important biospheric and atmospheric processes related to climate (adapted from Trenberth 1992).

To incorporate land surface processes to global climate studies we need to estimate sensible heat, latent heat, momentum, and fluxes of radiation. To perform these estimations we should gather measurements of the surface albedo, roughness length and moisture. The net radiation from the Earth surface to the atmosphere is $R_n = H + LE + G + P$, where P is the energy used for photosynthesis, G the heat flux into the ground, H the sensible heat flux, and LE. P and G are very small with respect to R_n, while H and LE are the important terms in the radiation balance equation. In some early works, Dorman and Sellers (1989), and Sato et al. (1989) showed how to improve the outputs of biophysical models and their applications for climate change prediction. They requested collecting more realistic measurements of albedo and roughness length, and to incorporate the spatial heterogeneity of land surface attributes like soil hydrology and evapotranspiration into the models.

The development of remote sensing techniques allowed for incorporating more realistic data of radiative transfer and related attributes in vegetation canopies, with a better spatial resolution, into biophysical models. For example, the incorporation of satellite-derived vegetation characteristics (type, density, phenology) simplifies the calculations and reduces the computational requirements. This top-down approach is specially useful for performing medium-range forecasts of climate change and related features, and to investigate low scale processes, like the impact of deforestation on climate.

The reverse, a bottom-up approach, is also necessary when we want to incorporate empirical data into climate studies and models, or to improve our knowledge about the biological mechanisms involved in climate change. Detailed studies of ecosystem structure and vegetation response to anthropogenic climate forcing (e.g., increased CO_2, increased temperature, and changes in precipitation) can be introduced in mesoscale models to improve the calculations of radiative transfer, turbulent transfer and transport of heat and water vapor (Sellers, 1992). The scaling-up of data is a serious difficulty for climate modelers, i.e., they require several equations for water and heat exchange when they model processes ranging from stomatal conductance to ecosystem-atmosphere interactions. Different models use diverse linear functions of environmental (temperature, leaf water potential and vapor pressure deficit stress) and biological variables (photosynthesis and respiration), to estimate canopy transpiration, one of the most important parameters in the study of land-climate processes (Sellers 1992).

In spite of the difficulties arising from the integration of biological mechanisms at the ecosystem level into the models (due to the high variability in their structural and physiological components), there are several advantages in this approach. A natural way to integrate processes and data into the models comes from the strong correlation between soil and vegetation properties. This relationship reflects most of the bulk characteristics of the ecosystems (vegetation type, photosynthetic capacity, and evapotranspiration), and allows, through large-scale field experiments for different types of land surfaces, to test and validate the parameterizations of the models (Dorman and Sellers 1989).

As a consequence, long-term experiments on the relationship between ecosystem dynamics and climate change are very useful in studying past global change and forecasting future climatic consequences of ecological disturbances (Katul et al. 1997). However, a full integration of terrestrial ecosystem dynamics into global climate models is still pending due to the difficulties arising from the lack of overlapping of the main driving processes, both in space and time (Fig. 1.6).

The use of climate model forecasts as forcing conditions to ecosystem spatial distribution and temporal dynamics, is very important in land use and cover change (LUCC) research (Nunes and Augé 1998). We should remember that LUCC is the second most important factor, after fossil fuel combustion, in explaining the current climate change. The coupling of AGCMs and vegetation functional models is also very useful for simulating the effects of climate change on terrestrial ecosystems. Using this approach, Betts et al. (1997) predicted a short-term enhancement of regional warming by vegetation physiology (a global leaf area index, LAI, increase leading to a global canopy conductance decrease) that will be balanced by a longer term modification of vegetation structure under equilibrium conditions at the time of CO_2 doubling. They suggested including dynamical changes in vegetation physiology and structure in the models, in order to cope with the wider range of timescales in which vegetation-climate feedbacks take place. This research line, however, is in the early stages of development. The use of vegetation dynamics to simulate the impact over terrestrial carbon stocks of climate shifts (doubling CO_2), increases the uncertainties in the calculations of future terrestrial carbon fluxes and, consequently, reduces the reliability of climate models forecasts (Kirilenko and Solomon 1998). Despite these difficulties, this is a very promising subject in climate modeling research.

1.5 Present certainties and uncertainties about the Earth climate

During the last two decades, controversial discussions about the fluctuations of the Earth's climate system during the last century have been maintained in scientific, social and political forums. The work of many researchers collected in the last report of the Intergovernmental Panel on Climatic Change (IPCC 2001) concludes that significant changes of the Earth climate are taking place since the late XIX century. Natural forcing alone is unlikely (estimation of confidence corresponding to 10-33% chance that the result is true, IPCC 2001) to explain the observed global warming and the changes in the vertical temperature structure of the atmosphere. In fact, a significant anthropogenic contribution is required to account for the climatic trends observed over the last 50 years. Different multi-signal models able to discriminate between the effects of different factors on climate, have been used to simulate the changes of the Earth global mean surface temperature over the last 140 years. Several simulations were performed using separately and combined natural and anthropogenic forcing factors. Although some uncertainties arose from sulphate aerosol and natural forcing, the best agreement between model simulations and observations is found when both anthropogenic and natural factors are considered.

The procedure established in the detection of global climate changes by the IPCC is reliable and highly conservative. The average temperature of the Earth surface has increased by 0.6 ± 0.2 °C since the late XIX century (Chap. 2, Fig. 2.1). Likely, (66-90% chance that the result is true) the XXth century warming rate and duration (0.15 °C per decade in the periods 1910-45 and 1976-2000) are larger than any others observed during the last 1 000 years. Sea surface temperature has increased on average 0.15 °C during the 1950-1993 period. The rate of increase of minimum air surface temperature doubled the warming rate of maximum temperatures (0.2 versus 0.1 °C per decade). Global ocean heat content has increased significantly since the late 1950s, mainly in the upper 300 m (0.04 °C per decade). Other major changes have been observed in the seasonal and spatial variability patterns of temperature and rainfall, that likely have caused significant impacts on terrestrial ecosystems and human related activities (Chap. 11). In general, the uncertain results of the previous IPCC report (1995) are now confirmed: the decrease of the snow pack and sea-ice extent in the Northern Hemisphere, the average increase of global precipitation, the faster warming in the land than in the ocean, and finally the increase of precipitation in tropical areas, the decrease in the Sub-tropics, and the relatively small increase in the high latitudes.

It is very likely (with a 90-99% probability of being true) that the sea level rise observed in tide gauges during the XX century (1-2 mm $year^{-1}$) resulted from the global Earth warming through the thermal expansion of sea water and the widespread loss of land and sea ice. However, the acceleration of sea level rise during the XXth century has been lower than that predicted by the models (-0.011 ± 0.012 mm yr^{-2}, Douglas 1995). There are also large uncertainties about the sea level rise forcing factors, especially the possible effect of ice volume variability on the interdecadal fluctuations of sea level. A further improvement of satellite observations and of tide-gauge density and geodetic measurements, both in the Arctic and Antarctic, is necessary to provide a better understanding and monitoring of sea level changes.

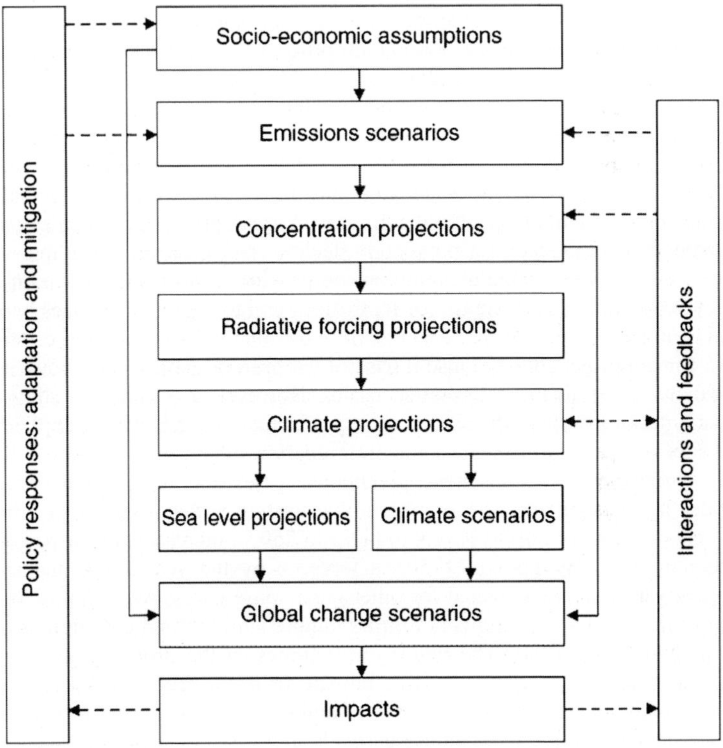

Fig. 1.7. The cascade of growing uncertainties in climate projections to be considered in the definition and development of climate scenarios for climate change impact, adaptation and mitigation assessment (adapted from Fig. 13.2 in "IPCC, Climate Change 2001: The Scientific Basis", available at http://www.grida.no/climate/ipcc_tar/wg1/477.htm)

However, some major uncertainties in the detection of global climate changes and causal attribution still persist. First of all, there are remarkable differences in the response of different models to the same forcing factors. There are also strong discrepancies between observations and model simulations of the trophospheric temperature change and of the internal climate variability. Some uncertainties remain due to the fact that the available solar and volcanic forcing data series are too short to reconstruct their contributions to climate change. Large uncertainties are also related to less known impacts of minor anthropogenic forcing factors (e.g., biomass aerosols and LUCC). To reduce these uncertainties, it is necessary to intensify the data gathering on land, oceans and atmosphere, and to refine the modeling techniques.

The global change projections of the future Earth climate provided by the AOGCMs are made using up to 40 different anthropogenic activity scenarios. Additional uncertainties related with the different emission scenarios can also arise at every single step of the overall simulation process (Fig. 1.7). The scenarios are a projection of future greenhouse gases and aerosol emissions under several socio-economic assumptions (Nakicenovic et al. 2000). The economic development assumptions range from a widespread fast economic growth leading to a homogeneous world without regional differences, to a highly heterogeneous economic development with outstanding local solutions to achieve economic, social and environmental sustainability. Population growth assumptions vary from a continuous exponential increase to a population decline after peaking in the mid-XXIst century. Similar wide-range assumptions are provided with respect to technological development to avoid the use of fossil fuels and to social differences between world countries. There is a lot of work still to be done from scenario description to future climate projections because it has not yet been possible to carry out model simulations using all the emission scenarios. However, a selection of illustrative emission scenarios encompassing a variety of situations, have been applied to a number of AOGCMs to provide some preliminary results.

The projections of atmospheric greenhouse gases and aerosol concentrations provided by the simulations (Fig. 1.8) vary according to the different scenarios. A CO_2 concentration of 540 to 970 ppm (a 90 to 250 % increase with respect to the concentration of 280 ppm in 1750) is foreseen by the year 2100, and similar ranges of variation are expected for other greenhouse gases. Accordingly, the anthropogenic radiative forcing also follows a quite similar trend although differing in magnitude (Fig. 1.9a). The major uncertainties in the projected atmospheric composition are the magnitude of the climate feedback from the terrestrial biosphere, the abundance of aerosols (both of anthropogenic and natural origin), and the emissions of indirect greenhouse gases. In any case, a general air quality degradation is expected in most scenarios due to the increasing background levels of tropospheric O_3.

The projections of future Earth climate also vary considerably among scenarios (Fig. 1.9b). Although there are some uncertainties arising from cloud modeling and other internal feedbacks, most models provide useful global and regional climate change predictions. The main temperature and sea-level projections in the IPCC reports (IPCC 1990, 1996, 2001) are the results of climate sensitivity experiments. The IPCC (1996) defines equilibrium climate sensitivity as the total change in global mean temperature corresponding to a continuous forcing of two times the pre-industrial atmospheric concentration of CO_2 ($2xCO_2$). Climate sensitivity has two components (IPCC 2001), the transient climate response (TCR) corresponding to the warming while the CO_2 atmospheric concentration is increasing (in most experiments the CO_2 concentration increases 1% yr^{-1}), and the additional warming due to the lag between the attainment of $2xCO_2$ conditions and the climate stabilization (fig 1.10). The different AOGCM yield climate sensitivities between 1.5 to 4.5°C and TCR between 1.1 to 3.1 °C (IPCC 2001), using a wide range of scenarios.

The IPCC projections in the period 1990 to 2100 also include forecasts for globally averaged evaporation, water vapor and precipitation. In spite of the wide range of greenhouse gases foreseen in the different emission scenarios, CO_2 and NO_2, and perhaps sulphate aerosols, are likely to be mainly main responsible for long-term climate change in precipitation. Precipitation will increase both in summer and winter in high-latitudes, in winter only in northern mid-latitudes, tropical Asia and Antarctica, and during summer in southern and eastern Asia, Australia and central America. Decreases in precipitation are expected in the southern African winter. Interestingly, a number of models predict a strong positive correlation between interannual variability and mean precipitation, which is opposite to the current relationship (Comín & Williams 1994).

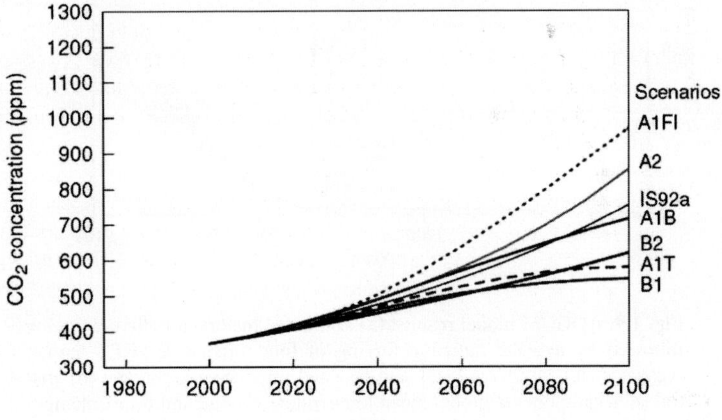

Fig. 1.8. Projected CO_2 concentrations resulting from six SRES scenarios (IPCC Special Report on Emissions Scenarios) and the IS92a scenario. The SRES scenarios represent the outcome of different assumptions about the future course of economic development, demography and technological change (adapted from Fig. 18 in "IPCC, Climate Change 2001: The Scientific Basis", available at http://www.grida.no/climate/ipcc_tar/wg1/030.htm)

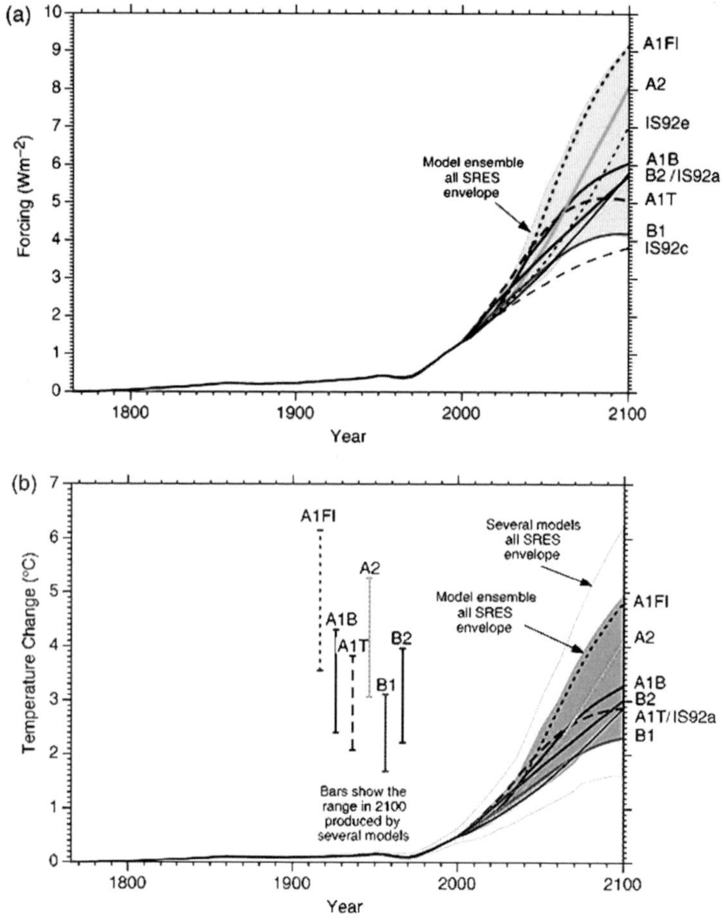

Fig. 1.9. AOGCM model results. (a) Estimated historical radiative forcing followed by average radiative forcing of four illustrative SRES marker scenarios (A1 family, A2, B1 and B2) and three IS92 scenarios. (b) Historical anthropogenic global mean temperature change and future changes for the same SRES scenarios and I292a. The scenarios in the A1 family illustrate different energy technology options. IS92 scenarios have different forcing factors. The darker shadow represents the average envelope of a full set with 35 SRES scenarios. The bars show the range of results produced by 7 different models applied to every scenario, and the light shadow represents the range of these results (adapted from Fig. 9.13 in "IPCC, Climate Change 2001: The Scientific Basis", available at http://www.grida.no/climate/ipcc_tar/wg1/353.htm)

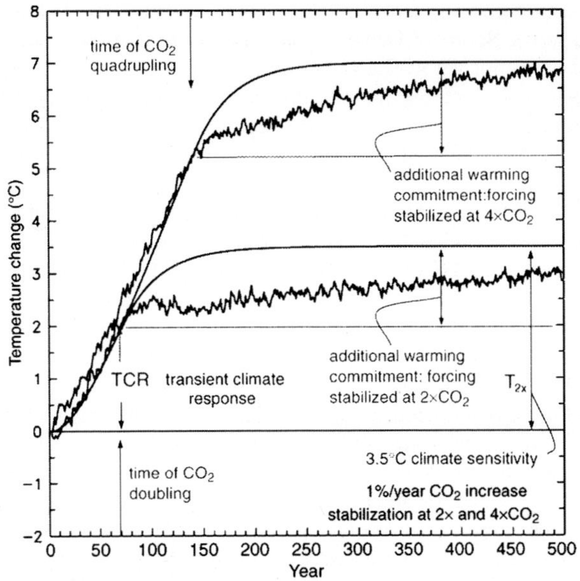

Fig. 1.10. Global mean temperature change for 1%yr^{-1} atmospheric CO_2 concentration increase with subsequent stabilization at $2xCO_2$ and $4xCO_2$. The noisy curves are from a coupled AOGCM simulation (GFDL_R15_a) while the smooth curves are from a simple illustrative model with no exchange of energy with the deep ocean. The transient climate response, TCR, is the temperature change at the time of CO_2 doubling and the "equilibrium climate sensitivity", T2x, is the temperature change after the system has reached a new equilibrium for doubled CO_2, i.e., after the "additional warming commitment" has been realised. (from Fig. 9.1, available at http://www.grida.no/climate/ipcc_tar/wg1/345.htm, in "IPCC, Climate Change 2001: The Scientific Basis").

The choice of model and scenario is important when studying regional differences in climate change projections. In fact, there are large uncertainties in the regional distribution of extreme events (maximum and minimum temperatures, precipitation intensification and drought risk). Fortunately, the global expected changes in extreme events for the XXIst century are in good agreement with those observed during the last part of the XXth century. However, an increase in the confidence of the models' regional projections will result from the future availability of additional long-term observations.

An average sea level rise of 0.11-0.77 m. is expected from the outputs of several AOGCMs simulations including sulphate aerosol forcing . Similar projection (0.09-0.88 m) are provided by the IPCC (2001) for different emission scenarios. Although most models disagree in the detailed regional distribution of sea level change, there is strong agreement in a larger than average rise in the Arctic and in a minor rise in the Southern Ocean. Uncertainties are large because the dynamics of grounded ice is not yet well understood and affects the confidence of sea-level change long-term projections. Despite this fact, the models predict an enlargement of the Antarctic ice sheet in response to a regional precipitation increase, and a severe ice-loss in Greenland where an excess of runoff with respect to precipitation is expected.

The IPCC (2001) also studies the climate change assuming stable CO_2 atmospheric concentrations other than 2xCO_2. They had especially analyzed the climatic consequences of 4xCO_2 conditions (approx. 1000 ppm) although the simulations have not been as intensive as those performed under the 2xCO_2 horizon. Under this assumption the TCR temperature and the climatic sensitivity are larger, as well as the additional warming after the stabilization of CO_2 concentrations (Fig. 1.10). The larger the horizon, the larger the temporal lag between greenhouse gas stabilization and temperature stabilization and the larger the additional warming. This is because sea-water, which plays a major buffering role in the climate system as a carbon sink, has a large residence time in the ocean basins. Similar projections are provided for other global changes commented before.

Impressive progress in the knowledge of the Earth system and its climate has taken place during the last two decades as a result of both the work of the scientific community and the development of new technologies. This work has been promoted by national and international organizations in an exemplary cooperative exercise that should be enlarged further in the near future. Results in past and future changes of the climate have been obtained with the best possible confidence and show good agreement with observations. Uncertainties are still present and new ones will arise, but they are constitutive element of scientific research and contribute to setting up new challenges. Some of them will be addressed in the next chapters of this volume, but in any case, uncertainties should be an excuse not to act. The number and confidence on present certainties overwhelms present uncertainties. We should decrease the stress on the Earth's climate system and we should rehabilitate its degraded components. We should employ this responsible strategy right now while the research continues!

References

Barnola, J.M., M. Anklin, J. Porcheron, D. Raynaud, J. Schwander and B. Stauffer (1995). CO_2 evolution during the last millennium as recorded by Antarctic and Greenland ice. Tellus 47 (B):264-272.

Berger, A., M.F. Loutre and C. Tricot (1993). Insolation and Earth's orbital periods. J. Geophys. Res. 98:10341-10362

Betts, R.A., P. M. Cox, S.E. Lee and F.I. Woodward (1997). Contrasting physiological and structural vegetation feedbacks in climate change simulations. Nature 387:796-799.

Bjerknes, J. (1969). Atmospheric teleconnections from the tropical Pacific. Monthly Weather Review 97:103-172.

Comín, F.A. and W.D. Williams (1994). Parched continents. Our common future?. In: ".). Limnology now. A paradigm of planetary problems", R. Margalef (ed.). Elsevier, Dordrecht, pp:473-527.

Dickinson, R.E. (1986). How will climate change?. The climate system and modelling of future climate. In J.E. Hanson and T. Takahashi (eds.). Climate processes and climate sensitivity. Geophysical Monograph 29, Maurice Erwing Volume 5, Amer. Geophys. Union, pags:58-72.

Dorman J.L. and P. J. Sellers (1989). A global climatology of albedo, roughness length and stomatal resistance for atmospheric general circulation models as represented by the Simple Biosphere model (SiB). J. Appl. Meteor. 28:833-855.

Douglas, B. C. (1995). Long-term sea-level variation. In: "Natural Climate Variability on Decadal-to-Century Time Scales", National Research Council (ed.). National Academy Press, Washington, pp: 264-272.

Frakes, L.A. (1980) Climate throughout geologic time. Elsevier, New York, 310 pp.

Glantz, M. H., R.W. Katz and N. Nicholls (1991). Teleconnections linking worldwide climate anomalies. Cambridge University Press, Cambridge, 535 pp.

Graedel, T.E. and P.J. Crutzen (1993) Atmospheric change. An Earth System Perspective. W.H. Freeman & Co., New York

Gribbin, J.R. (1986) El clima futuro. Biblioteca Científica Salvat, Barcelona, 240 pp

Houghton, J. T. and P. Morel (1984). The World Climate Research Programme. In: "The Global Climate" J.T. Houghton (ed.). Cambridge University Press, Cambridge, pp: 1-11.

IPCC (1990). Climate change: the IPCC scientific assessment. Houghton, J.T., G.J. Jenkins and J.J. Ephraums (eds.). Cambridge University Press, Cambridge, United Kingdom, 365pp

IPCC (1996) Climate Change 1995: The Science of Climate Change. Contribution of Working Group I to the Second Assessment Report of the Intergovernmental Panel on Climate Change. Houghton, J.T., L.G.M. Filho, B.A. Callandar, N. Harris, A. Kattenberg, and K. Maskell (eds). Cambridge University Press, New York, 572 pp.

IPCC (2001). Climate Change 2001: The Scientific Basis. Contribution of Working Group I to the Third Assessment Report of the Intergovernmental Panel on climate Change. Houghton, J. T., D. J. Ding, M. Griggs, M. Noguer, P. J. van der Linden, X. Dai, K. Maskell and C. A. Johnson (eds.). Cambridge University Press, Cambridge, 881 pp.

Katul, G.G., R. Oren, D. Ellsworth, C.I. Hsieh, N. Phillips, and K. Lewin (1997). A Lagrangian dispersion model for predicting CO2 sources, sinks and fluxes in a uniform Loblolly pine (Pinus taeda L.) stand. Journal of Geophysical Research 102, 9309-9321.

Kirilenko, A. P. and A. M. Solomon (1998). Modeling dynamic vegetation response to rapid climatic change using bioclimatic classification. Climatic Change 38(1):15-49.

Kutzsbach, J.E. (1995). Modeling large climatic changes of the past. In K. E. Trenberth (ed.). Climate System Modeling. Cambridge Univ. Press, New York, pp:669-701.

Lovelock, J.E. (1989a). Geophysiology: the science of Gaia. Reviews of Geophysics 27:215-222.

Lovelock, J.E. (1989b). The ages of Gaia. Norton, New York.

Nakicenovic, N. J., J. Alcamo, G. Davis, B. De Vries, J. Fenhann, S. Gaffin, K. Gregory, A. Grübler, T. Y. Jung, T. Kram, E. L. La Rovere, L. Michaelis, S. Mori, T. Morita, W. Pepper, H. Pitcher, L. Price, K. Raihi, A. Roehrl, H. H. Rogner, A. Sankovski, M. Schlesinger, P. Shukla, S. Smith, R. Swart, S. van Rooijen, N. Victor and Z. Dadi (2000). IPCC Special Report on Emissions Scenarios. Cambridge University Press, Cambridge, 599 pp.

Nunes, C. and J.L. Augé (1998). Land use and land cover change-LUCC. IGBP Report 48-IHDP Report 10, 125 pp.

Oerlemans, J. (1989). Glacier fluctuations and climatic change. Kluwer, Dordrecht, 417 pp.

Peixoto, J.P. and A.H. Oort (1992). Physics of climate. American Institute of Physics, New York, 520 pp.

Sato, N., P.J. Sellers, D.A. Randall, E.K. Schneider, J. Shukla, J.L. Kinter III, Y.Y. Hou and E. Albertazzi (1989). Effects of implementing the Simple Biosphere Model (SiB) in a general circulation model. J. Atmos. Sci. 46:2757-2782.

Schnaider, S.H. and R. Londer (1984) The coevolution of climate and life. Sierra Club Books, San Francisco, 563 pp

Tans, P.P., I.Y. Fang and T. Takahasi (1990). Observational constraints on the global atmospheric CO_2 budget. Science 247:1431-1438.

Taylor, J. A. and J. Lloyd (1992). Sources and sinks of atmospheric CO2. Australian Journal of Botany 40:407-418.

Trenberth, K.E. (1992) Climate System Modeling. Cambridge University Press, 788 pp

Turco, R. P. (1992). Atmospheric Chemistry. In: "Climate System Modeling", K. E. Trenberth (ed.). Cambridge Univ. Press, New York, pp:21 –23 .

Untersteiner, N. (1986). The Geophysics of Sea Ice. Plenum Press, 1196 pp.

Van der Veen, C.J. (1992). Land ice and climate. In: "Climate System Modeling", K. E. Trenberth (ed.). Cambridge Univ. Press, New York, pp.:437-450.

WMO (1975). The physical basis of climate and climate modeling. GARP Publications, Series No. 16, World Meteorological Organization, Geneva.

2 Why Global Warming Is A Controversial Issue

S. George Philander

Department of Geosciences, Princeton University. Guyot Hall, Washington Rd., Princeton, NJ 08544-1003, USA
gphlder@Princeton.edu

This article draws on S. George Philander's book *"Is the Temperature Rising? The Uncertain Science of Global Warming"* recently published (1998) by Princeton University Press.

2.1 The global warming debate

The debate about global warming is about the outcome of a gamble. We are betting that the benefits of our industrial and agricultural activities – increasing standards of living for the rich and poor alike –will outweigh possible adverse consequences of an unfortunate by-product of our activities, an increase in the atmospheric concentration of greenhouse gases that could lead to global warming and global climate changes. Some experts are warning that we are making poor bets, that global warming has started and that disasters are imminent. Others assure us that the chances of global warming are so remote that the outcome of our wager will definitely be in our favor. The impasse is disquieting because the issue is of vital importance to each of us; it concerns the habitability of our planet. How long will it be before the experts resolve their differences? How long before it is imperative that we take action?

Some people are under the false impression that global warming is a theory that still has to be confirmed. They do not realize that scientists are in agreement that a continual rise in the atmospheric concentration of greenhouse gases will inevitably lead to global warming and global climate changes. The disagreements are about the timing and amplitude of the expected warming. It is as if we are in a raft, gliding smoothly down a river, towards dangerous rapids and possibly a waterfall, and are uncertain of the distance to the waterfall. To avoid a disaster, we need to address two questions: how far is the waterfall? and when should we get out of the water? The first is a scientific question, the second is not. The first question, in principle, has a definite, unambiguous answer. The second is a difficult political question that probably will require compromises. This distinction between the science and politics becomes blurred should the scientific results have uncertainties.

Suppose that we have only approximate, not precise estimates of the distance to the waterfall. Rather than leave it at that -rather than accept that we can do no better than predict that we will arrive at the waterfall in thirty minutes plus or minus ten minutes- some people will minimize the distance and insist that we will arrive in twenty minutes or less, while others will maximize the distance, stating confidently that we won't be there for forty minutes or more. Do these people disagree for scientific reasons? (Some may have more confidence in their instruments and calculations than others do.) Or do their different opinions simply reflect the difference between optimists and pessimists?

To cope with this problem, we usually start by addressing the uncertainties in the scientific results. After all, everyone knows that science, in principle, can provide precise answers. One of the first scientist to be acclaimed by the public for his accurate predictions was Isaac Newton:

"Nature and Nature's law lay hid in night
God said, Let Newton be! and all was light".

Alexander Pope

Since Newton's accomplishments in the 17th century, scientists have continued to impress the public with remarkably accurate predictions that have led to inventions that continue to transform our daily lives. If, today, the results concerning a certain scientific problem has uncertainties then, surely, it is only a matter of time before scientists present us with more accurate results. It is therefore easy to agree on a postponement of difficult political decisions as regards certain environmental problems on the grounds that we will soon have more precise scientific information. This could prove disastrous should we suddenly find ourselves at the edge of the waterfall. We recently had such an experience.

The current fisheries crisis, which is most severe off the shores of New England and eastern Canada where many species of fish have practically disappeared, started a decade after scientists first warned that over-fishing could cause a dangerous reduction in fish stock. The scientists sounded a timely alert but poor judgment on the part of policy makers contributed to this disaster. That is not how policy makers view the matter. Some complain of the scientists' "penchant for speaking in terms of probabilities and confidence intervals" and propose that, in future, scientists make "more confident forecasts ... to catch the attention of regulators." As is often the case in environmental problems, we arrived at an impasse because of the reluctance of scientists to give definitive answers, and the unwillingness of policy makers to make difficult political decisions. Congressman George Brown, former chairman of the House Committee on Science, Space and Technology, wonders whether there is a conspiracy between these two groups, the scientists who are assured a continuation of funds to improve their predictions, and the politicians who avoid difficult decisions that can cost them their jobs.

The response of the Earth's climate to the perturbation we are introducing by injecting greenhouse gases into the atmosphere is such a complex problem that the scientific results concerning this matter have inevitable uncertainties. We nonetheless need to decide on appropriate actions to take in order to avoid disasters. Those opposed to any action claim that the uncertainties in the scientific results are too large to embark on policies that will ruin the economy. Implicit in such a statement is an appeal to a model of the economy that predicts the consequences of certain policies. Let us assume for a moment that the scientific results concerning large scale climate changes are free of uncertainties, and enquire about the uncertainties in the economic model. Climate models are based on the impersonal laws of physics and uncertainties in the results from those models are far, far smaller than uncertainties in the results from economic models which necessarily have to make assumptions about the behaviour of people. It is the difference between determining the distance to the waterfall, and deciding on the appropriate time to leave the water. Science, which addresses the first issue, can also guide us in our efforts to cope with the second. The spectacular successes of science over the past few centuries are attributable to a method of trial-and-error. Scientists continually subject any proposed solution to a problem to tests, and do not hesitate to modify (or even abandon) a solution should it prove inadequate. In our attempts to cope with our environmental problems, we should adopt a similar approach. Rather than implement comprehensive programs that decree a rigid course of action to reach grand, final solutions, we should promote adaptive programs whose evolution is determined by the results from those programs, and by new scientific results that become available. It will then be easier to take action when there is no scientific consensus, and it will be possible to correct mistakes at an early stage, before scarce resources have been wasted.

To decide on appropriate action as regards global warming, we need to familiarize ourselves with the nature of the uncertainties in the scientific results. In the debate about this matter, there is complete agreement amongst serious scientists that a continual increase in the atmospheric concentration of greenhouse gases will definitely result in global warming. (Our planet at present is comfortably warm and habitable because of global warming attributable to the greenhouse gases in its atmosphere. An increase in the atmospheric concentration of those gases will lead to too much of a good thing, too much warming.). The disagreements are about the amplitude and timing of the expected global warming and focus mainly on two issues: the temperature record of the past 150 years; and the accuracy of results from computer models of the Earth's climate.

2.1.1 Is global warming evident yet?

The atmospheric concentration of the greenhouse gas carbon dioxide has been rising rapidly since the industrial revolution some 150 years ago. Is there any evidence yet that global warming has started? The record of globally averaged temperatures at the Earth's surface is ambiguous. (Fig. 2.1) Between the turn of the century and the 1940s, temperatures increased but then the trend disappeared for a few decades. In the 1970's there was even a period of cooling that prompted some "experts" to warn that an ice age is imminent! Temperatures started to rise again in the early 1980's, at a pace that increased in the 1990's. Many scientists are now convinced that global warming is underway, but we can not be absolutely certain. To appreciate why, consider the annual global warming with which we are all familiar, the transition from winter to summer in response to the intensification of sunlight that starts on December 21, the winter solstice. In Princeton, New Jersey this past February was so warm that the forsythia started to bloom. The flowers were under the impression that spring had arrived! March, however, brought snow! Finally, in June, there was no doubt that summer had arrived. It clearly is a mistake to determine the transition from one season to the next by monitoring temperature changes on a daily basis.

The natural variability of our climate can mask the transition from one season to the next, and can similarly mask the onset of global warming. If we have complete information about that variability, we can estimate its effect. Unfortunately, our climate records are so short that they provide very limited information. We have no measurements, only historical records that tell us about the warm spell Europe enjoyed a thousand years ago when vineyards flourished in Britain, plagues of locusts descended on continental Europe, and Greenland could be colonized -there were enough souls for the pope to send an archbishop there. Then, in the 15th century, temperatures started to fall, forcing the abandonment of settlements in Greenland while, in Europe, cool, wet summers contributed to widespread famines. This cool period is exaggeratedly referred to as the Little Ice Age. Such climate fluctuations in the past -we do not know why they occurred but can rule out changes in the atmospheric concentration of carbon dioxide as a cause- suggest that we should not be surprised if the warm period that started in the 1980's soon comes to and end and is followed by a cold spell. In the same way that the flowering of forsythia in February is no guarantee that winter is over, so a few warm years in the 1990's may not be confirmation that global warming is underway. In due course there will be no doubt that we are experiencing global warming -by June there is no doubt that winter is over- but by then it will be too late to do much about it.

2.1.2 Models of the Earth's Climate

To make timely preparations for the changes of the seasons we consult a calendar. In the case of global warming, there is nothing comparable; we need to consult models that simulate the Earth's climate. By far the biggest contributor to uncertainties in the results from the models are clouds. Imagine trying to capture whimsical, capricious clouds by means of a few numbers in a computer model.

> *"HAMLET:* *Do you see yonder cloud that's almost in the shape of a camel?*
> *POLONIUS:* *By th' mass and 'tis, like a camel indeed.*
> *HAMLET:* *Methinks it is like a weasel.*
> *POLONIUS:* *It is backed like a weasel.*
> *HAMLET:* *Or like a whale.*
> *POLONIUS:* *Very like a whale"*
>
> *William Shakespeare*

Not only Hamlet but scientists too can use clouds to be mischievous. They can argue persuasively that clouds contribute to global cooling, and can then reverse themselves and argue equally persuasively that clouds cause global warming. This is possible because clouds are composed of a most remarkable chemical, water, that can change its appearance and properties dramatically in response to relatively small changes in temperature. A white cloud of transparent water droplets reflects sunlight, thus depriving the Earth of heat and contributing to global cooling. An increase in temperatures makes the cloud invisible when the visible droplets transform into the invisible gas water vapor. That gas is a powerful greenhouse gas and is a major contributor to global warming.

In the 1970's some experts argued that the rise in the atmospheric concentration of carbon dioxide could lead to global cooling. They granted that, because carbon dioxide is a greenhouse gas, an increase in its concentration will initially cause temperatures to rise, thus increasing evaporation from the oceans, and the amount of water vapor in the atmosphere. At this stage in the argument it is necessary to decide whether the water vapor simply accumulates in the atmosphere, or whether it condenses to form clouds. The experts of the 1970's chose the latter possibility and assumed that the clouds will produce snow on mountain peaks and in high latitudes. White snow, because it effectively deprives the Earth of heat by reflecting sunlight, causes temperatures to fall so that precipitation from the clouds continues in the form of snow rather than rain. As the snow-cover expands, it reflects even more sunlight, causing even lower temperatures, even more extensive snow-cover, and in due course we have an ice age.

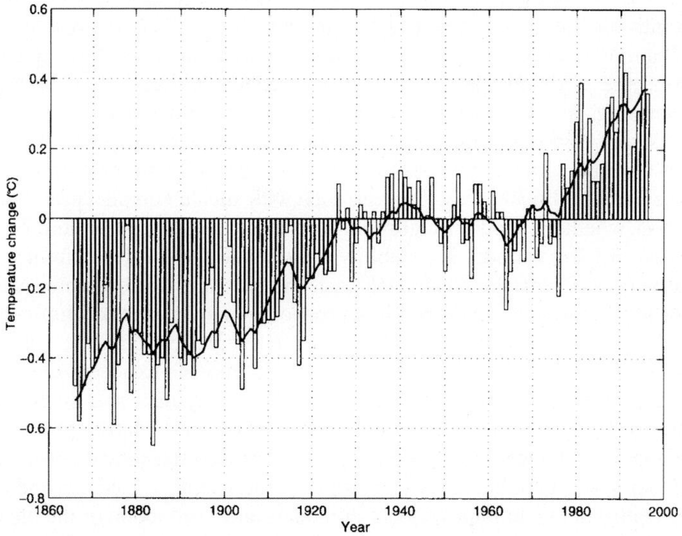

Fig. 2.1. Globally averaged land and sea surface temperatures, relative to the average for the period 1951 to 1980, over the past century.

To argue in favor of global warming rather than an ice age, assume that evaporation from the oceans results in an accumulation of water vapor in the atmosphere, without any clouds forming. The greenhouse effect of water vapor increases temperatures, thus causing even more evaporation from the oceans so that more water vapor enters the atmosphere, resulting in further increases in temperature, and so on; the atmosphere grows continually hotter until the oceans have evaporated entirely. (The planet Venus which, at the birth of the solar system, had a composition similar to that of the Earth, may have suffered such a runaway greenhouse effect; it has no water today)

These different arguments, for global warming and global cooling respectively, both invoke an escalating tit-for-tat (or positive feedback). For example, in one model, an increase in snow-cover lowers temperatures, hence causes a further increase in snow-cover that further decreases temperature, and so on. By favoring a different feedback, it is possible to argue the very opposite, that temperatures will rise steadily. It is possible to make a case for any eventuality by emphasizing some feedbacks more than others. The Earth's climate happens to depend on a great many feedbacks. The debate about global warming is primarily about the relative importance of different feedbacks, many of which involve water in one form or another. Realistic climate models (known as General Circulation Models or GCM) cope with numerous feedbacks simultaneously. Some critics identify specific feedbacks that are absent from GCM, show that those feedbacks, in isolation, can produce results different from those in the GCM, and conclude that the GCM are seriously flawed. This is sophistical reasoning because any feedback must be evaluated, not in isolation, but in the context of all the others and only the GCM are capable of doing that. Any newly identified feedback needs to be incorporated into the GCM in order to determine whether it alters the results significantly. If a model that predicts global warming is found to have a flaw, it is logical to question the prediction, but it is illogical to conclude that there is no threat of global warming. It is possible that, after the flaw in the model had been corrected, the results stay essentially the same. For an objective assessment of the relative importance of the various feedbacks we are obliged to rely on the computer models of the Earth's climate; only they can cope with such a complex problem. How do we know whether a model weighs the various feedbacks appropriately?

The best test for a model is its ability to simulate the Earth's current climate, and also very different climates in the past, the Ice Ages for example. Although the GCM are remarkably successful in reproducing many aspects of those climates realistically, and are improving rapidly, they have imperfections. The vast majority of scientists are sufficiently impressed with the models to accept their predictions as regards future climate changes. A small minority disagree. Such a lack of unanimity is the rule when it comes to results concerning scientific problems that are so complex that uncertainties are inevitable. The skeptics play an essential role because they force a continual reexamination of the scientific methods and results, and thus contribute to an improvement of models and a reduction of the uncertainties. Skeptics are vocal even when the issues being debated are of strictly academic interest, and concern another planet, Mars for example. That global warming has both academic and political aspects complicates matters enormously. The role of the skeptics changes dramatically because they become the focus of attention for reasons unrelated to the scientific merits of their arguments. The public gains the impression that there is little agreement in the scientific community even though the number of dissenters is very small.

2.1.3
The Geological Record

The geological record of past climates is of interest not only because it provides tests for climate models, but also because it provides information about the sensitivity of our planet to disturbances. (We are introducing a disturbance by altering the composition of the atmosphere). The dramatic and recurrent Ice Ages over the past two million years are of special interest for two reasons at least. First of all, as is evident in Fig. 2.2, the records show a remarkably high correlation between fluctuations in the Earth's temperature, and in the atmospheric levels of carbon dioxide and other greenhouse gases. Whenever temperatures are high, those levels are high; whenever it is cool, those levels are low. The records provide convincing empirical evidence that the current increase in the atmospheric concentration of greenhouse gases will lead to global warming. The second reason for paying close attention to the records is related to the cause of the Ice Ages.

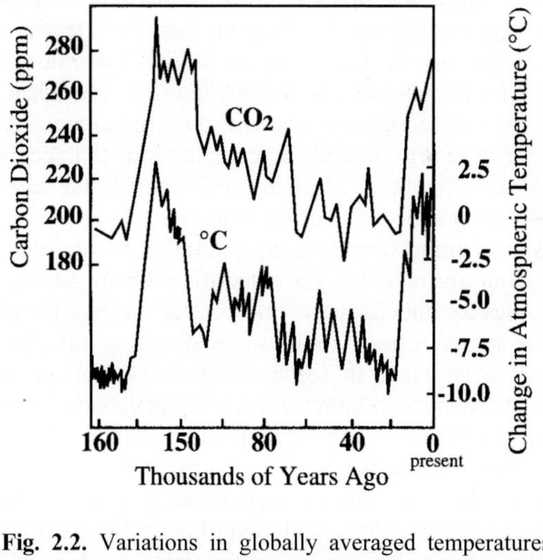

Fig. 2.2. Variations in globally averaged temperatures and in the atmospheric concentration of carbon dioxide over the past 160,000 years. The carbon dioxide data come from air bubbles trapped in ice cores extracted from glaciers over Antarctica. The temperature was estimated by measuring the oxygen isotopic ratio of the frozen water.

Our planet spins about its tilted axis each day, and orbits the sun in an ellipse each year. Over the millenia the tilt of the axis changes slightly -today the pole star is Ursa Polaris Minor but 11 000 years ago it was a different star, Vega- and there are also moderate changes in the eccentricity -at times the orbit is practically a circle. These very slight changes are periodic and are associated with moderate changes in the distribution and intensity of sunlight at the Earth's surface. The Earth's dramatic response to those modest variations in sunlight includes the Ice Ages. This is clear evidence that the Earth's climate is very sensitive to modest disturbances. The Earth's response to a small change in sunlight is not only a directly forced, small change in temperatures, but includes additional factors such as a variation in the atmospheric concentration of greenhouse gases. -this is evident from the geological record as mentioned above- which amplifies the change in temperature. Presumably a positive feedback comes into play and continues the amplification of the initially modest response. At this time scientists do not understand exactly how a modest change in sunlight translates into an enormous climate change, but they have clear evidence that it does happen. This is disturbing information because we are in the process of introducing a perturbation that amounts, not to a few percent, but to 100%. We are in the process of doubling the atmospheric concentration of carbon dioxide, at a rate that is accelerating. Our situation is similar to that of the gardener who finds that his pond has one lily pad on a certain day, two the next day, four the subsequent day and so on. If, after 100 days, the pond is completely filled with lily pads, on what day was the pond half full? The answer, day 99, illustrates how any problem involving explosive growth requires action at a very early stage, long before there are clear indications of impending trouble. In the case of the debate about global warming, in which some people insist that we are close to day one while others are adamant that we are close to day 100, the riddle indicates that, far more important than a precise answer that brings the debate to an end, is recognition of the special nature of the problem, its geometric growth. With such problems it is far wiser to act sooner rather than later. To defer action is to court disaster.

In the debate about global warming, uncertainties in the scientific results are invoked as justification for inaction. The scientific problem, although it is so complex that uncertainties are inevitable, is nonetheless far simpler than the political question of how we should respond to this impending calamity. Whether a particular policy will ruin or stimulate the economy is a matter of considerable uncertainty. Given the uncertainties in both the science and economics, it is wise, in our efforts to limit the growth in the atmospheric concentration of greenhouse gases, to avoid comprehensive programs that decree a rigid course of action to reach a grand, final solution. The other extreme, doing nothing at all, is also unacceptable. A sensible compromise is the implementation of adaptive programs whose evolution is determined by the results from those programs. Such a method of trial and error makes it easier to take action when there is no scientific consensus, and to correct mistakes at an early stage before scarce resources have been wasted. We are courting a disaster and need to accept that uncertainties do not justify inaction. Rather than wish for another Isaac Newton to provide precise answers, we need to accept that

"Between the idea
And the reality
Between the motion
And the act
Falls the Shadow"

T.S.Eliot

3 El Niño: A Predictable Climate Fluctuation

S. George Philander

Department of Geosciences, Princeton University. Guyot Hall, Washington Rd., Princeton, NJ 08544-1003, USA
gphlder@Princeton.edu

3.1 El Niño

El Niño is so versatile and ubiquitous -he causes torrential rains in Peru and Ecuador, droughts and fires in Indonesia, and abnormal weather globally- that the term is now part of everyone's vocabulary; it designates a mischievous gremlin. Hence, if the stock-market in New York is erratic, or the traffic jams in London are exceptionally bad, it must be El Niño. This is consistent with our practice of using meteorological phenomenon as a metaphors in our daily speech: the president is under a cloud, the examination was a breeze. We know exactly what these statements mean because we have a life-long familiarity with clouds and breezes. Despite all the publicity it receives, El Niño, Spanish for Child Jesus, remains a puzzle. Why is the rascal named after Child Jesus? How can scientists claim that they are able to predict it months in advance when they are unable to predict the weather more than a few days ahead?

3.1.1 Why is this disaster named after Child Jesus?

Originally El Niño was the apposite name given to the warm, southward, seasonal current that appears along the barren coasts of Peru and Ecuador around Christmas when it provides a respite from the very cold, northward current that otherwise prevails. Every few years the southward current is exceptionally warm and intense, penetrates far south, and bears gifts. A visitor to Peru described one such occasion, in the year 1891, as follows: "the sea is full of wonders, the land even more so. First of all the desert becomes a garden. The soil is soaked by heavy downpour, and within a few weeks the whole country is covered by abundant pasture. The natural increase of flocks is practically doubled and cotton can be grown in places where in other years vegetation seems impossible". The "wonders" in the sea can include long yellow and black water snakes, alligators, bananas, and coconuts (Philander 1998).

With time, we reserved the use of the term El Niño, not for the annual, coastal current, but for the more spectacular, interannual occurrences that affect much of the globe. Not only our terminology, but also our perceptions changed; we now have a pejorative view of El Niño, not because its character has changed, but because we have changed. Heavy rains still transform the desert into a garden, but they also wash away homes, bridges and roads, products of economic development and of a huge increase in population. As our economy and numbers continue to grow, the damage inflicted by natural phenomena such as El Niño, hurricanes and severe storms are likely to continue increasing, even in the absence of global climate changes.

3.1.2 How is El Niño related to the Southern Oscillation?

Towards the end of the 19th century scientists discovered that when atmospheric pressure is high over the western tropical Pacific, it is low over the eastern tropical Pacific and *vice versa* (Walker and Bliss 1932). This see-saw in pressure, known as the Southern Oscillation, has a period of approximately four years, and is associated with fluctuations in a number of other variables.

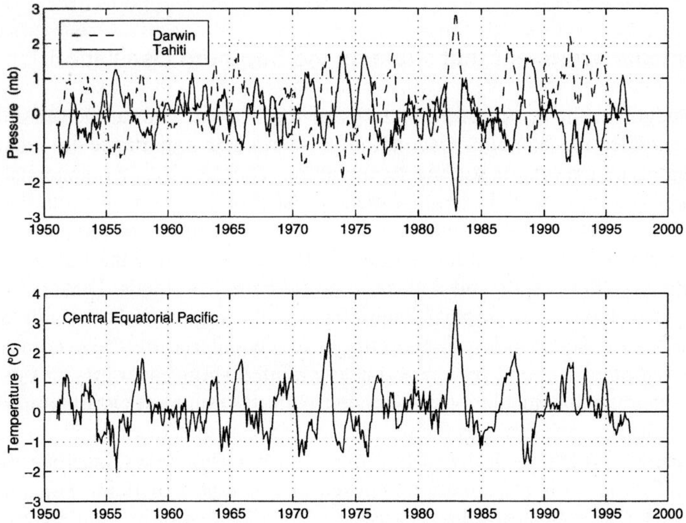

Fig. 3.1. The Southern Oscillation, the interannual fluctuations in pressure that are out of phase at Darwin (Australia) and Tahiti. These variations are highly coherent with those in sea surface temperature in the central equatorial Pacific. The lower panel shows the departure of sea surface temperature from 22 °C.

For example, during one phase of the Southern Oscillation the trade winds are intense, and rainfall is heavy over the western Pacific but light over the eastern Pacific. During the complementary phase, the trades are weak while rainfall is light over the western and heavy over the eastern tropical Pacific.The reason for this interannual climate fluctuation became apparent in 1957 when, for the first time during the occurrence of El Niño, sea surface temperatures across the entire Pacific were available. Those measurements revealed that the warming along the shores of Peru associated with El Niño is not confined to a narrow coastal zone but extends thousands of kilometers offshore and influences an area so enormous that the global atmospheric circulation is affected. This led Bjerknes (1969) to realize that El Niño is one phase of the Southern Oscillation as is evident in Fig. 3.1. El Niño is a period of high sea surface temperatures and heavy rainfall in the eastern Pacific, of droughts in the "maritime" continent of southeastern Asia and northern Australia. During El Niño of 1997-98 the droughts in the west contributed to devastating forest fires, while in the east tropical storms spawned off Mexico, and heavy downpours drenched Chile and Peru. Usually El Niño is followed by its complement, known as La Niña, a period of low sea surface temperatures and dry conditions in the eastern tropical Pacific, and plentiful rainfall in the western tropical Pacific. Fig. 3.2 depicts El Niño and La Niña which, together, constitute the Southern Oscillation.

3.1.3 The circular argument that explains the Southern Oscillation

The changes in the temperature patterns of the tropical Pacific Ocean shown in Fig. 3.2 induce the Southern Oscillation. Of particular importance are changes in the regions of maximum surface temperature. The air rises spontaneously over such regions, creating tall cumulus towers that provide plentiful rainfall locally. To sustain the rising motion, winds in the lower atmosphere converge onto the regions of maximum temperature. Those winds, the trades over the Pacific, harvest moisture over the oceans and deposit it in the cumulus clouds. During La Niña, the warm waters of the tropical Pacific (under the region of heavy rainfall) are confined to the west. During El Niño the eastern equatorial Pacific becomes warm and the region of cumulus towers and heavy rainfall shifts eastward. This explanation for the climate fluctuations associated with the Southern Oscillation leads to another question: what causes the sea surface temperatures to change?

Temperature patterns at the surface of the earth reflect subsurface oceanic conditions. The ocean is effectively composed of two layers of fluid: a shallow surface layer of warm water, some 100 meters deep, above a deep, cold layer that extends to depths in excess of 4 km. The thermocline, a thin region of large vertical temperature gradients, separates the two layers. In the absence of any winds, the thermocline, which is the interface between the warm and cold water, is horizontal. Under such conditions, the warm surface waters are spread uniformly over the cold layer. There is a tendency towards this state during El Niño when the trades are weak. The intensification of those winds, during La Niña, drives the warm surface waters westward, causing the thermocline to tilt downwards to the west so that the cold water becomes exposed to the surface in the east (see Fig. 3.2.) Hence, from an oceanographic perspective, the Southern Oscillation corresponds to a sloshing, back and forth across the Pacific Ocean, of the warm surface waters in response to trade winds that fluctuate from weak (during El Niño) to strong (during La Niña).

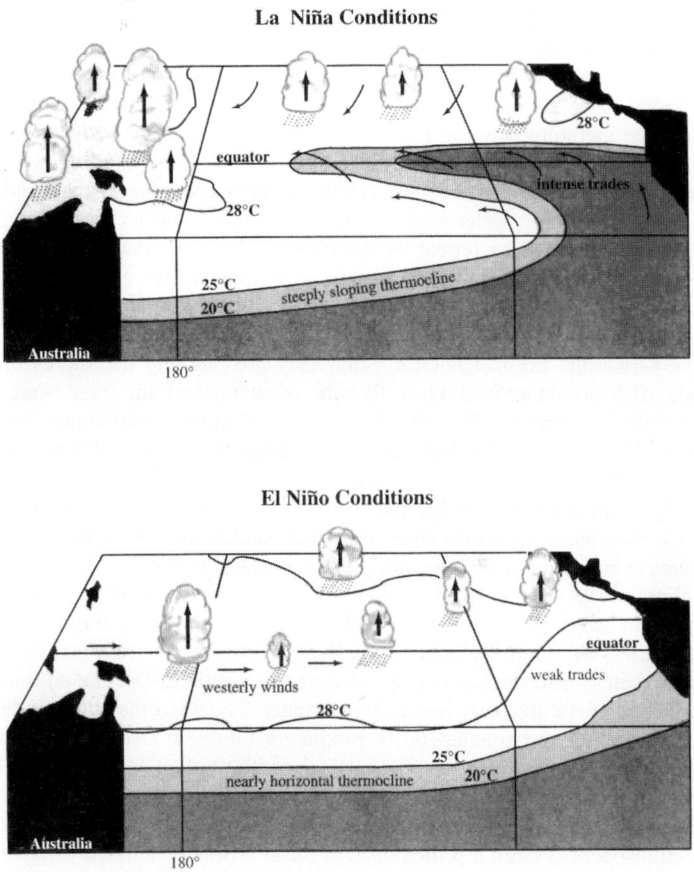

Fig. 3.2. A schematic view of La Niña (top) and El Niño (bottom). During La Niña, intense trade winds cause the thermocline to have a pronounced slope from east to west so that the equatorial Pacific is cold in the east, warm in the west where moist air rises in cumulus towers. The air subsides in the east, a region with little rainfall except in the doldrums where the southeast and northeast trades converge. During El Niño the trades along the equator relax, as does the slope of the thermocline when the warm surface waters flow eastward.

From an atmospheric perspective, changes in sea surface temperatures cause the Southern Oscillation, including changes in the winds over the Pacific. From an oceanic perspective, those changes in the winds cause the sea surface temperature changes. In response to this circular argument, it is tempting to ask which changed first, the winds or the temperatures? This is a futile approach, equivalent to asking whether the chicken or the egg came first. It is far more rewarding to explore the behaviour of the coupled ocean-atmosphere system Consider a random disturbance which, at the height of La Niña, causes a slight relaxation of the trades. Those winds drive the warm surface waters westward and cause the region of high temperatures, rising air, and heavy rainfall to be confined to the western Pacific. A weakening of those winds therefore causes some of the warm water to start flowing back eastward, thus expanding the region of high surface. The difference in temperature between the eastern and western Pacific, which drives the winds, is now smaller, so that the winds become weaker. In other words, the change in ocean temperatures reinforces the initial weakening of the wind. As a result, even more warm water flows eastward, causing a further weakening of the winds. An initial, cautious retreat by the trades induces a tentative eastward step by the warm surface waters, which hastens the retreat, which emboldens the pursuit. The interactions between the ocean and atmosphere amount to an escalating tit-for-tat (a positive feedback) that causes the warm surface waters and humid air to surge across the tropical Pacific. Soon they are hugging the shores of Latin America. El Niño has arrived. Once El Niño is established, the stage is set for La Niña to make its entrance. This new phase of the Southern Oscillation is an inversion, a mirror image, of the first part of this tango for ocean and atmosphere. A slight intensification of the winds drive some of the warm water westward thus increasing the temperature difference between the eastern and western Pacific. That increase makes the winds stronger, which causes the temperature difference to become even bigger, and so on until La Niña is established.

El Niño and la Niña are beautiful examples of how the whole can be greater than the sum of the parts. The ocean, by itself, can not produce the Southern Oscillation unless the winds over the ocean fluctuate in a certain manner. The atmosphere by itself is also incapable of producing the Southern Oscillation unless the sea surface temperatures vary in a certain manner. Together, the ocean and atmosphere can interact and spontaneously produce a Southern Oscillation with wind and temperature fluctuations that are perfectly orchestrated. To ask why El Niño or La Niña occurs is equivalent to asking why a bell rings or a taut violin string vibrates. The Southern Oscillation is a natural mode of oscillation of the coupled ocean-atmosphere system; it is the music of the atmosphere and hydrosphere.

The atmosphere and ocean are partners in a dance. But who leads? Which one initiates the eastward surge of warm water that ends La Niña and starts El Niño? Though intimately coupled, the ocean and atmosphere do not form a perfectly symmetrical pair. Whereas the atmosphere is quick and agile and responds nimbly to hints from the ocean, the ocean is ponderous and cumbersome and takes a long time to adjust to a change in the winds. The atmosphere responds to altered sea surface temperature patterns within a matter of days or weeks; the ocean has far more inertia and takes months to reach a new equilibrium. The state of the ocean at any time is not simply determined by the winds at that time because the ocean is still adjusting to and has a memory of earlier winds, a memory in the form of waves below the ocean surface. They propagate along the thermocline, the interface that separates warm surface waters from the cold water at depth, elevating it in some places, deepening it in others. These vertical displacements of the thermocline effect the transition from one phase of the Southern Oscillation to the next so that it is of critical importance to monitor oceanic conditions in order to anticipate future developments. That is why oceanographers developed and now maintain the measurement array shown in Fig. 3.3. On the basis of those oceanic measurements, which are available on the world wide web a few days after the measurements had been made, scientists were able to anticipate the arrival of El Niño of 1997-98 months in advance, and could alert the public of the impending changes in weather and climate.

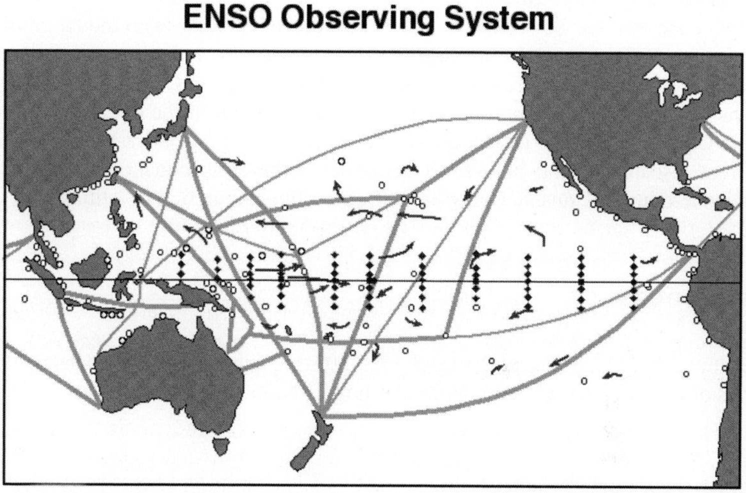

Fig. 3.3. The array of instruments that monitor oceanic conditions. All these measurements are relayed to stations on land, in "real time". (The web site www.pmel.noaa.gov/toga-tao/ftp.html displays some of the measurements.) The grey lines are the tracks of commercial ships that deploy instruments that measure temperature to a depth of a few hundred meters. The arrows are drifting buoys that measure temperature and the winds, and whose movements yield information about surface currents. The dots are tide gauges that measure sea level which depends on the average temperature of a water column. The diamonds, buoys moored to the ocean floor, are locations where temperature is measured over the upper few hundred meters of the ocean. The squares indicate where oceanic currents are measured.

3.1.4
The Predictability of El Niño

A bell, to ring, must be struck whereafter its oscillations last for a while before dying out. It appears that there are periods during which the Southern Oscillation "rings" by itself, and periods where a disturbance (blow) is needed to start it "ringing" again. During the 1980's it seemed to be ringing -- El Niño of 1982 and of 1987 were both followed by La Niña episodes but El Niño of 1992 persisted for several years, as if a new disturbances was needed for the ocean-atmosphere system to start ringing again. Disturbances that very effectively can excite El Niño, because their surface winds have a spatial structure that coincides with those of the Southern Oscillation, are with two-week bursts of westerly winds that sporadically occur over the far western equatorial Pacific. Such wind bursts were influential in initiating El Niño of 1997-98. Will the ringing now continue, with another El Niño making its appearance in the year 2002?[1]

Evidence that the Southern Oscillation is subject to long-term modulations, so that it is prominent and energetic during some decades, less so during others, is available from coral records that cover a century or more. One of the factors responsible for this modulation is the time-averaged depth of the equatorial thermocline, which depends on exchanges between the tropical and extra-tropical oceans. Current research on the Southern Oscillation is therefore concerned with ocean-atmosphere interactions, not only in the tropics, but also in higher latitudes.

References

The Journal of Geophysical Research, vol 103, C7, June 29, 1998, is devoted to a series of excellent review articles concerning El Niño and related topics.

Bjerknes J. (1969) Atmospheric teleconnections from the equatorial Pacific Mon. Weather Rev. 97, 163-172.

Philander S. G. H. (1998) *Is the Temperature Rising? The Uncertain Science of Global Warming*. 258 pp Princeton University Press.

Walker G.T. and E. W.Bliss (1932) World Weather V *Mem. Royal Meterol. Soc.* 4, 53-84.

[1] This chapter was written in early 2000 (the editors)

4 Climate Information Systems and Their Applications

Chet F. Ropelewski[1] and Bradfield Lyon[2]

International Research Institute for Climate Prediction, Columbia University. 61 Rt. 9W, Monell Building, P.O. Box 1000, Palisades, NY 10964-8000, USA
[1]chet@iri.columbia.edu
[2]blyon@iri.columbia.edu

Abstract. Over the past two decades considerable advances have been made in understanding climate variability on the seasonal time scale. This scientific understanding has led to the development of techniques that demonstrate some modest skill in seasonal climate prediction. Having developed this skill the crucial question is how can this scientific achievement be utilized in the real world of practical decision-making. This brief chapter outlines one attempt to provide a framework for such a comprehensive Climate Information System. The elements of this System are: 1) reliable and complete climate observations in real time, 2) complete and error-free climate data archives to provide a basis for placing the observations into historical context, 3) complete real time analyses of current climate data, 4) access to current climate forecasts from a variety of sources, 5) easy-to-understand forecast products, 6) complete records and analyses of past forecast performance, 7) routine methods to disseminate climate information to user groups and sectors and 8) active collaboration and feedback from the user community. It is hoped that this outline will provide the basis for further discussion and the development of such Systems to promote the use of climate information.

4.1 Introduction

Advances in our scientific understanding of seasonal climate variability and parallel advances in the ability to predict important aspects of the climate system have served to underscore a long-recognized need for climate information (Goddard et al. 2001). Climate information includes all aspects of climate, from observations and forecasts to application of this information for relevant segments of society. This brief Chapter outlines the essential components of an envisioned Climate Information System that would lead to the effective use of our current understanding and predictive capabilities in seasonal climate. In outline, an effective Climate Information System must include:

- Reliable and complete climate observations in real time,
- Complete and error-free climate data archives to provide a basis for placing the observations into historical context,
- Complete real time analyses of current climate data,
- Access to current climate forecasts from a variety of sources,
- Easy-to-understand forecast products,
- Complete records and analyses of past forecast performance,
- Routine methods to disseminate climate information to user groups and sectors and
- Active collaboration and feedback from the user community.

The components of this system and their interactions are shown schematically in Fig. 4.1. Each of these essential features is discussed in detail below. Implicit in these requirements is that there is an adequate number of scientific and professional staff that have the training and capacity to support the various components of the Climate Information System. This requirement will be discussed further at the end of this Chapter.

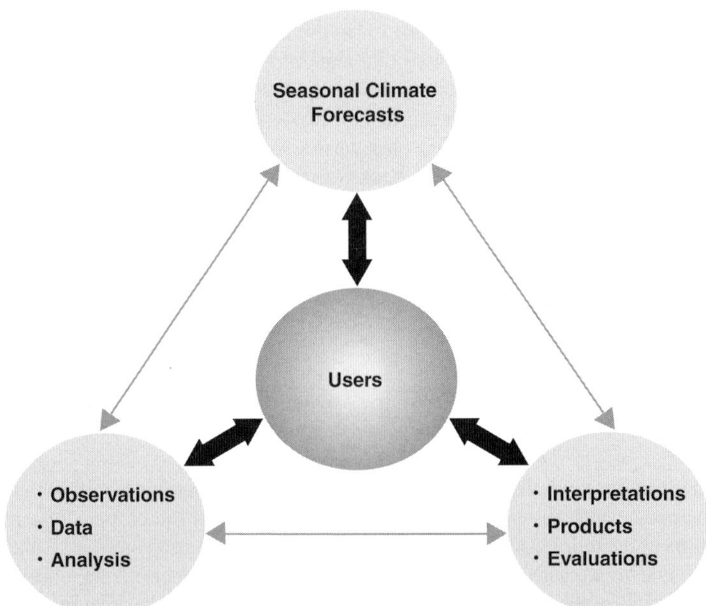

Fig. 4.1. Schematic of Climate Information System components

4.2 Monitoring Climate

Many years ago a famed hurricane forecaster and Director of the National Hurricane Center in the United States was asked what accounted for the significant increase in hurricane forecast skill. He replied that, in fact, hurricane forecast skill had only improved slightly over the past several years. "What had improved", he said, was the early identification and observations of tropical storms and hurricanes. Our current situation in climate prediction is very much analogous with respect to ENSO forecasting in particular, and seasonal climate forecasting, in general. That is, great improvements in our understanding of the climate system have occurred because of increased attention to observations of the climate system and the enhanced ability to interpret those observations. For instance, much of the potential to prepare for ENSO-related climate impacts comes from an ability to observe and recognize the early stages of ENSO. Much of our understanding of ENSO itself was made possible by observations from a system of moored-buoys, called the Tropical Atmosphere Ocean (TAO) array, across the equatorial Pacific, originated in the Tropical Ocean Global Atmosphere (TOGA) Program in the mid 1980's (Hayes et al. 1991). Since then the TOGA-TAO Array has become a multinational program and a mainstay of ENSO observations. Researchers are building a complementary observational array, called PIRATA, in the Tropical Atlantic with hopes that it will lead to comparable understanding of climate variability especially in Africa and South America (Servain et al. 1998).

Further improvements in the application of climate information for greater benefit to society could be envisioned with further improvements in climate observations. Very often such improvements require only that existing weather observations be reliably continued, that any changes in observational instruments or procedures be quantified and documented, and that the data be readily available. However, if there is to be an increase in the efficient application of climate information it will require improvements in the ground-based observational networks themselves. That is, many applications, and potential applications, of climate information require increases in the number of observations in both space and time. Users and providers of climate information will have to work together to evaluate the need and provide justification for increased observations. An effective and complete Climate Information System will require close collaboration and interaction among agencies and organizations that are responsible for making the observations and the users of the climate information such as farming groups, public health officials, civil defense and disaster preparedness officials and others for whom climate variability has a direct or indirect impact. Indeed, many of these user groups already maintain their own observational networks. An effective Climate Information System would develop the methods to foster exchange of data from many diverse sources and develop efficient methods to integrate observations from these sources.

The same considerations apply to the growing number of satellite-based observations. For example, there is a need for continuity in satellite observations of sea surface temperature as well as satellite estimates of rainfall, vegetation, and land surface state (i.e., soil moisture or soil wetness) (Reynolds and Smith 1995, Xie and Arkin 1995). Satellite observations of weather as well as estimates of the land and ocean surface are generally available from a number of national sources, e.g., METEOSAT (European), GOES and POES (NOAA, USA), GMS (Japan) and INSAT (India). An effective Climate Information System requires reliable and complete access to the climate data and analyses from these satellites, at least on a regional basis, at spatial resolutions high enough to be useful to satisfy the requirements from local user groups. In addition, the Climate Information System should include mechanisms for the receipt, analysis and archive of relevant satellite data in the context of user requirements.

4.3 Climate Data

There are at least two compelling reasons to justify efforts to maintain and improve the historical climate records. The historical data are necessary 1) to place current observations of climate into historical context and 2) to document the "general" conditions associated with various climate states, for example conditions typical of El Niño or La Niña. The most commonly available meteorological data records, or archives, contain monthly mean values of precipitation, temperature and other climate variables (Vose et al. 1992). For many years considerable national and international efforts have focused on obtaining, and archiving this type of data. For example, the WMO-sponsored DARE project has "rescued" data from several locations around the world including the Caribbean, Latin America and Africa. In many instances, however, these data have been transferred to more permanent storage media e.g., microfilm versus high acid paper, and are not available in digitized form making them unavailable for the types of historical analyses required for an effective Climate Information System. These and other sources of historical data e.g., from National Meteorological Services and Agencies, need to be made available in computer-compatible form if they are to live up to their potential usefulness. This is especially urgent since the long-term stability of microfilm has been recently called into question (Baker 2001).

Another practical obstacle to the efficient use of climate data is that many of the organizations that gather meteorological and other data, including national agencies, do not allow the free and open exchange of data. Many of these agencies attempt to partially recover the costs of making and maintaining the observations through data access fees. Very often such data access carries restrictions to its further use and dissemination. An effective Climate Information System will need to work with such agencies to develop mechanisms for the access and analysis of climate-relevant data. This requires development of policies that are in harmony with cost-based requirements while providing for climate analyses useful to the society at large that supports these observations in the first place. To this end some preliminary efforts have been initiated and need to be encouraged and expanded (VMO 1996).

Historical monthly-mean data are the backbone of traditional climate analysis but they are not sufficient, in themselves, to support all users needs. Historical records of daily weather data need to be developed and maintained since many user communities require more than monthly-mean values of weather variables. Modern data archives must be able to support analyses of the statistics of daily weather on both monthly and seasonal time scales. The statistics of weather includes things like "the number of days with rainfall" and "the starting date of the rainy season", and could include any number of additional user-driven summaries. While daily weather observations are performed by virtually all National Meteorological Services and often by other national agencies, long historical records of daily data are not generally available or easily accessible. Considerable efforts are thus needed to both develop and maintain these archives. As with the monthly mean data, appropriate mechanisms must be developed between various nations and agencies to allow use and analysis of these data in a Climate Information System. Because climate variations typically span large regions and national boundaries the potential benefits of an open data exchange to a nation or a region far outweigh the incremental cost recovery that agencies receive by selling their data. An effective Climate Information System will need to develop cooperative agreements with appropriate national and regional agencies to facilitate the equitable and cost effective uses of those data.

To be useful for climate analyses historical data sets should extend back in time as far as possible. The World Meteorological Organization suggests a minimum of 30 years for historical monthly data. The requirements for daily data should meet, or exceed, this guideline where possible. Climate research suggests that, beyond seasonal to interannual timescales, there is significant climate variability for periods of several years to decades or longer. At some locations there is a tendency for the climate system to tend towards a particular climate state or mode, "dry" or "wet" for instance, for several years in a row. To understand and interpret this kind of climate variability will require several decades of data. As the use of climate information grows so will the requirements for increased understanding of the climate system at these longer time scales. Such understanding will require analysis of long data sets, e.g. several decades. An effective Climate Information System will necessitate the development of such data sets. Some international efforts along these lines have been initiated under the Global Climate Observing System (GCOS) program (Peterson et al. 1997).

4.4 Climate Analysis

Gathering real-time global climate observations and maintenance of global climate data archives sets the stage for the diagnosis and further analysis of these data. In the context of a Climate Information System some of the most basic information relates to the current state of ENSO, e.g., "Is an El Niño starting? Or, "Has the La Niña ended?" The monitoring of the global climate including the status of ENSO is performed at several national and international centers, e.g., the Bureau of Meteorology (Australia), CPTEC (Brazil), the Climate Prediction Center/NOAA (USA), the Japanese Meteorological Agency as well as the IRI.

It is essential that regional and user-specific climate monitoring activities have access to the latest information from these sources and have the capabilities to interpret these observations and analyses in the national and regional context. For instance, given the latest information about the current state of ENSO, a Climate Information System must be capable of:

- Evaluating and interpreting statements about the current state of the equatorial Pacific
- Documenting the current state of the climate,
- Comparing the current climate to the past data record and noting differences and similarities and
- Evaluating the current and potential impacts of ENSO on the various user communities, agencies and groups.

The Climate Information System should include analyses of precipitation, temperature and other user-relevant quantities along with similar historical analyses (for example, in the case of ENSO, analyses of its mean evolution and past impacts). While these analyses are valuable in themselves they are not sufficient for many applications. Methods must also be developed to evaluate probable climate impacts in the context of recent climate anomalies. For instance, if the climate record shows increased regional rainfall associated with El Niño the appropriate user response to a current ENSO episode may differ if the previous season was very dry or very wet. A useful Climate Information System should be capable of placing current climate information, including forecasts, into the context of recent and current climate anomalies. The System will often require climate information in greater detail than provided by global observation networks as well as close and continuing coordination with user groups to identify evolving needs for specific, tailored, climate information.

4.5 Seasonal Climate Forecasts

Seasonal climate forecasts are primarily based on the influence of ocean sea surface temperature (SST) on the atmospheric circulation. The best-documented and most important oceanic influence on the atmosphere is ENSO. Since the evolution of a typical ENSO episode takes place over several months, the single most important type of seasonal climate information relates to understanding the current state of the equatorial Pacific Ocean. The next most important information relates to understanding and interpreting the future state of this ocean basin through the latest ENSO and SST forecasts and outlooks. Thus a Climate Information System must have reliable access to the latest SST forecast information from a variety of sources. The System should include procedures that not only access the latest SST forecasts themselves but also provide interpretation of these forecasts while making both available to appropriate users. In general, SST forecasts generated by the various research and operational groups are very technical in nature and thus require interpretation and evaluation to be useful to the user community. Currently, sources of seasonal SST forecasts include: coupled model SST forecasts for the equatorial Pacific from NCEP/NOAA, global coupled model SST forecasts from ECMWF and global hybrid SST forecasts from the IRI. In addition there are readily available monthly summaries of equatorial Pacific SST forecasts complied by the Australian Bureau of Meteorology and the Forecast Bulletin edited and distributed by quarterly COLA.

In addition to SST forecasts, seasonal forecasts of global rainfall and temperature patterns associated with these predictions of SST are available, for example, from the ECMWF, the IRI and Forecasts from appropriate Seasonal Climate Outlook Forums (the Climate Outlook Forums are a series of regional meetings that bring climate experts and users together to discuss the current seasonal climate forecast for a region.) It is essential that a Climate Information System access several sources of seasonal climate forecasts since experience has suggested that each seasonal prediction system performs poorly at different times as initial conditions vary (Graham et al. 2000).

4.6 Evaluation of Seasonal Climate Forecasts

Many computer models and statistical techniques of varying complexity have shown skill in predicting the evolution of the equatorial Pacific sea surface temperature associated with ENSO for both El Niño and La Niña (Goddard et al. 2001). However, virtually all of these studies also suggest that ENSO prediction techniques have some difficulties in predicting the details of the onset, magnitude, and demise of ENSO episodes. Thus a useful Climate Information System should include mechanisms for monitoring and evaluating the skill of sea surface temperature forecasts produced by a variety of methods. This may be as simple as having the System access the evaluations performed by various research and academic institutions. Information on forecast model skill is essential to safeguard against occurrences of "false alarms", i.e., forecasts of El Niño or La Niña based on models that have poor historical performance under certain initial conditions.

In addition to evaluation of the SST forecasts, the Climate Information System should also be able to evaluate associated precipitation and temperature forecasts from various computer models and statistical techniques. Comparisons of several seasonal climate forecast models have shown considerable spatial and temporal variation in the skill of such models. For instance, some models have difficulty in predicting the ENSO related rainfall in northern Australia while other models perform quite well. On the other hand, all models seem to have relatively less skill for predictions made for the Northern Hemisphere in the springtime (Latif et al. 1998). No current model outperforms every other model for all areas and all seasons indicating that all models have low predictive skill for particular seasons and areas. Again, documentation of model performance is essential for forecast interpretation and the successful operation of a Climate Information System.

Another important consideration is that of spatial scale. Forecast evaluations for precipitation and temperature are typically made on spatial scales of the computer models. These spatial scales may or may not be relevant to various users. Much of the effort in building an effective Climate Information System will need to be focused on the development of forecast products and forecast evaluations on spatial (and temporal) scales appropriate to user needs.

4.7 Climate Product Development

The production of global seasonal climate predictions requires significant resources, including a highly trained professional staff as well as state of the art computers and communications systems. Thus, it is likely that routine production of global seasonal climate forecasts will be performed routinely at only at a few Centers. However, the development of useful products based on these forecasts needs to involve a much wider community. We emphasize that the production of seasonal forecasts should not be considered the end result of a climate forecast system but only one of the early links in a long chain of tailored climate information and forecast products that should feed into a Climate Information System.

If society at large is to benefit from climate science's modest abilities in seasonal climate prediction then methods, mechanisms and management structures must be developed to build and maintain a strong, sustained and dynamic interaction between the end users of climate information and those involved in the production of such information. This puts a burden on the climate community to engage and encourage collaboration with users. Very often the practical use of climate information can only be achieved at the "regional", "local", or "sector" level. The practical application of climate information must be accomplished through collaborative efforts by those with local expertise, not only in climate, but also in each of the sectors that can profit from the expanded use of climate information. For example, the development of specialized climate products in agriculture requires close collaboration among climate scientists (who need to be made aware of the needs of the farming community) and those involved in managing, guiding and advising farmers as well as the farmers themselves (who need to be made aware of the potential advantages and pitfalls in climate prediction). Climate scientists acting in isolation from the user community can emphasize a particular climate feature or climate forecast (e.g., probability of the minimum temperature falling below (or above) some value) but to be useful the determination of what constitutes a critical climate feature must come from collaborative efforts with users. To reiterate, the determination of critical climate features and the applied research to see whether these features are predictable can only be done in collaboration with the end users of climate information.

Thus a Climate Information System will be most effective if intrinsic to the System are mechanisms to insure close collaboration, feedbacks and interaction with the actual and potential users of climate information. A major part of such a System should be identification of institutional, social, educational and infrastructure impediments to the use of available climate information and the development of ways to remove these impediments. This includes the requirement to identify and train potential users as to what constitutes a scientifically defensible climate forecast i.e., what can and cannot be forecast, as well as the careful determination of the potential utility of the forecasts themselves.

4.8 Dissemination of Climate Information

To be successful a Climate Information System must insure that the tailored climate information discussed above gets into the hands of appropriate users in a timely fashion in order to have some influence on practical decisions. It will do society no good if the results of climate applications research remain in the academic or research arena. Considerable efforts are required to develop the appropriate means for the effective dissemination of the climate information. The producers of climate information generally do not have easy and direct access to the end users of that information and in practice the mechanisms for dissemination need to be developed through a range of organizations and structures. These could include National Meteorological Services, who generally have a well-developed set of users, to representatives of the ministries, or government organizations, who advise user communities e.g., public health, agriculture, water resources, risk/hazard management, and appropriate Non-Government Organizations (NGOs).

It is likely that the mix of users of climate information will vary by country and application sector requiring an ideal System to take advantage of regional organizations, social structures and methods for disseminating information. In the end, a Climate Information System must provide managers with enough information to influence informed decisions.

The user communities are essential to facilitating the means to disseminate climate information. As such, the dissemination of climate information may have some very sophisticated components, such as Internet and/or direct satellite links but it is also likely to have some very effective components that are less dependent on advanced technology. Among these are radio, newsletters, local civic meetings, meeting reports, and educational materials. An effective Climate Information System will have mechanisms to develop, maintain and improve these types of user-links and dissemination methods. Once again a large part of the work in building a Climate Information System is the close collaboration with the various user sectors. This further requires clear formal (and informal) agreements be in place among the various links in a dissemination system. For example, suppose a National Weather Service, perhaps in collaboration with international or regional climate centers, has determined that there is a high probability of ENSO occurrence in the coming season(s). The mechanisms to disseminate this information to the user communities, in formats compatible with user requirements, should have been agreed upon in advance. The mechanisms for moving the information from the various government agencies, NGOs, and representatives of the user communities to the end users should also have been addressed.

The building of these mechanisms and agreements is an integral and vital part of any Climate Information System. In addition, the System design should contain provisions for ensuring that these mechanisms and agreements are continually developed and evaluated.

4.9 Feedback

For a Climate Information System to be effective requires regular, organized interactions among all of the participants in the System, from the providers of the climate information through to the users of this information. The development of criteria to measure "success" and "effectiveness" of the System is a major challenge. It is essential that criteria for evaluating the System be developed and agreed upon in advance. It is virtually certain that for any given forecast for a climate phenomenon, e.g., ENSO, there will a spectrum of impacts, some good, some bad, and possibly, some disastrous. Even an ideal Climate Information System will not be capable of predicting, much less eliminating, all of the negative effects of an ENSO but the System must at least aid in mitigating these effects and should support planners in the identification of sectors and regions of climate vulnerability.

The evaluation of the Climate Information System must also deal directly with the inherent probabilistic nature of climate forecasts. For instance, a seasonal probability forecast of 50% wet, 35% average, and 15% dry conditions is not "wrong" if a particular season turns out dry but would be "wrong" if, in a large number of such forecasts the observed probabilities were substantially different from those forecast, say 35% wet, 35% average, and 30% dry. Such evaluations require a large data set, i.e. several dozen seasonal forecasts over several years. In any case, it is important that the evaluation of the Climate Information System be considered separately from the evaluation of the seasonal forecasts themselves.

4.10 Training

To a large extent the elements of an ideal Climate Information System depend on the existence of well-trained and informed personnel who can advance and strengthen each link in the System. The required training and skills are diverse. They include a sufficient number of formally educated climate scientists and interpreters of climate information as well as various managers and users of climate information who may have attended and participated in any number of training or certificate courses in climate or courses related to "climate affairs". Educational materials may consist of booklets, pamphlets and information sheets as well as radio or television presentations by climate experts and users of climate information.

Acknowledgements

This paper was based on work partially supported by the Inter-American Development Bank Project IDB ATN/JF-65-69-RG. The IRI was established as a cooperative agreement between the U.S. NOAA Office of Global Programs and Columbia University. Taiwan joined the IRI in 2000.

References

Baker, N. (2001) Double Fold: Libraries and the Assault on Paper. Random House, Pub., First Edition 2001, 288 pp.

Goddard, L., S.J. Mason, S.E. Zebiak, C.F. Ropelewski, R. Basher, and M.A. Cane (2001). Current Approaches to Seasonal-to-Interannual Climate Predictions. *International Journal of Climatology:* 21, pp. 1111-1152.

Graham, N.E., A.D.L. Evans, K.R. Mylne, M.S.J. Harrison, and K.B. Robertson (2000) An Assessment of seasonal predictability using atmospheric general circulation models. *Quarterly Journal of the Royal Meteorological Society:* 126, pp. 2211-2240.

Hayes, S. P., L.J. Mangum, J. Picaut, A. Sumi, and K. Takeuchi (1991) TOGA-TAO: A Moored Array for Real-time Measurements in the Tropical Pacific Ocean. *Bulletin of the American Meteorological Society:* 72, No. 3, pp. 339-347.

Latif M., Anderson D. L. T., Barnett T. P., Cane M. A., Kleeman R., Leetmaa A., O'Brien J., Rosati A., Schneider E. (1998) A review of the predictability and prediction of ENSO. *Journal of Geophysical Research* 103: pp.14375-14393.

Peterson, T., H. Daan, and P. Jones (1997). Initial Selection of a GCOS Surface Network. *Bulletin of the American Meteorological Society:* 78, No. 10, pp. 2145-2152.

Reynolds, R. W. and T. M. Smith (1995) A High-Resolution Global Sea Surface Temperature Climatology. *Journal of Climate:* 8, No. 6, pp. 1571-1583.

Servain, J., A. J. Busalacchi, M. J. McPhaden, A. D. Moura, G. Reverdin, M. Vianna, and S. E. Zebiak (1998) A Pilot Research Moored Array in the Tropical Atlantic (PIRATA). *Bulletin of the American Meteorological Society:* 79, No. 10, pp. 2019-2032.

Vose, R. S., R. L. Schmoyer, P. M. Steurer, T. C. Peterson, R. Heim, T. R. Karl, and J. Eisheid (1992) The Global Historical Climatology Network: Long-term monthly temperature, precipitation, sea level pressure, and station pressure data. ORNL/CDIAC-53, NDP-041, 325 pp. [Available from Carbon Dioxide Information Analysis Center, Oak Ridge National Laboratory, PO Box 2008, Oak Ridge, TN 37831].

WMO (1996) Exchanging Meteorological Data: Guidelines on Relationships in Commercial Meteorological Activities, WMO Policy and Practices. WMO/TD No. 837, World Meteorological Organization, Geneva. Booklet 24 pp.

Xie, P. and P. A. Arkin (1995) An Intercomparison of Gauge Observations and Satellite Estimates of Monthly Precipitation. *Journal of Applied Meteorology:* 34, No. 5, pp. 1143-1160.

Part II

El núvol blanc que ara passa, pel blau del cel molt suaument, no ha estat posat, ni poc ni massa, per ser mirat, amb els ulls meus.

(The white cloud that's now passing, through the blue of the sky very softly, has not been put there at all, to be looked at with my eyes).

Raimon,
Del blanc I el blau, 1987

5 An Introduction to Ocean Climate Modeling

Stephen M. Griffies

Geophysical Fluid Dynamics Laboratory, NOAA. Princeton Forrestal Campus Rte. 1, P.O. Box 308, Princeton, NJ 08542-0308, USA
smg@gfdl.noaa.gov

5.1 Introduction and motivation.

This chapter introduces some of the science of numerical ocean climate modeling. The discussion pedagogical and self-contained, thus requiring little previous knowledge of numerical ocean modeling. It is geared towards the science, engineering, and/or mathematics student or researcher who wishes to garner a sense of the goals, methods, and some details of ocean climate modeling. Notably, it only presents a small hint at the large body of literature which supports the field of climate modeling. To do more justice would require many textbooks.

Climate models are a core element in the quantitative toolbox of climate research. Furthermore, results from climate models are being consulted with increasing frequency by non-scientists, such as environmental managers and policy makers. Therefore, it is essential that the scientific foundations of climate models be articulated in a thorough and sound fashion. Only by doing so will the models achieve a level of "intelligent use" so that their methods may be improved and their output more critically interpreted.

It is important to emphasize that although ocean climate models have reached a reasonably high level of respect amongst the traditional oceanography and climate communities, the models are far from perfect. Their future improvement will necessarily involve a broad range of scientists whose expertise ranges from the computational to phenomenological. Such efforts are ongoing and continue to grow at an exciting pace.

The problems of climate science necessitate a sound intellectual and rational approach using a hierarchy of tools and methods, both quantitative and qualitative. Amongst the ensemble of methods currently available are the numerical models described here, as well as an impressive array of observational methods, both *in situ* and remote. Indeed, the near future promises to be a very exciting time in ocean climate studies due to the increasing ability of both modeling and observational methods to provide direct and useful feedbacks for each other. The result will hopefully be a steadily improving ability to measure the ocean in both space and time, and to interpret the measurements according to scientific theories. Only by doing so will the core problems of climate be addressed in a manner consistent with scientific methods.

The bulk of climate science problems take on an air of immediacy due to the often dire predictions of anthropogenic climate change. Hence, they warrant focused attention for societal reasons. Additionally, these problems are deeply interesting and intellectually challenging. Thus, they have captured the obsession of many scientists.

5.1.1 Some key questions

There are many fundamental questions that climate science is trying to address, with the following list providing examples of interest to both science and society:

- What are the physical mechanisms accounting for the mean climate state, variability about this mean, and its stability?
- Do different climate phenomenon interact? For example, do variations in the Atlantic's thermohaline circulation, which occur on multi-decadal time scales, have any effect on Atlantic tropical storm frequency or inter-annual fluctuations of the North Atlantic Oscillation?
- Can we rationalize past climate behavior and provide useful predictions for its future?
- Is the observed warming of the planet due to human effects, and how much more warming could occur in the 21st century and beyond?

These questions are clearly of relevance for human society, and they are extremely interesting and complex scientifically. Indeed, they may never lend themselves to wards a "clean" answer so familiar to those brought up on the exactitude of other branches of post-World War II physics. Instead, as emphasized in Philander's book (Philander 1998), we are dealing which questions that can only be answered in a probabilistic manner, similar to the probabilities inherent in the insurance industry.

5.1.2 Climate phenomena where the ocean is fundamental

The ocean is a fundamental component of the earth's climate. Most simply, it is the flywheel which largely determines the climate mean, variations about this mean, and stability of climate regimes. That is not to say that the ocean is everything. Instead, the ocean is a fundamental part, along with the atmosphere, cryosphere, and biosphere, of the earth's climate. The large-scale changes in the ocean occur on monthly to multi-decadal time scales, rather than the daily to weekly changes occurring in the atmosphere. The ocean carries a large amount of heat poleward within its massive currents, thence playing a role, roughly equal to the atmosphere, in keeping the low latitudes from scorching under its net input of solar radiation, and the higher latitudes from freezing solid. The ocean interacts with the other components of the climate system. The El Niño phenomenon is perhaps the most beautiful illustration of this interaction, hence making the study of ocean dynamics less an esoteric endeavor, and of more direct relevance to human cultures (see, for example, the book by Philander 1990 and his chapters in this volume, Chap. 2 and Chap. 3, for discussions of El Niño).

Most studies of climate start with observations which are lacking physical explanations. It is the job of the climate scientist to associate physical mechanisms with the phenomenon. An observation often involves a time series of a climate index, such as mean temperature over a particular spatial region. In addition to time series, it is useful to determine spatial patterns associated with the time series. For example, did the temperature change coherently over a region, or did it rise in some parts and fall in others?

There are two time series, with associated patterns, that currently occupy a great deal of research in climate dynamics. One is the El Niño-Southern Oscillation (ENSO) phenomenon (Fig. 5.1), which affects large parts of the tropical and mid-latitude areas in and around the Pacific, and another is the observed rise in global mean surface temperature over the past century (Fig. 5.3).

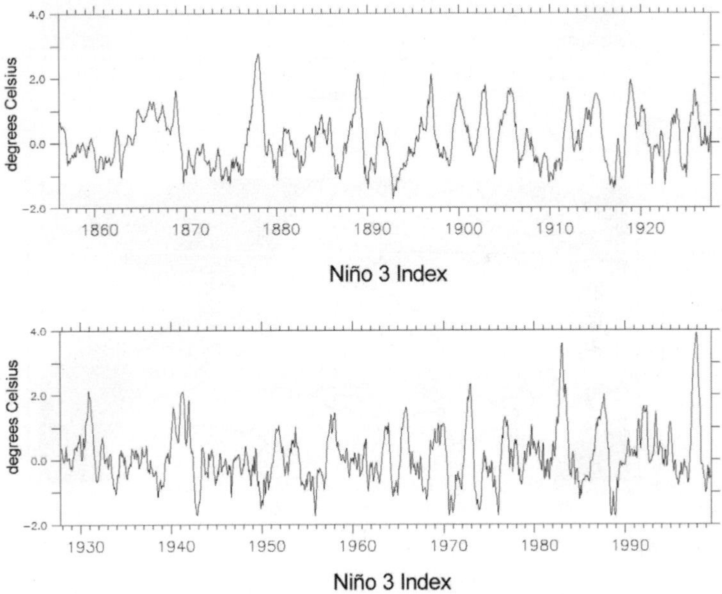

Fig. 5.1. Time series of SST for the Niño-3 region of the Tropical Pacific. This is a region between 5°N and 5°S latitude, and 150°W and 90°W in longitude; i.e., the eastern equatorial Pacific. Note the two largest events in the 20th century occurred during 1982 and 1998. This time series was compiled by researchers at the Lamont-Doherty Earth Observatory of Columbia University. It is available at their website at (http://ingrid.ldgo.columbia.edu/).

Fig. 5.1 shows a time series for the Niño 3 index, which represents the anomalous sea surface temperature (SST) for a region in the eastern tropical Pacific. The fluctuations in this temperature range from nearly –2 °C to 4 °C. Note that the two largest fluctuations for the 20th century occurred in 1982 and 1998. Associated with the 1998 event is the SST pattern for February 1998 shown in Fig. 5.2. The huge region of very warm water in the central east Pacific is characteristic of an El Niño "event." In contrast, the same region in February 1999 shows much cooler temperatures in the eastern Pacific, with warmer temperatures in the west.

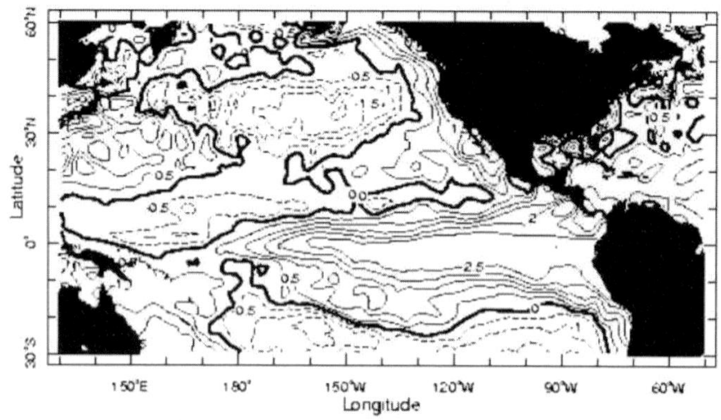

Feb 1998

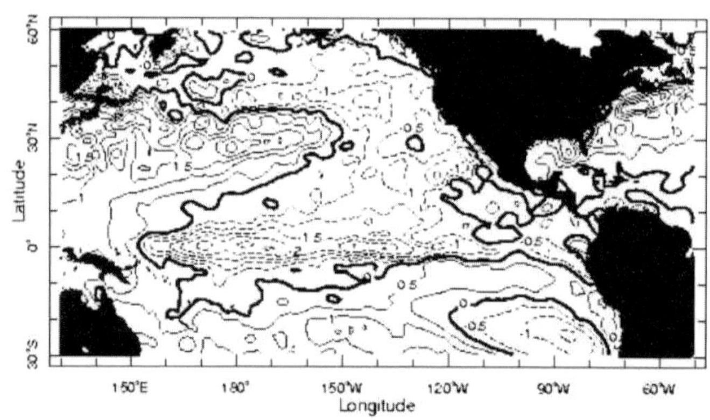

Feb 1999

Fig. 5.2. Top panel: SST anomaly (degrees Celsius) in the Tropical Pacific for February 1998. The large warming is associated with a strong El Niño event. Lower panel: SST anomaly for February 1999, showing much cooler temperatures across the Pacific recurring after the departure of the previous year's El Niño event. This observational data is taken from the web-site at Lamont-Doherty Earth Observatory of Columbia University (http://ingrid.ldgo.columbia.edu/).

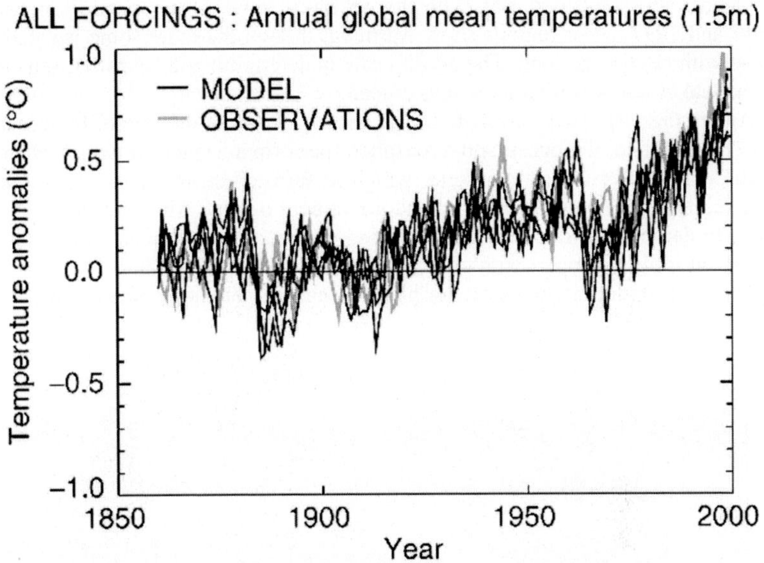

Fig. 5.3. Time series for annual mean global surface temperature anomaly with respect to the values averaged over 1880-1920 from the instrumental record compared with ensembles of four simulations with a coupled ocean-atmosphere climate model (from Stott et al., 2000; Tett et al., 2000) forced with natural forcing (solar and volcanic) and with anthropogenic forcing (well mixed greenhouse gases, changes in stratospheric and tropospheric ozone and the direct and indirect effects of sulphate aerosols). The thick line shows the instrumental data while the thin lines show the individual model simulations in the ensemble of four members. Note that the data are annual mean values. The model data are only sampled at the locations where there are observations. The changes in sulphate aerosol are calculated interactively, and changes in tropospheric ozone were calculated offline using a chemical transport model. Changes in cloud brightness (the first indirect effect of sulphate aerosols) were calculated by an offline simulation (Jones et al., 1999a) and included in the model. The changes in stratospheric ozone were based on observations. The volcanic forcing was based on the data of Sato et al. (1993) and the solar forcing on Lean et al. (1995), updated to 1997. The net anthropogenic forcing at 1990 was 1.0 Wm^{-2} including a net cooling of 1.0 Wm^{-2} due to sulphate aerosols. The net natural forcing for 1990 relative to 1860 was 0.5 Wm^{-2} , and for 1992 was a net cooling of 2.0 Wm^{-2} due to Mt. Pinatubo. Note the rough agreement of the time series for the gradual warming occurring in the late 20th century. Such comparisons between observations and numerical models provide an important avenue for validating the numerical simulations and lending some confidence to their predictions (from IPCC, 2001).

Fig. 5.3 shows a time series of anomalous annual mean surface temperature over the latter half of the 19th and all of the 20th centuries. The rising temperature trend is clear, although note the large amount of fluctuations, especially on the short time scales. This figure shows both model simulated and observed data. Fig.

5.4 shows the associated map with the observed trends in temperature between 1949 and 1997. Most regions show warming, though there are some isolated regions with relative cooling. The ocean's role in determining these patterns, as well as the rate of temperature increase, is crucial.

Again, these are two forms of climate variations where the role of the ocean is key. For El Niño, the ocean and atmosphere perform a "dance" in which changes in the ocean affect the atmosphere, which in turn affect the ocean. For climate change, the ocean sequesters a tremendous amount of heat, which acts in some regions to dampen the rate at which warming occurs. Ocean climate models are an important tool, in concert with observations and theory, whose intelligent use can enhance our understanding of these phenomena and perhaps improve our predictive capabilities.

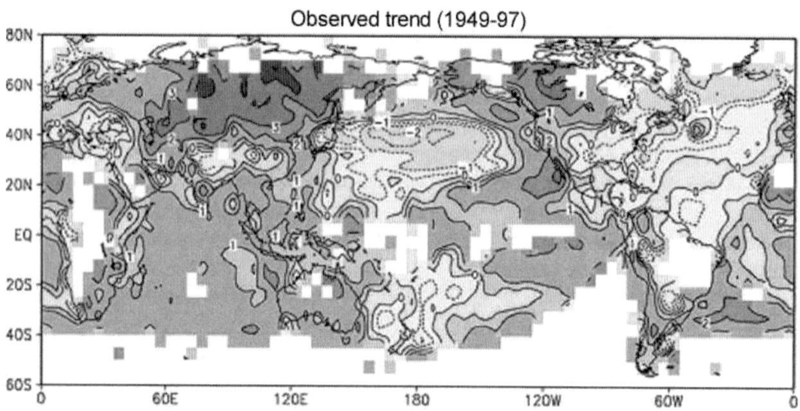

Fig. 5.4. Map of the observed trend in air surface temperature between 1949 and 1997 (IPCC, 2001). Note the general warming over most of the planet, with some localized exceptions. Blank regions denote areas with insufficient data. The observed temperature changes are discussed in the review article by Jones at al. (1999b)

5.1.3 Some recommended review articles and books

There is much in this chapter biased towards the author's personal research. To partially remedy this shortcoming, a fair number of references are cited so that the reader can consult the fundamental literature for more thorough and/or detailed presentations. In particular, the following review articles and texts are recommended for those wishing to pursue further study of ocean climate modeling:

- Bryan (1975), Pond and Bryan (1976), Bryan (1979), Bryan and Sarmiento (1982), Bryan (1989, 1991a,b). When read in temporal order, these papers provide an interesting perspective on the evolution of ocean climate modeling.

- Haltiner and Williams (1980): This text is a classic in the field of numerical atmospheric modeling. It contains very good discussions of the basic dynamical and numerical issues which must be considered when constructing models, and many issues relevant for atmospheric modeling are also relevant for oceanic modeling.
- O'Brien (1986): This book contains technical chapters written in a pedagogical fashion by various ocean modeling experts.
- Haidvogel and F. Bryan (1992): This is a well written review of how ocean models are perceived in the context of climate modeling, as of 1989. It is contained within a highly recommended book on climate system modeling.
- McWilliams (1996): This review summarizes the state-of-the-art in ocean climate models as of 1995.
- The book edited Chassignet and Verron (1998) provides a valuable source for pedagogical articles on various fundamental and applied aspects of ocean climate modeling and parameterization.
- Haidvogel and Beckmann (1999): This is a textbook devoted to ocean circulation modeling. It will hopefully signal the writing of further texts to help solidify the scientific foundations of this field.
- The two books by Kantha and Clayson (2000a, 2000b) provide a significant and valuable contribution to the field of ocean modeling.
- The paper by Griffies et al. (2001), provides a review, with many references, of recent advances in ocean climate modeling.

5.1.4 Remainder of this chapter

The remainder of this chapter consists of the following sections. Sect. 5.2 presents some fundamental ideas which form the key components of an ocean model. This discussion is given in words and pictures, and it is followed up in Sect. 5.3 with more formal mathematical statements. Sect. 5.4 discusses some future areas of research in ocean climate modeling. Sect. 5.5 finishes the chapter with closing remarks.

5.2 Some basics of ocean dynamics.

The equations of an ocean model constitute a discrete version of the continuum equations of fluid dynamics along with the laws of linear irreversible thermodynamics. The dynamical equations, generally known as the Navier-Stokes equations, are an expression for a continuous media of Newton's second law of motion. That is, the acceleration experienced by a fluid parcel is due to the force per unit mass exerted on that parcel. The thermodynamic laws express the basic ideas of energy conservation and increase in entropy. The remainder of this section introduces, in words and pictures, more details concerning these dynamical ideas as they relate to ocean modeling.

5.2.1 Rotation and stratification

The study of atmospheric and oceanic fluid dynamics constitutes the subject matter of geophysical fluid dynamics (GFD). Two fundamental aspects of GFD are 1) the fluid motion occurs on a rotating sphere and 2) the fluid is stratified according to density. From a pedagogical perspective, it is useful to isolate these two features in order to understand motions, say, of a uniform fluid on a rotating sphere, or a stratified fluid on a resting plane. Individually and together, one will necessarily encounter processes of relevance for the ocean. For those interested in such a dynamical characterization of GFD, the book by Gill (1982) is highly recommended. Other recommended books discussing GFD with an ocean perspective, given in increasing order of sophistication, are those by Cushman-Roisin (1994), Apel (1987), Pedlosky (1987), and Salmon (1998).

For regions isolated away from the equator, the large-scale horizontal motions in the earth's atmosphere and ocean occur in a dynamical regime in which the Coriolis force and pressure gradients are the two leading order terms affecting the dynamics. For the atmosphere, the relevant length scale is on the order of 1000's of kilometers and larger (e.g., from synoptic weather patterns to the mid-latitude jet stream), whereas in the ocean it is order 100's of kilometers and larger (e.g., from Gulf Stream eddies and rings to basin-wide gyre currents). These different scales are a result of the difference in vertical density stratification of the fluids, with the atmosphere being about 10-100 times more stratified than the ocean.

The balance between pressure gradients and rotational effects, embodied in the Coriolis force, is known as geostrophic balance. For example, mid-latitude atmospheric weather charts show areas of high and low pressure. Due to the Coriolis force, winds in the northern hemisphere circulate counter-clockwise around a low pressure center and clockwise around a high. The opposite circulation orientation occurs in the southern hemisphere. Without rotation, the winds would simply flow from high to low pressure as water does in a pipe. Similar effects are relevant for the ocean, in which the height of the ocean, relative to a resting geoid, provides the oceanic analogue for the atmospheric pressure patterns. Such surface elevations can now be measured quite accurately through methods of satellite altimetry (e.g., Fu and Smith 1996).

An important dynamical balance relevant for vertical motions is the hydrostatic balance. This balance says that the pressure at a point in the fluid is due to the weight of the overlying fluid. It is the result of disregarding vertical accelerations of fluid parcels. The degree to which the hydrostatic balance is respected is largely determined by the ratio of the vertical length scale to the horizontal length scale. For large-scale oceanography, relevant vertical heights are on the order of at most a few thousand meters (the ocean depth is typically some 5-6 x 10^3 m), whereas the horizontal distance is some few hundred kilometers (one degree in latitude is roughly 10^5 m), hence making the hydrostatic balance relevant. Processes where small horizontal length scales are relevant, such as convection, breaking internal waves, salt fingers, etc., introduce non-hydrostatic effects which are excluded, or filtered, from models which assume a hydrostatic balance.

Through the hydrostatic balance, the ocean's pressure field is directly linked to the density field. The equation of state for ocean water relates density to temperature, salinity, and pressure. Hence, oceanographers can deduce the geostrophic velocity field through analysis of water mass properties. Such analysis has formed a major part of observational oceanography over the last 50 years. With an ocean model, the water mass, or hydrographic, properties of the simulation are computed along with the velocity fields. Consequently, there is no need to perform the "inverse calculations" needed by the observationalist.

As mentioned earlier, the ocean is stratified into different layers of density, with the denser waters in the deep ocean. For much of the ocean, the deviation of density from the surface to the abyss is less than 2% from 1035 kg m^{-3} (Gill, 1982). Consequently, for some purposes, the vertical stratification of the ocean can be ignored. It is through the *Boussinesq Approximation* that one systematically ignores such stratification when it is reasonable to do so, but keeps its essential effects, such as those related to buoyancy forces.

5.2.2
Ocean tracers

The ocean is a wonderful mixer of various properties. For example, great clouds of phytoplankton are mixed over many hundreds of kilometers through the action of the ocean currents, especially through the turbulent eddying patterns created by hydrodynamic instabilities. Fig. 5.5 shows a striking satellite image of the northwest Atlantic, including part of the Gulf Stream. Instabilities in the Gulf Stream produces eddies, or rings, which stir less biologically active waters from the south into more active northern waters. This figure is representative of many splendid pictures of ocean currents now available from satellites.

As the clouds of phytoplankton are transported by the ocean currents, they become stretched and strained into filamentary structures. These structures are characteristic of turbulent transport, a process which interests a broad range of scientists and mathematicians. To leading order, phytoplankton does not affect the ocean's density field, and so will not affect the ocean currents. Instead, phytoplankton is passively carried around by the currents, at the whim of variations in the ocean dynamics, as well as sources and sinks which affect its population.

Other important passive tracers include radioactive isotopes which were deposited into the ocean during the 1950's and 1960's from atmospheric atomic bomb tests. These tracers, although produced under rather unfortunate circumstances, have proven to be of great value for diagnosing pathways of ocean currents (e.g., Broecker and Peng, 1982). This method is directly analogous to how medical doctors introduce weakly radioactive tracers into humans for diagnosing certain ailments.

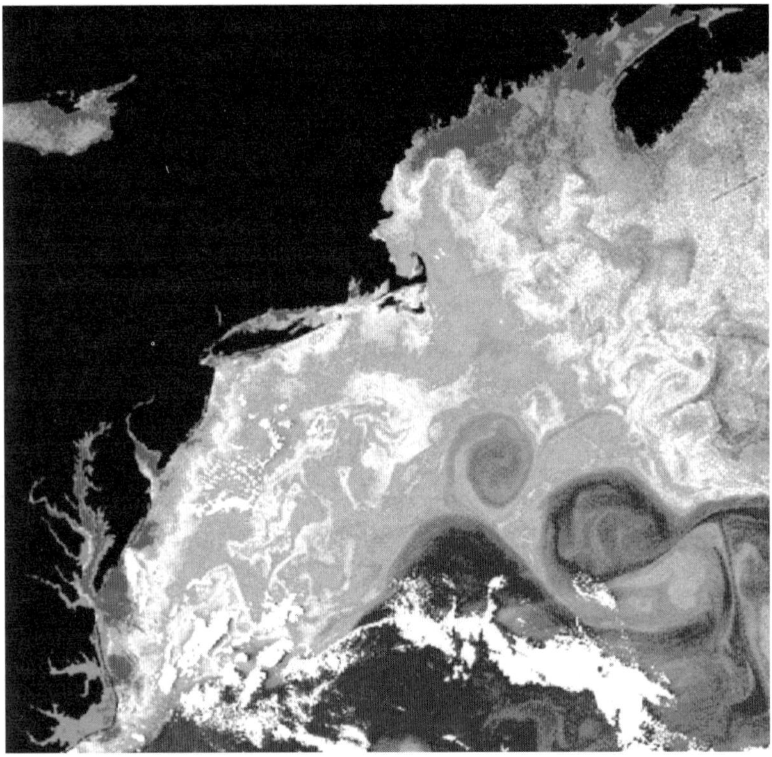

Fig. 5.5. Satellite image of biological production in the northwest Atlantic near the North American coast (Cape Hatteras, North Carolina is at the bottom left corner and Nova Scotia is at the top right). The lighter gray colors near the coast are associated with active phytoplankton production, typically present in regions of cooler waters found north of the Gulf Stream. The darker gray colors to the south are regions of less biological activity. The large dark gray rings represent eddies of low productive waters shed from the unstable Gulf Stream jet. These eddies are the ocean's equivalent of the mid-latitude storms familiar from daily atmospheric weather. The white represents atmospheric clouds. Satellite views of biological fields are important for studying the ocean's biological activity. Additionally, biological fields provide the ocean fluid dynamicist with beautiful images of the ocean's turbulent eddy fields with its many filamentary scales. In contrast, such fine scale patterns in sea surface temperature are more rapidly washed out by strong interactions with the atmosphere. This picture was taken from the website at NASA's Goddard Space Flight Center (http://daac.gsfc.nasa.gov)

Heat is another tracer which is transported by ocean currents. As mentioned earlier, heat, or temperature, is a key tracer for purposes of understanding the ocean's role in the earth's climate system. In general, heat is carried from the equatorial region, where there is a net input of solar radiation, to the higher latitudes, where there is a net sink of heat. Since the temperature of ocean water affects the density field, temperature is termed an active tracer. That is, through the hydrostatic balance, the effects of temperature on density will alter the pressure field, which in turn affects the forces acting on the fluid and thence affects the velocity field. In this manner, the transport of heat in the ocean is a nonlinear process.

There is yet another tracer, known as potential vorticity, whose diagnosis allows for the deduction of many dynamical properties of the fluid. Indeed, under certain scalings, such as the quasi-geostrophic approximation (QG) (e.g., Gill 1982, Pedlosky 1987), the potential vorticity (PV) field completely specifies the dynamics. It can be considered a generalization, to rotating stratified fluids, of the scalar potential useful for ideal two-dimensional fluid flow. The conservation of PV is related through symmetry principles to the arbitrariness of labeling fluid parcels (e.g., see Müller (1995) or Salmon (1998) for discussions of symmetry principles applied to fluid mechanics). In general, PV has proven to be a very important field for studying ocean dynamics. As this tracer is a function of the flow, its transport is a nonlinear process.

As the above examples may indicate, tracer transport has captured the attention of many oceanographers. In particular, because many tracers can be measured in the real ocean, they provide a useful means to evaluate the fidelity of ocean model simulations. Indeed, due to the strong geostrophic balance in the ocean, if the model successfully reproduces the observed temperature and salinity distributions (e.g., as found in an atlas such as that of Levitus 1982), then the model will necessarily do a good job with the simulated currents. Refining the simulations of ocean climate models so that they remain close to the Levitus Atlas is a difficult challenge. Recent advances have enhanced the solution integrity to the point of being quite good in capturing many large-scale features (see, e.g., the review by McWilliams 1996).

5.2.3 Forces acting on the ocean

The ocean is forced at its boundaries through interactions with the atmosphere, sea ice, rivers, and solid earth. These are the forces which impart an acceleration to the oceanic fluid. In this section we review some of the more important forces.

The atmosphere transfers momentum to the ocean through wind stress acting on the sea surface. This stress is a central driving force for the large-scale ocean currents. An interesting means of viewing how stress acts on the surface ocean is to look at maps of surface wave heights, as shown in Fig. 5.6. As the surface waters are forced by the winds, they transfer momentum downward through the fluid column, thus affecting the full ocean. The highest waves, which correspond to the strongest winds, are seen in the Southern Ocean. The strong westerly winds at this latitude drive the Antarctic Circumpolar Current (ACC), which is the ocean's most powerful current. It carries a volume of water equivalent to roughly 100-200 times that of all the rivers in the world combined.

Buoyancy forces are also a crucial means of altering the ocean currents as they alter the ocean's density structure. For example, through strong cooling from bitter cold winter storms, high latitude waters increase their density (cold water is denser than warm water), which means the water looses buoyancy and thence sinks to deeper levels.

Salinity changes also affect buoyancy, since fresh water is less dense than salty water. Consequently, the transfer of moisture through evaporation and precipitation plays an important role in changing the ocean density. The Mediterranean Sea is a region experiencing a net loss of fresh water through evaporation (as residents of South Spain are quite aware). Indeed, due to evaporative losses, roughly 5% more water enters the Mediterranean through the Straits of Gilbraltar than leaves. Thus, although the water is relatively warm as it leaves through the Straits, its high salt concentration makes it more dense than water entering the Mediterranean. As a result, water leaving the Mediterranean tends to sink when entering the Atlantic Ocean (e.g., Robinson and Golnaraghi 1994, Bryden et al. 1994).

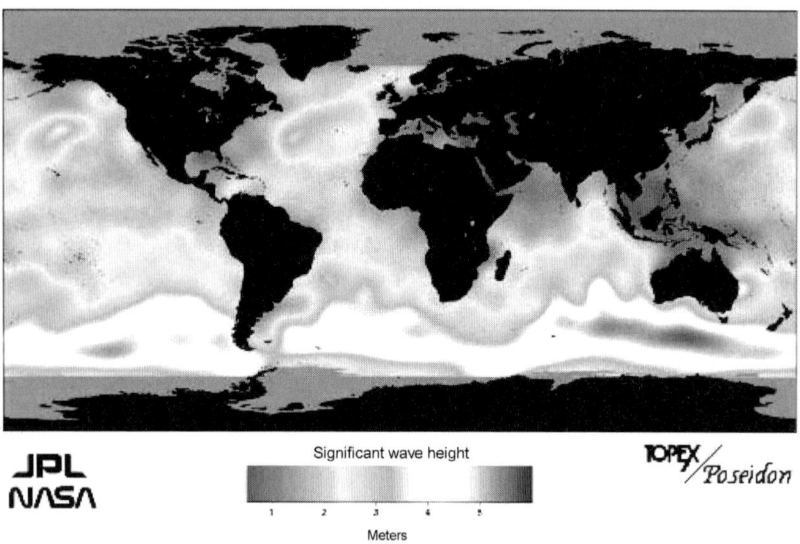

Fig. 5.6. Satellite image of significant surface ocean wave height for May 1996. The data are from the Topex/Poseidon satellite, as prepared by scientists as NASA's Jet Propulsion Laboratory (JPL). Note the very large waves in the Southern Ocean. The homogeneous regions in the high latitudes are areas where the satellite does not sample. The image is taken from the JPL website http://podaac.jpl.nasa.gov/topex/

Rivers transport fresh water into the ocean, which adds buoyancy to the ocean waters. Such effects are most important near the mouths of large rivers, such as the Amazon. They also play an important role in the dynamics of the Arctic Ocean due to the input of water from Siberian rivers.

Sea ice exchanges heat and water with the ocean, hence also affecting the ocean buoyancy. For example, as sea ice melts, it freshens the nearby ocean water, thereby increasing the ocean buoyancy. Especially in the high latitudes of the Atlantic, and in the Southern Ocean near Antarctica, the interaction between sea ice and the ocean waters is essential to understanding ocean water mass properties.

The ocean bottom is quite varied in its degree of smoothness. In some regions, such as the southeast Pacific, one finds large regions of reasonably smooth bottom relief. In other regions, such as the mid-Atlantic ridge, one finds very rough topography. Other features include the continental shelves, in which the ocean bottom rises to within a few hundred meters from the surface, and ocean trenches, which can extend to more than 10^4 m depth. Each of these features are important for determining the ocean dynamics. The obvious effects include the routing of deep currents through trenches, sills, and passage-ways. Other effects include the interaction of bottom relief and the ocean's potential vorticity field. These interactions can lead to effects which are somewhat counter-intuitive, such as the acceleration of mean currents (e.g., Holloway 1992 or Salmon 1998) rather than the more expected deceleration or drag.

Yet another effect of topography, especially rough topography, is related to processes in which the fluid becomes intensely turbulent, hence increasing the amount of mixing which occurs near these regions. This process, although fundamentally quite small (order meters or less), has profound effects on the large-scale stratification of the ocean. The central idea is that if one places a light parcel of fluid above a heavy parcel, and then mixes the parcels, the resulting homogeneous fluid will have more potential energy than the two unmixed parcels. One source for this increase in potential energy is thought to be the ocean's lunar tides (Munk and Wunsch 1998), which continually move the abyssal waters across the rough topography, thus causing parcels to mix due to enhanced turbulence in this region. The cumulative effect of this mixing, occurring over many years, can have as profound an effect on the ocean's stratification as the surface forcing. These ideas are only now being solidified with ocean measurements (e.g., Polzin et al. 1997), and future research will likely see more exciting advances.

5.2.4 General overview of an ocean model

Quite generally, a numerical ocean model is a method for taking a representation of the ocean state at one instance of time into a future instance of time. There are three basic, and nontrivial, requirements necessary to bring about a numerical representation of the ocean's evolution: the initial state, the forcing fields, and the evolution equations.

Means for garnering information regarding the present ocean state combine methods of observational oceanography with mathematical methods for sensibly assimilating observational information into a numerical model. The job is technically difficult partly because of the huge size of the oceanic system which admits numerous degrees of freedom, and partly because of the paucity of ocean data to constrain the degrees of freedom onto Nature's "phase space". In general, the earth is a very under-sampled system, both in space and time. Such sampling problems pervade all of climate science and lead to many active debates regarding past and future climate.

Once a representation of the ocean is obtained, which will in practice be quite rough and incomplete, it is necessary to obtain information regarding the forces acting on the ocean. The forces, as described above, are largely those imparted by the atmosphere, ice, and solid earth. Again, obtaining accurate and complete forcing data is nontrivial due to sampling limitations.

Finally, given a representation of the present ocean state and forces acting on this state, it then falls on the basic laws of physics, chemistry, and biology to bring about an evolution of the ocean to a new updated state. An ocean model is a discrete numerical representation of these laws, in combination with the initial and forcing data, allowing the climate scientist an ability to rationalize a massive amount of information and to project that information into the future.

5.3 The dynamical equations of an ocean model.

The purpose of this section is to present the continuum form of the dynamical equations solved by the GFDL Modular Ocean Model (MOM). These equations share much in common with those solved by many other ocean models. Hence, they provide a useful avenue for introducing some details of ocean models. Although necessary, knowledge of the continuum equations is not sufficient for building an ocean model. The essential, and often quite difficult, next step is to take these equations and apply the principles of finite or discrete differential equations to render the dynamics suitable for computer calculations. Details of these methods are beyond the scope of this chapter, yet may be found in Haltiner and Williams (1980), O'Brien (1986), Bryan (1989), Haidvogel and Beckmann (1999), and Pacanowski and Griffies (1999), Kantha and Clayson (2000), amongst others.

5.3.1 Conservation of volume

A fluid may be idealized as a collection of very small parcels whose size is macroscopically infinitesimal yet microscopically huge. That is, we may employ the general rules of differential calculus without worrying about the molecular or atomic nature of the fluid. This is the realm of classical fluid dynamics. A fundamental constraint on the dynamics of fluid parcels is that without sources or sinks, the mass of the parcel must be conserved. Mathematically, one requires $dM = \rho\, dV$ to be a constant of the motion, where dM is the mass of the infinitesimal fluid parcel, dV is its volume, and ρ is its mass per unit volume. For a Boussinesq fluid, mass conservation translates into volume conservation, which means that the volume dV of a parcel is constant. In Cartesian coordinates, $dV = dx\, dy\, dz$, with z the vertical coordinate and x,y the horizontal coordinates. Setting $dV/dt = 0$ leads to the continuity equation

$$\nabla \cdot \mathbf{u} = 0, \tag{5.1}$$

where $\mathbf{u} = (u, v, w) = (dx/dt, dy/dt, dz/dt)$ is the three-dimensional velocity of the fluid parcel. For those with a physics training, it is interesting to note that a divergence-free velocity field is analogous to the constraint respected by the magnetic field in classical electrodynamics (e.g., Jackson 1975).

5.3.2 Conservation of momentum

Newton's second law is an expression of momentum conservation. On an infinite at plane, in the absence of external forces, the linear momentum $\mathbf{u}\, \rho\, dV$ of a fluid parcel is conserved. For a Boussinesq fluid, the linear momentum takes the form $\mathbf{u}\, \rho_o\, dV$, where ρ_o is a constant reference density. On a rotating sphere, it is the total angular momentum of the parcel about the axis of rotation which is conserved. The ocean primitive equations typically are not written in terms of angular momentum, but instead are expressions of the linear momentum of the parcels in the longitudinal and latitudinal directions. As linear momentum is not conserved on a sphere, these equations will not take the form of traditional conservation laws.

5.3.3 Stress acting on a parcel

We already discussed the forces acting on the ocean through its boundaries. More formally, the forces acting on an element of a continuous media are of two kinds. External or body forces, such as gravitation, Coriolis, or electromagnetic forces, act throughout the media. Internal or contact forces, such as pressure forces, act on an element of volume through its bounding surface. When the bounding surface of the parcel contacts the ocean-atmosphere or ocean-solid earth boundary, then the boundary forces acting on the parcel are also called contact forces.

The stresses acting within a continuous media can be organized into a second order stress tensor with generally 3 x 3 independent elements. The divergence of these stresses gives rise to the internal forces acting in the media. A proper account of the angular momentum budget implies that the stress tensor is symmetric, which brings the number of independent stress elements down to six. Further restrictions related to the hydrostatic approximation and basic symmetry requirements lead to a stress tensor with two independent degrees of freedom. For a general discussion of stresses in a fluid, one may consult most fluid mechanics textbooks, such as Aris (1962), Batchelor (1967), or Landau and Lifshitz (1987). For discussion of stress as employed in ocean models, the papers by Smagorinsky (1993), Wajsowicz (1993), and Griffies and Hallberg (2000) should be consulted.

The two most important internal or contact forces for large-scale ocean modeling are the pressure and frictional forces. As discussed earlier, the pressure force is a fundamental driving force for fluid motion. Frictional forces represent the fact that as real fluid parcels move past one another, they exchange momentum and ultimately produce heat and increase entropy. For example, fast fluid moving next to slow fluid will speed up the slow fluid and slow down the fast fluid.

This exchange of momentum occurs at very small scales, much below that resolved by an ocean model. As such, the process must be parameterized. The traditional form used to parameterize this exchange is motivated by ideas from the kinetic theory of gases, for which a precise form is derivable from first principles (e.g., Salmon 1998). Since the kinetic theory involves molecular sized objects, and the typical grid boxes in an ocean model are often hundreds of kilometers across, the relevance of kinetic theory for setting the form of friction in ocean models is questionable.

Instead, a more pragmatic perspective is necessitated based on fundamental constraints placed on finite difference equations. Basically, it is impossible to integrate the finite difference equations of fluid flow without introducing some form of a grid filter. Without such filters, the solution will soon degenerate into a "sea of grid scale noise" and so possess little physical relevance. The filters used in ocean models generally take the form of a dissipation operator which can be interpreted as a form of friction. Often, this friction is simply a Laplacian operator multiplied by a viscosity coefficient: $A\,\nabla^2\,\mathbf{u}$, while other times it takes the form of higher order operators such as a biharmonic operator $B\,\nabla^4\,\mathbf{u}$. These forms are schematic; the precise form, relevant for a sphere, are discussed by Griffies and Hallberg (2000) and references therein.

5.3.4 The equations of motion

The previous ideas ultimately lead to the following prognostic equations of motion

$$\frac{Du_h}{Dt} = (f + (u/a)\tan\phi)\,(u_h \wedge \hat{z}) - \nabla_h(\rho/\rho_0) + F \tag{5.2}$$

In this expression, $D\mathbf{u}_h/Dt$ represents the acceleration of a parcel in the frame of reference of that parcel. In a frame fixed with respect to the earth, one requires the transformation from the total or Lagrangian time derivative to an Eulerian frame

$$\frac{Du_h}{Dt} = \frac{\partial u_h}{\partial t} + \nabla\cdot(u\,u_h), \tag{5.3}$$

where the first term is the usual partial time derivative taken at a point fixed on the earth. The second term in Eq. 5.3, which is nonlinear, is associated with advection, where $\mathbf{u} = (\mathbf{u}_h, w)$ is the three-dimensional velocity field. As the currents flow, they carry, or advect, momentum, which in turn alters the currents through the presence of the nonlinear advection term.

Eq. 5.2 provides a means for updating in time the horizontal ocean currents $\mathbf{u}_h$, thus motivating the term prognostic equation. To compute the time derivative, it is necessary to know the horizontal pressure gradient $\nabla_h p$, the horizontal forces from friction and boundary forces $\mathbf{F}$, and the Coriolis force $\mathbf{u}_h \wedge f\,\hat{z}$. The term $(\mathbf{u}/a) \tan\phi\,(\mathbf{u}_h \wedge \hat{z})$, where a is the earth's radius, arises from motion on the sphere. The parameter $f = 2\,\Omega\,sin\,\phi$, where $\Omega = 7.292 \times 10^{-5}\ \mathrm{s}^{-1}$ is the earth's radial rotation rate and ϕ is the latitude, is known as the Coriolis parameter. It measures the strength of the rotation, and is maximum at the poles and vanishes at the equator. Its non-vanishing latitudinal gradient $\partial_y f = 2(\Omega/a) \cos\phi$ turns out to be quite important for many forms of motion in GFD (Gill 1982, Pedlosky 1987). Note that the cross product term does not perform work on the fluid parcel, since these forces are always orthogonal to the parcel's trajectory. In this way, the Coriolis force is directly analogous to the Lorentz force encountered when a charged particle moves in a magnetic field (e.g., Jackson 1975).

In addition to the prognostic equation for the horizontal motion of the fluid parcel, there are three diagnostic equations

$$\partial_z p = -g\rho \tag{5.4}$$

$$\nabla\,\mathbf{u} = 0 \tag{5.5}$$

$$\rho = \rho\,(T, s, p) \tag{5.6}$$

The term diagnostic refers to the absence of time derivatives, which is in contrast to the prognostic equation (Eq. 5.2). Diagnostic equations represent constraints satisfied by the dynamical fields at each time step. The first diagnostic equation is the hydrostatic balance, which says that the vertical gradients of pressure are due to buoyancy forces. Or equivalently, the pressure at a point equals the weight of the overlying fluid. The second diagnostic equation is the continuity equation discussed earlier, which expresses a constraint on the fluid's velocity field due to conservation of volume. The final diagnostic equation is the equation of state, which expresses the density in terms of temperature, salinity, and pressure. The equation of state for most liquids is quite complicated, and the ocean is no exception (see Appendix III of Gill 1982, for discussion).

As the ocean currents move, they transport scalar quantities, such as temperature, salinity, and passive tracers. This transport takes the form of a conservation law

$$\frac{DT}{Dt} = \psi_T \tag{5.7}$$

where T is an arbitrary tracer concentration and ψ_T represents a tracer dependent source or sink. The update of tracer fields closes the set of ocean equations of motion. As described here, with the hydrostatic and Boussinesq approximations, this set of equations is termed the ocean primitive equations.

5.3.5
Splitting between fast and slow dynamics

Disturbances represented in a linearized version of the ocean primitive equations can be partitioned into an infinite number of orthogonal eigenmodes, each with a different vertical structure (e.g., Sec. 6.11 of Gill 1982). Because of the weak compressibility of the ocean, motions associated with the zeroth eigenmode, also called the external or barotropic mode, are weakly depth dependent, and so correspond to elevations of the sea surface. These barotropic disturbances constitute the fast dynamics of the ocean's primitive equations. They are associated with such motions as Tsunami's, which can travel at speeds upwards of 200 ms^{-1} in the deep ocean. The infinity of higher eigenmodes, known as baroclinic or internal modes, have disturbances associated with undulations of internal density surfaces. Advection, baroclinic planetary waves, and baroclinic gravity waves constitute the slow dynamics, which can be 100 times slower than the barotropic waves.

For computational efficiency, large-scale ocean models aim to exploit the split between the fast barotropic and slow baroclinic dynamics. Doing so in an efficient and complete manner has been the subject of many studies; e.g., Bryan (1969), Killworth et al. (1991), Dukowicz and Smith (1994), Higdon and Bennett (1996), Hallberg (1997), and Griffies et al., (2001). Although performing such a split complicates the numerical algorithm, the computational savings are tremendous relative to an un-split algorithm.

5.3.6
Subgrid-scale tracer parameterizations

As mentioned earlier, the typical finite grid spacing in ocean models is on the order of hundreds of kilometers. For example, present models being constructed for studies of global warming are approaching the one degree resolution, which requires 360 longitude points and 180 latitude points. In the vertical, it is necessary to resolve the density stratification with some 20-50 levels, with 50 levels much preferred. This amounts to a model with $N = 360 \times 180 \times 50 = 3.24 \times 10^6$ grid points. In contrast, resolving most of the spectrum of oceanic variability requires resolutions on the order of 10^{-3} m in all three directions. Assuming an ocean with an average depth of D = 5000m, an ocean model with $\Delta x = \Delta y = \Delta z = 10^{-3}$ m requires $N \approx 4 \pi a^2 D/(\Delta x \Delta y \Delta z) \approx 2.6 \times 10^{27}$ grid points. This is a huge number which will likely remain well beyond the capabilities of computers for many years to come. As a result, it is necessary to provide physically based parameterizations of the effects of the subgrid scales (SGS) on the resolved scales.

The problems of SGS parameterization are not unique to ocean modeling. Indeed, they play a central role in almost all areas of computational fluid dynamics. Furthermore, they generally touch on elements of the turbulence closure problem, and parameterizations which correctly summarize SGS physics do not come without some level of fundamental understanding. For an illuminating discussion of the SGS parameterization problem in oceanography, with comments on its relation to turbulence theories, see the review by Holloway (1989).

For ocean modeling, the 1990's has seen a focus on the problem of parameterizing subgrid scale effects in the tracer equation (Eq. 5.7). The central reason is due to the importance of tracer transport, especially that of heat, for determining the effects of the ocean on the global climate. Additionally, due to the geostrophic balance, the proper distribution of tracers in an ocean model will yield a reasonable current structure.

Mathematically, the fundamental problem of tracer transport can be boiled down to the following. First, write the source-free tracer equation in an Eulerian reference frame

$$\partial_t T = -\nabla \cdot (uT) \tag{5.8}$$

Now assume that one has access only to discrete points in space and time, perhaps because of a finite sized computer model. Such a limitation of information can be formally associated with a space-time filtering process, which leads to the modified tracer equation

$$\partial_t \overline{T} = -\nabla \cdot (\overline{u}\overline{T}) - \nabla \cdot \overline{u'T'} \tag{5.9}$$

In this equation, $T = \overline{T} + T'$ and $u = \overline{u} + u'$ are representations of the tracer and velocity fields in terms of their filtered, or mean, values and fluctuations about this mean. The filter is assumed to annihilate the fluctuations $\overline{T'} = \overline{u'} = 0$, whereas the filtered version of the fields $\overline{T}$ and $\overline{u}$ is all that is available to the modeler. Hence, the correlation $\overline{u'T'}$ must be parameterized. The form of the parameterization depends on the particular physical process contributing to the generally nonzero correlations. It also depends on the range over which the filter is applied, which is determined by the model's space-time grid resolution.

Mesoscale eddies are the most energetic motions in the oceans. These eddies feed off of the potential energy associated with tilted density surfaces. Unfortunately, the eddies are difficult to explicitly resolve in ocean models until reaching grid resolutions of a few tens of kilometers. Hence, their effects have to be parameterized.

A common assumption has been that the subgrid mesoscale processes act to mix the tracer down its mean gradient, thence leading to

$$\overline{u'T'} = -\kappa \nabla \overline{T} \tag{5.10}$$

where κ is a diffusivity with dimensions of squared length per time. As pointed out by Redi (1982), it is essential that this diffusion occur along surfaces of constant density rather than surfaces of constant depth. The reason is that the mesoscale eddies which perform much of the large-scale mixing preferentially mix along, rather than across, constant density surfaces. The tools for performing the necessary rotation of the diffusion tensor in MOM are described by Griffies et al. (1998).

The diffusion envisioned by Redi does not affect the underlying density field. In 1990, Gent and McWilliams (Gent and McWilliams 1990 (GM90), see also Gent et al. 1995) pointed out that another important effect of mesoscale eddies is to reduce the ocean's potential energy. Hence, the GM90 parameterization is based on the introduction of a means to extract potential energy from the model solution. Importantly, this potential energy sink will not mix the layers of density, but will rotate them to a horizontally uniform state.

As reviewed by McWilliams (1996), the Redi and GM90 parameterizations are popular with many ocean climate modelers since the schemes provide for an improved simulation of the water mass properties, as gauged by their comparison to the Levitus Atlas (Levitus 1982). They are also very simple to implement in tandem, as discussed by Griffies (1998). Given such results, the work of Redi and GM90 have prompted a flurry of activity which aims to clarify, refine, and/or correct, depending on the author's perspective, the original ideas.

5.4 Some future directions

One of the most challenging areas of large-scale climate modeling relates to the problem of constructing models which faithfully represent ENSO as well as the large-scale thermohaline circulation (THC) (see Chap. 8, by Griffies in this volume for more on the THC), all while providing a sensible projection of future climate over a decadal to century time scale. To reach this goal, it is necessary to add model resolution to admit better representations of the ocean currents, waves, and boundary forcing. Doing so requires increased computer power, which is becoming more readily available.

Additionally, more physically relevant simulations require better formulations of the SGS physics. That is, as we progress in our computational abilities and gather more measurements from the real ocean, we find that to improve the models does not necessarily mean simply increasing the resolution. It also means improving our numerical representations of ocean mixing and turbulence on all scales. Consequently, this area should continue to attract much research activity over the next decade.

Another area that shows promise of occupying many researchers during the next decade concerns the choice of vertical coordinate. Bryan (1969) chose the conceptually simple z-coordinate, in which the vertical levels are discretized according to the distance from a resting ocean surface. All versions of the GFDL Modular Ocean Model have maintained this choice. There are two other approaches that have become popular during the last decade. Sigma modelers (e.g., Blumberg and Mellor 1987, Haidvogel et al 1991) choose to discretize the vertical coordinate according to the bottom relief: $\sigma = z/H(x, y)$, where z is the distance from a resting ocean surface and $H(x, y)$ is the ocean depth. These models are well suited for studies of coastal processes as they provide for a natural means to incorporate bottom boundary layer mixing, which is important in these areas. Another approach is to discretize the vertical coordinate according to the density (e.g., Bleck et al. 1992, Oberhuber 1993, Hallberg 1995). This approach is useful when aiming to simulate the ocean interior, where most of the dynamics occurs without mixing properties across the stably stratified density layers.

Each choice of vertical coordinate has its pluses and minuses, some of which are highlighted in the DYNAMO project (The DYNAMO Group, 1997), which recently compared the results of three models, each using different vertical coordinates. The conventional thinking is that no one vertical coordinate has a clear advantage. Hence, many ocean model designers are thinking about how to build the next generation of models in a manner to incorporate only the best of each vertical coordinate choice. Such hybrid coordinate modeling is difficult both conceptually and technically, and so likely will account for much research into the next decade.

5.5 Closing remarks

In closing this chapter, it is worth reflecting on why the field of climate modeling has blossomed to the extent that it has, with literally thousands of researchers employing numerical climate models of one form or another.

One reason is related to the huge increase in computational power realized over the past few decades. What was once possible only on the largest supercomputers is now possible on a desktop personal computer. This increase in power has removed a substantial barrier that once separated the bulk of university climate researchers from those working in the few laboratories possessing the necessary computers.

Another reason is related to there being only one sample of the earth's climate. This situation is in contrast to some other fields of science, such as the fundamental branches of physics, chemistry, and biology, where multiple controlled and reproducible laboratory experiments can be performed to directly investigate and isolate a particular phenomena. Yet there are other fields, mostly dealing with complex systems such as those in ecology and cosmology, where it is difficult to directly perform controlled experiments.

As a result of there being only a single climate system, numerical climate modeling has evolved to the point of being a key tool for use in scientific investigations of climate. A well designed numerical model can be employed in a series of "controlled and reproducible experiments", where various parameters are easily altered and thousands of years of simulated data generated. These experiments have been found to be useful for initiating and testing various ideas, summarizing observational data, and, more controversially, forecasting future aspects of the climate system. Consequently, numerical models have become the climate scientist's laboratory. Just like the conventional laboratory, it is the responsibility of the experimentalist to intensely scrutinize the design of the numerical experiment and to determine the integrity of the tools and methods. Without doing so, it will be unclear how relevant the simulated results are for understanding natural phenomenon.

As obvious as these points may be today, they are largely the result of the continued evolution of the models. The early models of the 1960's and 1970's produced simulations which were quite far from Nature, so much so that many scientists could not foresee the day when the models would be more than a mere curiosity. It is largely due to the persistence, leadership, and vision of the early pioneers that the models continued to improve throughout those years. Today, although far from perfect, the models provide a reasonably faithful representation of the observed climate system. Hence, many "hard-core" oceanographers, some who employ relatively conservative or "old-fashioned" methods, find themselves adding the numerical ocean models to their toolbox. This is clearly a healthy state for the field, since those who directly view the real ocean are those most able to critique the simulations.

Acknowledgments

This chapter is based on the first of two talks given during a conference on Global Climate held March 1999 at Museu de la Ciència de la Fundació "La Caixa" in Barcelona, Catalunya, Spain. Many warm thanks go to the organizers of this conference for holding a wonderful meeting in this splendid city. In particular, thanks go to Francisco Comín, Paquita Ciller, Xavier Rodó, and Jorge Wagensberg. Thanks also go to the participants, and to the translators who brilliantly kept up with the lectures. Further thanks go to my ocean modeling mentors and collaborators, including Kirk Bryan, Bob Hallberg, Ron Pacanowski, Tony Rosati, Eli Tziperman, and many others.

References

Apel, J.R. (1987) Principles of Ocean Physics. International Geophysics Series, Volume 38. Academic Press. 634 pages.

Aris, R. (1962) Vectors, Tensors and the Basic Equations of Fluid Mechanics, Dover publishing.

Batchelor, G. K. (1967) An Introduction to Fluid Dynamics, Cambridge University Press. 615 pages.

Bleck, R., C. Rooth, D. Hu, and L. T. Smith (1992) Salinity-driven thermocline transients in a wind and thermohaline forced isopycnic coordinate model of the North Atlantic. Journal of Physical Oceanography, 22, 1486-1505.

Blumberg, A. F., and G. L. Mellor (1987) A description of a three-dimensional coastal ocean circulation model. Three-Dimensional Coastal Ocean Models. Vol. 4, N. Heaps, Ed., American Geophysical Union. 208 pp.

Broecker, W.S., and T.-H. Peng, (1982) Tracers in the Sea. Lamont-Doherty Earth Observatory Publishers. 690 pages.
Bryan, K. (1975) Three-dimensional numerical models of the ocean circulation. In Numerical Models of Ocean Circulation, National Academy of Sciences. Washington, D.C.
Bryan, K. (1989) The design of numerical models of the ocean circulation. In Oceanic Circulation Models: Combining Data and Dynamics, edited by D.L.T. Anderson and J. Willebrand. pages 465-511.
Bryan, K. (1991a) Michael Cox (1941-1989): His pioneering contributions to ocean circulation modeling. Journal of Physical Oceanography, 21, 1259-1270.
Bryan, K. (1991b) Poleward heat transport in the ocean: a review of a hierarchy of models of increasing resolution. Tellus, 43AB, 104-115.
Bryan, K. and J. L. Sarmiento (1985) Modeling ocean circulation. Advances in Geophysics, Volume 28A, 433-459.
Bryden, H. L., J. Candela, and T. H. Kinder (1994) Exchange through the Strait of Gibraltar, Progress in Oceanography, 33, 201-248.
Chassignet, E.P., and J. Verron, editors (1998) Ocean modeling and parameterization, NATO Advanced Study Institute, Kluwer Academic Publishers, 451 pages.
Cushman-Roisin, B. (1994) Introduction to Geophysical Fluid Dynamics, Prentice-Hall Publishers. 320 pages.
Dukowicz, J. K. and R. D. Smith (1994) Implicit free-surface method for the Bryan-Cox-Semtner ocean model. Journal of Geophysical Research, 99, 7991-8014.
DYNAMO Group (1997) DYNAMO: Dynamics of North Atlantic models: simulation and assimilation with high resolution models. Institut Für Meereskunde an der Christian-Albrechts-Universität. Copies of this report are available from Institut Für Meereskunde an der Universität Kiel. Abt. Theoretische Ozeanographie. Düsternbrooker Weg 20. D-24105 Kiel, Germany.
Fu, L.-L. and R. D. Smith (1996) Global ocean circulation from satellite altimetry and high-resolution computer simulation. Bulletin of the American Meteorological Society, 77, 2625-2636.
Gent, P. R., and J. C. McWilliams (1990) Isopycnal mixing in ocean circulation models. Journal of Physical Oceanography, 20, 150-155.
Gent, P. R., J. Willebrand, T. McDougall, and J. C. McWilliams (1995) Parameterizing eddy-induced tracer transports in ocean circulation models. Journal of Physical Oceanography, 25, 463-474.
Gill, A. E. (1982) Atmosphere-Ocean Dynamics. Academic Press Inc. 662 pages.
Griffies, S.M. (1998) The Gent-McWilliams skew-flux. Journal of Physical Oceanography, 28, 831-841.
Griffies, S.M., A. Gnanadesikan, R. C. Pacanowski, V. Larichev, J. K. Dukowicz, and R. D. Smith (1998) Isoneutral diffusion in a z-coordinate ocean model. Journal of Physical Oceanography 28, 805-830.
Griffies, S.M. and R.W. Hallberg (2000) Biharmonic friction with a Smagorinsky viscosity for use in large-scale eddy-permitting ocean models. Monthly Weather Review, 128, 2935-2946.
Griffies, S. M., R.C. Pacanowski, M. Schmidt, and V. Balaji (2000) Improved tracer conservation with a new explicit free surface method for z-coordinate ocean models. Monthly Weather Review in press 2001. Available from www.gfdl.noaa.gov/~smg.html.
Griffies, S.M., C. Böning, F.O. Bryan, E.P. Chassignet, R. Gerdes, H. Hasumi, A. Hirst, A.-M. Treguier, D. Webb (2001) Developments in ocean climate modeling. Ocean Modeling, 2, 123-192.
Haidvogel, D.B., and A. Beckmann (1999) Numerical Ocean Circulation Modeling. Imperial College Press. 300 pages.
Haidvogel, D.B., and F. Bryan (1992) Ocean general circulation modeling. In Climate System Modeling, pages 371{412. Edited by K.E. Trenberth. Cambridge University Press, 788 pages.

Haidvogel, D. B., J. L. Wilkin, and R. E. Young (1991) A semi-spectral primitive equation ocean circulation model using vertical sigma and orthogonal curvilinear horizontal coordinates. Journal of Computational Physics, 94, 151-185.

Hallberg, R. (1995) Some aspects of the circulation in ocean basins with isopycnals intersecting the sloping boundaries, Ph.D thesis, University of Washington, Seattle, 244 pp.

Hallberg, R. W. (1997) Stable split time stepping schemes for large-scale ocean modeling. Journal of Computational Physics, 135, 54-65.

Haltiner, G. J. and R. T. Williams (1980) Numerical Prediction and Dynamic Meteorology, Wiley.

Haywood, J. M., R. J. Stoufer, R. T. Wetherald, S. Manabe, and V. Ramaswamy (1997) Transient response of a coupled model to estimated changes in greenhouse gas and sulfate concentrations. Geophysical Research Letters, 24, 1335-1338.

Higdon, R. L. and A. F. Bennett (1996) Stability analysis of operator splitting for large-scale ocean modeling. Journal of Computational Physics, 123, 311-329.

Holloway, G. (1989) Subgridscale representation. In D. L. T. Anderson and J. Willebrand (eds.), Oceanic Circulation Models: Combining Data and Dynamics. pages 513-593. Kluwer Academic Publishers.

Holloway, G. (1992) Representing Topographic Stress for Large-Scale Ocean Models. Journal of Physical Oceanography, 22, 1033-1046.

IPCC (2001). Climate Change 2001: The Scientific Basis. Contribution of Working Group I to the Third Assessment Report of the Intergovernmental Panel on Climate Change, Cambridge University Press

Jackson, J.D. (1975) Classical Electrodynamics, Second edition. Wiley publishers.

Jones, A., D.L. Roberts and M.J. Woodage (1999a) The indirect effects of anthropogenic aerosol simulated using a climate model with an interactive sulphur cycle. Hadley Centre Tech. Note 14. Hadley Center for Climate Prediction and Research, Meteorological Office, Bracknell RG12 2SY UK, pp39.

Jones, P.D., M. New, D.E. Parker, S. Martin, and I.G. Rigor (1999b). Surface air temperature and its changes over the past 150 years. Review of Geophysics, 37: 173-200.

Kantha, L.H., and C.A. Clayson (2000a) Numerical Models of Oceans and Oceanic Processes, International Geophysics Series, vol. 66. Academic Press. 936 pages.

Kantha, L.H., and C.A. Clayson (2000b) Small Scale Processes in Geophysical Fluid Flows, International Geophysics Series, vol. 67. Academic Press. 883 pages.

Killworth, P. D., D. Stainforth, D. J. Webb, and S. M. Paterson (1991) The development of a free-surface Bryan-Cox-Semtner ocean model. Journal of Physical Oceanography, 21, 1333-1348.

Landau, L. D., Lifshitz, E. M. (1987) Fluid Mechanics. Course of Theoretical Physics, Volume 6. Pergamon Press, Oxford.

Lean, J., J.Beer and R. Bradley (1995). Reconstruction of solar irradiance since 1600: Implications For climate change. Geophys. Res. Lett., 22, 3195-3198.

Levitus, S. (1982) Climatological atlas of the world ocean. NOAA Prof. Paper 13. 173 pages. U. S. Government Printing Office, Washington, D. C.

McWilliams, J.C. (1996) Modeling the oceanic general circulation. Annual Review of Fluid Mechanics, 28, 215-248.

Müller, P. (1995) Ertel's potential vorticity theorem in physical oceanography. Reviews of Geophysics, 33, 67-97.

Munk, W.H., and C. Wunsch (1998) The moon and mixing: abyssal recipes II. Deep Sea Research, 45, 1977-2010.

Oberhuber, J. M. (1993) Simulation of the Atlantic circulation with a coupled sea ice-mixed layer-isopycnal general circulation model. Journal of Physical Oceanography, 23, 808-829.

O'Brien, J.J. (1986) Advanced Physical Oceanographic Numerical Modeling. D. Reidel Publishing Company.

Pacanowski, R. C., and S. M. Griffies (1999) MOM 3.0 Manual, NOAA/Geophysical Fluid Dynamics Laboratory, Princeton, USA 08542. 680 pages. Available from http://www.gfdl.noaa.gov/MOM/MOM.html.

Pedlosky, J. (1987) Geophysical Fluid Dynamics, 2nd edition. Springer-Verlag Publishers. 710 pages.

Philander, S.G. (1990) El Niño, La Niña, and the Southern Oscillation. Academic Press. 293 pages.

Philander, S.G. (1998) Is the Temperature Rising? The Uncertain Science of Global Warming. Princeton University Press. 262 pages.

Polzin, K. L., J. M. Toole, J. R. Ledwell, and R. W. Schmidt (1997) Spatial variability of turbulent mixing in the abyssal ocean. Science, 276, 93-96.

Pond, S., and K. Bryan (1976) Numerical models of the ocean circulation. Reviews of Geophysics and Space Physics, 14, 243-263.

Redi, M. H. (1982) Oceanic isopycnal mixing by coordinate rotation. Journal of Physical Oceanography, 12, 1154-1158.

Richardson, L. F. (1922) Weather Prediction by Numerical Process. Cambridge University Press, reprinted by Dover, 1965. 236 pages.

Robinson, A. R., and M. Golnaraghi (1994) The Physical and Dynamical Oceanography of the Mediterranean Sea, in Ocean Processes in Climate Dynamics, P. Malanotte-Rizzoli and A. R. Robinson (editors), pp. 255-306, Kluwer Academic.

Salmon, R. (1998) Lectures on Geophysical Fluid Dynamics, Oxford University Press. 378.

Sato, M., J.E. Hansen, M.P. McCormick and J. Pollack (1993) Statospheric aerosol optical depths (1850-1990). J. Geophys. Res. 98, 22987-22994

Semtner, Jr., A. J. (1974) An oceanic general circulation model with bottom topography. In Numerical Simulation of Weather and Climate, Technical Report No. 9, UCLA Department of Meteorology.

Smagorinsky, J. (1993) Some historical remarks on the use of nonlinear viscosities, in Large Eddy Simulation of Complex Engineering and Geophysical Flows, edited by B. Galperin and S. A. Orszag. Cambridge University Press.

Stott, P.A., S.F.B. Tett, G.S. Jones, M.R. Allen, J.F.B.Mitchell and G.J. Jenkins (2000b) External control of twentieth century temperature variations by natural and anthropogenic forcings. Science, 15, 2133-2137.

Tett, S.F.B., G.S. Jones, P.A. Stott, D.C. Hill, J.F.B. Mitchell, M.R. Allen, W.J. Ingram, T.C. Johns, C.E. Johnson, A. Jones, D.L. Roberts, D.M.H. Sexton and M.J. Woodage (2000) Estimation of natural and anthropogenic contributions to 20th century. Hadley Center Tech Note 19, Hadley Center for Climate Prediction and Response, Meteorological Office, RG12 2SY, UK pp52.

Wajsowicz, R. C. (1993) A consistent formulation of the anisotropic stress tensor for use in models of the large-scale ocean circulation. Journal of Computational Physics, 105, 333-338.

6 An Introduction to Linear Predictability Analysis

Stephen M. Griffies

Geophysical Fluid Dynamics Laboratory, NOAA. Princeton Forrestal Campus Rte. 1, P.O. Box 308, Princeton, NJ 08542-0308, USA
smg@gfdl.noaa.gov

6.1
Introduction and motivation.

This chapter introduces some basic notions of climate predictability. The tools used are those from linear stochastic processes. As an application, we analyze simulations of the Atlantic thermohaline circulation.

6.1.1
Linear systems

The relevance of linear models for climate studies is largely based on the empirical finding that many forms of climate variability, when considered on decadal to centennial time scales, are uni-modal and often close to Gaussian. This is not to say that multi-modality or non-Gaussian behavior are irrelevant for climate studies (e.g., Ghil and Childress 1987, Manabe and Stouffer 1988). Instead, it indicates the relevance of linear analysis as a general "null hypothesis." That is, only by ruling out the possibility of a linear stochastic interpretation should one consider more sophisticated, and often more data demanding and theoretically speculative, means of analysis. This perspective is not new, and can be found in various forms in Hasselmann (1976), Wunsch (1992), Sarachik et al (1996), and Schneider and Griffies (1999).

The discussion of linear stochastic systems will be given from the perspective of stochastic differential equations. This approach lends itself to connections with well studied systems in statistical physics, most notably Brownian motion. Thorough treatments of these topics can be found in many places, with the following familiar to the author: Chandrasekhar (1943), MacDonald (1962), Reif (1965), Gardiner (1985), and Kubo et al (1985) provide discussions of stochastic processes from a mathematical physics perspective; Jenkins and Watts (1968), Chatfield (1989), and Lütkepohl (1993) provide more statistical perspectives, and Honerkamp (1994) combines the approaches.

6.1.2 Ensemble predictability experiments

Predictability studies are concerned with the feasibility of making predictions for dynamical systems on the basis of incomplete or "imperfect" initial conditions. That is, predictability is concerned with the length of time for which a prediction can be carried out before errors in the initial conditions rise to the levels of ambient variability. From a climate modeling perspective, predictability studies are a first step towards assessing the accuracy of forecasts. In the early stages of such studies, validation of the forecast's accuracy against reality is not considered. Rather, a characterization of the model's precision is made in order to determine the statistical significance of a potential forecast.

A tool for addressing precision is the ensemble experiment. Here, a dynamical model is integrated a number of times starting from slightly different independent initial conditions. The model is exhibiting predictable behavior if the ensemble statistics at a particular time are distinguishable from the statistics of a climatology. A climatology refers to the statistics (e.g., mean, standard deviation, probability distribution function) from a long realization of the system. For linear Gaussian processes, which form the basis of the analysis in this chapter, a complete comparison of the ensemble and climatology is afforded by comparison of their mean and variance. The work of Anderson and Stern (1996), and references therein, provide additional discussion of these points.

6.1.3 Time scale separation and a stochastic perspective

The climate system, through its various sub-systems, evolves on a broad range of time scales which can be characterized by representative auto-correlation times. In the context of the coupled ocean-atmosphere system, the limits on the time scales are roughly those defined by the short time scale atmospheric processes and the long time scale oceanic processes. Of interest here is the long time (yearly to decadal) and large space (basin-wide to global) scale responses attributed to the ocean's thermohaline circulation. This large separation of time scales between the synoptic atmospheric phenomenon and the thermohaline circulation is fundamental to the considerations in this chapter.

Atmospheric predictability studies indicate a loss of deterministic predictability of synoptic scale motions (i.e., motions associated with daily weather patterns) at the 10 day time scale (e.g., Lorenz 1969, 1973, 1975, Palmer 1996). For times longer than the synoptic predictability time, the atmosphere is effectively random. Consequently, through interactions between the atmosphere and the ocean, the coupled ocean-atmosphere system may be considered a stochastic or non-deterministic system at these longer time scales. This loss of determinism does not equate to the absence of useful statistical predictions. Rather, through exploiting the memory of the slower time scale processes, namely the ocean circulation, and its interactions with the atmosphere, useful statistical predictions regarding the complete system are conceivable (see, for example, the lecture about El Niño by Philander in this volume, Chap. 3).

Motivated by this time scale separation, considerations of the coupled ocean-atmosphere system at time scales longer than atmospheric synoptic times motivate a simplification in which the atmosphere is considered a source of white noise forcing acting on the ocean. Technically, this approximation equates the time auto-correlation of the atmospheric processes to a delta function. In the frequency domain, this equates to equal power at all frequencies, hence the appellation "white". In this context, the ocean's evolution is thought of as a particular realization of a stochastic process. This "atmosphere as a noise-box" approach, albeit simple-minded, can serve as a first step in a more complete characterization of climate predictability. This approach may not be useful for all forms of climate variability, yet it has been useful for the multi-decadal variability of the thermohaline circulation studied by Griffies and Bryan (1997a,b).

6.1.4 Remainder of this chapter

The remainder of this chapter consists of the following sections. Sect. 6.2 discusses some aspects of the thermohaline circulation and its variability. Sect. 6.3 and an appendix describe predictability seen in two linear stochastic processes: red noise and a noise-driven damped oscillator. Sect. 6.4 discusses some methods of pattern analysis which aim to partition variability and predictability into their dominant modes. Sect. 6.5 finishes with closing remarks.

6.2 The thermohaline circulation.

As discussed in the lecture on ocean climate models in this volume (Chap. 5), a great deal of the ocean's water mass structure, and hence its large-scale currents, are determined by the flux of buoyancy across the ocean surface. As these fluxes affect the temperature and salinity of the ocean, the resulting circulation is termed the thermohaline circulation (THC).

There are special regions of the ocean where buoyancy forcing plays a key role in forming the deep ocean water masses. One is the Weddell and Ross Seas in the Southern Ocean, and another is the Labrador and Nordic Seas of the North Atlantic. In these parts of the world, winter storms cause huge losses of buoyancy and hence make the water quite dense. Additionally, the Atlantic waters are generally more saline than the Pacific, thereby further reducing the buoyancy of surface waters in the high North Atlantic. Furthermore, sea ice plays a crucial part because of participation in the salt and heat budgets of these regions.

Upon losing enough buoyancy, surface waters undergo vertical convection, which results in a tremendous mixing of properties throughout a vertical column of water. This process largely determines the structure of the abyssal ocean. It also provides a conduit for incorporating the effects of atmospheric forcing into the deep ocean. These deep water masses are then transported to other parts of the globe, eventually reaching the surface after some years to centuries. It is interesting to note that absent the effects of sea ice, many of the same processes active in the high latitude Atlantic are also seen in parts of the Mediterranean Sea (e.g., Robinson and Golnaraghi 1994). Since the climate of the Mediterranean is more suitable for mid-winter cruises than, say, the Greenland Sea, the Mediterranean has seen many important studies of ocean convection. For a lucid and thorough review of ocean convection, including studies of the Mediterranean, see Marshall and Schott (1999). For a discussion of the climate variability seen in the North Atlantic convective activity, see the review of Dickson et al. (1996).

Currently, there are studies indicating the added importance of the Antarctic Circumpolar Current (ACC) for determining the structure of the ocean's water masses. Crucially, the ACC is the only ocean current which has no meridional boundaries interrupting its zonal flow. Hence, the dynamical balances active for the ACC are quite distinct from those relevant for, say, the North Atlantic, which is enclosed by the American and European/African continents. These differences are key to understanding the ACC's role in the World Ocean. Much present research is focused on clarifying these ideas, many of which have been prompted by some compelling modeling studies (e.g., Toggweiler and Samuels 1995, Gnanadesikan 1999, and Vallis 1999). Although these topics are quite interesting, the following discussion will focus on just the North Atlantic's overturning circulation, and in particular its variability.

6.2.1 North Atlantic climate variability

About the mean ocean circulation state, there are possibilities for nontrivial modes of variability. Variability associated with the rates of deep water formation occurring in the North Atlantic forms the focus for the following.

Historical data for sea surface temperatures (SST) indicate that the North Atlantic has experienced significant multidecadal climate variability superimposed on a positive trend (e.g., Folland et al 1986, Mann and Park 1994). Further analysis by Hansen and Bezdek (1996) and Sutton and Allen (1997) shows that a yearly to decadal time scale propagation of SST anomalies may be traced in the surface circulation around the North Atlantic sub-polar gyre. Earlier work, such as that of Levitus (1990), documents the large scale multidecadal changes that took place in the Atlantic sea surface height.

These studies suggest that variability in mid-depth to surface Atlantic Ocean properties contains an intriguing amount of coherence over multiple years to decades, thus suggesting the possibility for nontrivial predictability of such climatic fluctuations. A primary objective of several international climate research programs is to assess the possibilities for projecting climate over decadal and longer time scales. For this purpose, it is essential to quantify the predictability of such variability as it manifests itself in the ocean. The central reason for this interest is that the ocean represents the source of memory upon which useful long-term climate forecasts could be based.

6.2.2 Conceptual models of THC variability

Perhaps the simplest conceptual model of Atlantic variability is Hasselmann's (1976) random walk model of air-sea interaction. In this model, the ocean is taken to be a homogeneous reservoir of water with random heat fluxes forcing it at the upper boundary. The random nature of the forcing is meant to represent the high frequency (with period on the order of days) atmospheric storms crossing the North Atlantic. As mentioned earlier, since there is little temporal coherence of this forcing beyond a week or two, it can be roughly approximated as random noise.

In the Hasselmann model, long term effects of the surface heat balance are parameterized as a linear negative feedback which prevents the reservoir from moving too far away from a stable climate equilibrium. The combined effect of the random heating and cooling, and the negative feedback, is to allow random excursions of SST anomalies whose magnitude are determined by the strength of the feedback. The temporal variability arising from this mechanism displays a red noise power spectrum, which decays as ω^{-2} for high frequency ω and becomes a constant for low frequency (see Eq. 6.14, in the appendix).

The relevance of the Hasselmann model to the problem of North Atlantic variability was suggested by Bryan and Hansen (1995), who considered a stochastically forced Stommel (1961) two-box model of the thermohaline circulation. Building on the ideas of Bryan and Hansen (1995), Griffies and Tziperman (1995) introduced another conceptual model based on the stochastic forcing of a four-box model originally considered by Huang, Luyten, and Stommel (1992).

In the THC box models, circulation is driven by latitudinal buoyancy gradients set up by surface fluxes of heat and salt. In general, the fluxes have a time-mean tendency to increase density of northern waters in the North Atlantic, hence causing water to sink. This high latitude sinking sets up an overturning circulation, with water moving northward at the surface, sinking as it reaches the northern boundary, moving southward at depth, and rising at the southern boundary. Fig. 6.1 illustrates this situation for the Griffies and Tziperman four box model.

In addition to a time-mean component, the buoyancy fluxes have a large stochastic component associated with atmospheric disturbances similar to those considered in the Hasselmann model. When introduced into the box model of Griffies and Tziperman, the effect provides a driving for the THC on all time scales. The model's thermohaline dynamics provide a response which has power concentrated in the multidecadal time scale. In particular, the feedbacks from temperature anomalies (acting as a negative feedback) and salinity anomalies (acting as a positive feedback) contribute to the decay and growth, respectively, of the THC circulation. A time delay in the THC response to north-south pressure gradients allows for damped oscillatory changes in the ocean circulation. The mechanical analog is a noise-driven, damped linear oscillator.

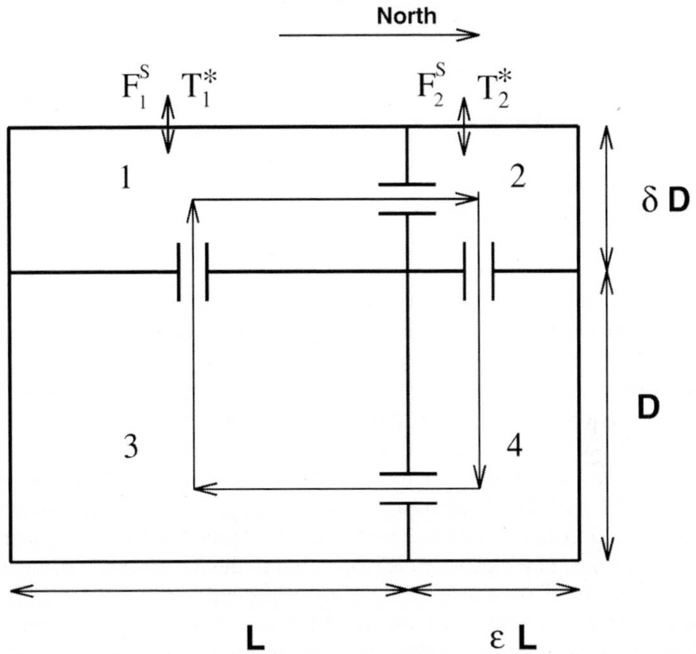

Fig. 6.1. Geometry of the Griffies and Tziperman four box model. The mean circulation, with sinking in the north and rising in the south, is indicated. The boxes have homogeneous temperature and salinity. The surface boxes are forced with fluxes of salt (F^S) and damped to the temperatures (T^*).

6.3 Predictability in linear noise-driven systems.

In this section and the appendix we provide a mathematical analysis of two linear stochastic systems: the univariate red noise process, and noise-driven damped harmonic oscillator. These systems are simple enough that analytic expressions are available for various diagnostics quantifying their variability and predictability. Hence, they are useful as a means to pedagogically introduce predictability concepts. Additionally, they find much use in more complicated systems, since any stable linear system can be reduced to a system of coupled damped linear oscillators (e.g., see Hasselmann 1988 and Penland 1989, 1996 for multivariate examples). The oscillator will find particular use when interpreting the climate model variability of the Griffies and Bryan (1997a,b) papers. Some of the material in this section, as well as the appendix, formed an appendix to Griffies and Bryan (1997b).

6.3.1 The red noise process

The first example of a linear noise-driven process is the red noise or Brownian process $v(t)$, which is described by

$$\dot{v}(t) = -\alpha v(t) + \xi(t), \tag{6.1}$$

where the overdot indicates a time derivative. The parameter $\alpha > 0$ represents the effects of frictional forces, which are dissipative and act to relax $v(t)$ back to its equilibrium value $v_{eq}(t) = 0$; i.e., it provides a linear negative feedback. $\xi(t)$ represents the rapidly de-correlating forces, and is the fluctuation term. It will be modeled as a Gaussian white noise process with zero mean and auto-covariance $\langle \xi(t)\xi(s) \rangle = \sigma^2_\xi\, \delta(t-s)$, with $\delta(t)$ the Dirac delta function and σ^2_ξ determining the power of the noise. As traditionally treated in statistical physics, the expectation operator $\langle\rangle$ can be considered a probability average, or an average over an extremely large, or formally infinite, ensemble. The ensemble consists of identically prepared elements, in which only the realization of the white noise forcing differs.

In the context of Brownian motion, $v(t)$ represents the velocity dx/dt of the Brownian particle (see e.g., Reif 1965). This is also the process proposed by Hasselmann (1976) as a means for describing the integrative response of a slow subsystem of the climate, such as the ocean, to a shorter time scale forcing, such as that provided by the synoptic scale atmosphere. Some further details regarding the stationary statistics of this process are given in the appendix.

Forecasting red noise

The Brownian process is stochastic and hence non-deterministic. Therefore, its future cannot be precisely determined. Given observations of a particular realization of the red noise process, what is its most probable future state?

The simplest answer is to say that the future remains the same as the present. This trivial forecast is called persistence: $v_{persist}(\tau) = v(0)$. The persistence forecast uses no previous information regarding the system, only the last measured value. Indeed, without any previous temporal information, persistence is all that is available.

When given an extended temporal sequence of measurements, one can estimate the correlation function discussed in the appendix (see Eq. 6.11). Given such climatological information, as well as the last measurement $v(0)$, it is reasonable to expect that one can produce a forecast of greater utility than the persistence forecast. Indeed, this is the case. Since the process is Gaussian, the most probable future state is also that state which minimizes the least squares difference from the actual realization (see, e.g., Penland 1989). For red noise, this criteria defines the damped persistence forecast (Lorenz 1973)

$$v_{dp}(\tau) = v(0)e^{-\alpha\tau} \tag{6.2}$$

That is, the damped persistence forecast is simply a damping of a persistence forecast $v_{persist}(\tau) = v(0)$ back to a zero anomaly. The relaxation time for this damping is given by the system's auto-correlation time α^{-1}.

In order to show how useful a forecast is, one needs to define an error metric. For this purpose, we first obtain an expression for the exact solution of the red noise equation (Eq. 6.2). That is, a formal expression for a particular element from an ensemble of red noise processes can be written

$$v(\tau) = v_{dp}(\tau) + \int_0^\tau e^{-\alpha(\tau-u)}\xi(u)du \tag{6.3}$$

This expression is a function of the white noise process $\xi(u)$, which differs for each ensemble element. Upon taking expectations of both sides, we see that $\langle v(\tau)\rangle = v_{dp}(\tau)$, since $\langle\xi(u)\rangle = 0$, and the expectation operator commutes with integration. That is, the damped persistence forecast is the same as the mean of an infinite sized ensemble of red noise processes, where each ensemble element starts at the initial state $v(0)$ yet experiences different realizations of the white noise forcing.

Using the last result (Eq. 6.3), we can compute the mean square difference between the damped persistence forecast $v_{dp}(\tau)$ and an infinite number of realizations of red noise processes, each of which starts at $v(0)$. This squared difference is a natural definition of the forecast error, and for red noise it takes the form

$$\sigma^2_{dp}(\tau) \equiv \langle(v(\tau) - v_{dp}(\tau))^2\rangle = \langle v^2\rangle\,(1 - e^{-2\alpha\tau}), \tag{6.4}$$

where $\langle v^2\rangle = \sigma^2_\xi/2\alpha$ is the climatological variance of the process (see Eq. 6.12 in the appendix). The error can also be interpreted as the variance of an infinite sized ensemble about its mean state, or equivalently as the covariance between the damped persistence forecast and an infinite number of realizations of red noise, each of which starts at $v(0)$.

The squared forecast error σ^2_{dp} approaches the climatological variance with an e-folding time $1/(2\alpha)$. The larger the feedback coefficient α of the red noise process, the faster the forecast error saturates to that defined by the climatology, and hence the less useful the forecast becomes. The same error calculation for the persistence forecast $v_{persist}(\tau) = v(0)$ yields the mean squared error

$$\sigma^2_{persist}(\tau) = \langle(v(\tau) - v(0))^2\rangle = 2\,\langle v^2\rangle\,(1 - e^{-\alpha\tau}), \tag{6.5}$$

which saturates to twice the climatological variance yet at a slower rate than the damped persistence forecast. The error in the persistence forecast, however, is always larger than that of damped persistence, and so it is less useful. Again, since the persistence forecast utilizes only the initial condition, it is reasonable to expect it to be less useful than the damped persistence forecast, which utilizes both the initial condition and the correlation time.

A numerical example

Fig. 6.2a shows a 200 year realization of a red noise process. The parameters used for generating this signal are indicated in the caption. The structure of this time series is not dissimilar from many time series of geophysical fields (this point is emphasized by Wunsch 1992). The sample auto-correlation function (*acf*) computed from 400 years of the process is shown in Fig. 6.2b. Fig. 6.3a then shows a nine element ensemble of red noise processes. This ensemble size, though mathematically quite small, is not uncommon in practice for predictability studies which use realistic climate models (e.g., Griffies and Bryan 1997a,b).

The noise forcing is identical for all the ensemble elements up to around year 33. Afterwards, the noise differs and so the trajectories diverge. The ensemble mean is shown in Fig. 6.3b. The smoother dashed line is the mean of the infinite sized ensemble, i.e., the damped persistence forecast given by Eq. 6.2. Fig. 6.3c shows the ensemble variance, which measures the spread of the ensemble about its mean. The smoother dashed line is the variance which results from an infinite sized ensemble, which is identical to the optimal forecast error (Eq. 6.4). Note that we follow the discussion of Penland (1989) when numerically realizing the stochastic processes considered here (see also Kloeden and Platen 1992 for a more complete treatment).

6.3.2 The harmonic Brownian process

This section presents the analogous calculations of the previous section for the damped harmonic oscillator driven by Gaussian white noise. This process will be referred to here as the harmonic Brownian process, and it is governed by

$$x(t)+\omega_0^2(t)=-2\beta\dot{x}(t)+\xi(t) \tag{6.6}$$

The parameter $\beta \geq 0$ represents the contribution of frictional or viscous dissipation, the frequency ω_0 is the natural frequency of the noise free and dissipation free system, and the white noise forcing $\xi(t)$ is like that in the red noise example. The second section in the appendix discusses the stationary statistics for this process. The remainder of this section describes the problem of forecasting and characterizing the predictability of this system when it is underdamped; i.e., $\Omega^2 \equiv \omega_0^2 - \beta^2 > 0$.

Forecasting the noise-driven oscillator

As in the last subsection, since the process is linear and Gaussian, the optimal forecast (also the most probable future state) is the same as the mean of an infinite ensemble; i.e., it is the noise free solution of Eq. 6.6. The forecast satisfying these conditions is

$$x_{dhp}(\tau) = \frac{e^{-\beta\tau}}{\Omega}\left(x(0)\Omega\cos(\Omega\tau) + \left[\dot{x}(0) + \beta x(0)\right]\sin(\Omega\tau)\right) \tag{6.7}$$

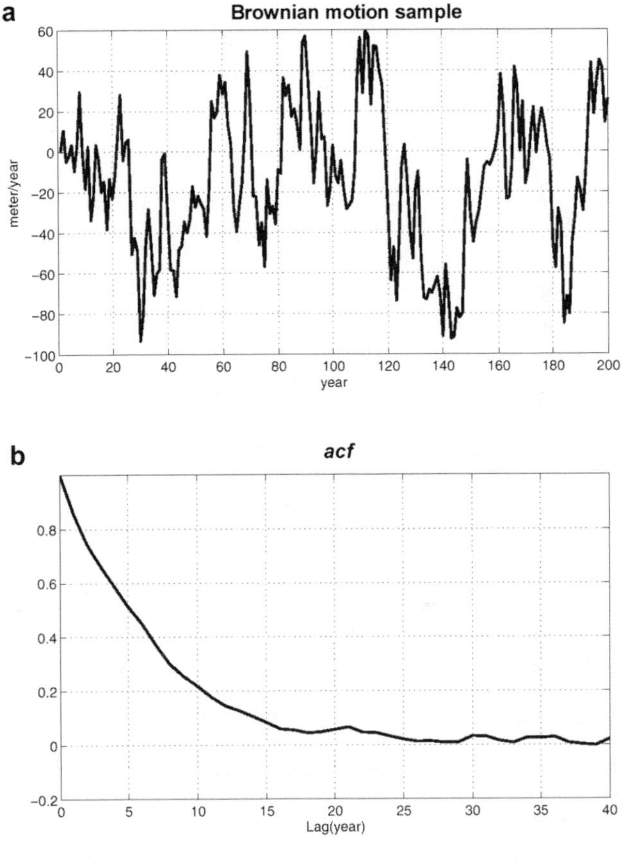

Fig. 6.2. a) A 200 year realization of the red noise or Brownian process (the continuous time process is described by Eq. 6.1) using an inverse damping time $\alpha^{-1} = 7$ years and noise variance 4 meter^2/years^3. b) The *acf* computed from 400 years of the process, including the 200 years in (a). This function approximates that from the infinite sized ensemble, which is a pure exponential decay, given by Eq. 6.11 in the appendix. An approximate 7 year e-folding time is apparent. Note that for illustrative purposes, the dimensions of the process *v(t)* are chosen to be those of a velocity, as is relevant for Brownian motion.

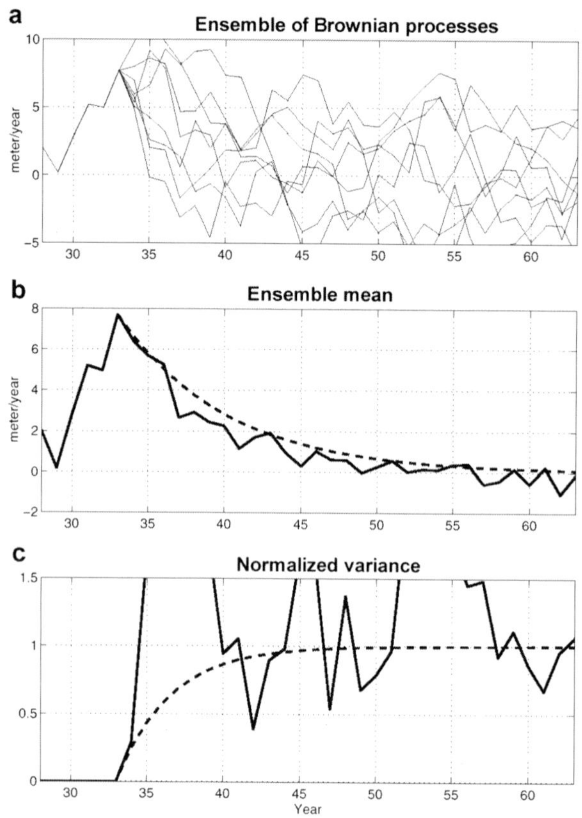

Fig. 6.3. a) A nine element ensemble of red noise processes, each element of which has a different realization of the white noise process after starting from the same initial state near year 33. b) The ensemble mean (solid line) and fit of the damped persistence (dashed line; Eq. 6.2). Note the different vertical axis relative to (a). c) The ensemble variance (solid line) and the error function (dashed line; Eq. 6.4) of the damped persistence forecast.

This optimal forecast (Eq. 6.7) will be called damped harmonic persistence in analogy with the damped persistence forecast given by Eq. 6.2. Note that for certain initial conditions relevant for the growing phase of an oscillation, the damped harmonic persistence forecast results in an initially growing anomaly. This growth is in contrast to the damped persistence forecast (Eq. 6.2) which always results in a damping of anomalies. The damped harmonic persistence forecast requires knowledge of the initial state $x(0)$, the initial time tendency $dx/dt\ (0)$, the damping coefficient β, and the natural frequency ω_0.

The mean squared error in the damped harmonic persistence forecast $\sigma^2_{dhp} = \langle (x(\tau) - x_{dhp}(\tau))^2 \rangle$ is given by

$$\sigma^2_{dhp}(\tau) = \langle x^2 \rangle \left(1 - \frac{e^{-2\beta\tau}}{\Omega^2} \left[\omega_0 - \beta^2 \cos(2\Omega\tau) + \beta\Omega \sin(2\Omega\tau) \right] \right) \tag{6.8}$$

The error σ^2_{dhp} exponentially saturates to the climatological variance $\langle x^2 \rangle = \sigma^2_\xi / (4\beta\omega^2_0)$. The corresponding error for the persistence forecast $x_{persist} = x(0)$ is

$$\sigma^2_{persist}(\tau) = 2\langle x^2 \rangle \left(1 - e^{-\beta\tau} \left[\cos(\Omega\tau) + \frac{\beta}{\Omega} \sin(\Omega\tau) \right] \right) \tag{6.9}$$

which saturates to twice the climatological variance, and is always larger than the damped harmonic persistence forecast error. Figures 4 and 5 show examples of this process.

6.4 Elements of pattern analysis.

When a geophysical field is sampled, it is often of interest to summarize its spatio-temporal variability using just a few dominant or canonical patterns. The method for doing so has long been part of climate analysis (e.g., Wilks 1995, von Storch and Navarra 1995) and statistics (e.g., Jolliffe 1986), where is it commonly known as principle component analysis, and the dominant modes are called empirical orthogonal functions (EOFs). Relatedly, it can be of interest to partition a field into its dominant patterns of predictability. The latter approach was recently introduced by Schneider and Griffies (1999). Some aspects of these methods are introduced in this section.

6.4.1 A climate model dataset

The predictability studies of Griffies and Bryan (1997a,b) employed a coupled atmosphere-ocean-land-ice model developed at GFDL to study climate (Manabe et al. 1991). Of interest is the model's multidecadal North Atlantic variability seen in the 1000 year control experiment analyzed by Delworth et al (1993). This variability is associated with variations in the model's thermohaline circulation in which there is a strong correlation between fluctuations in the oceanic overturning and changes in the thermocline density distribution and sea surface properties.

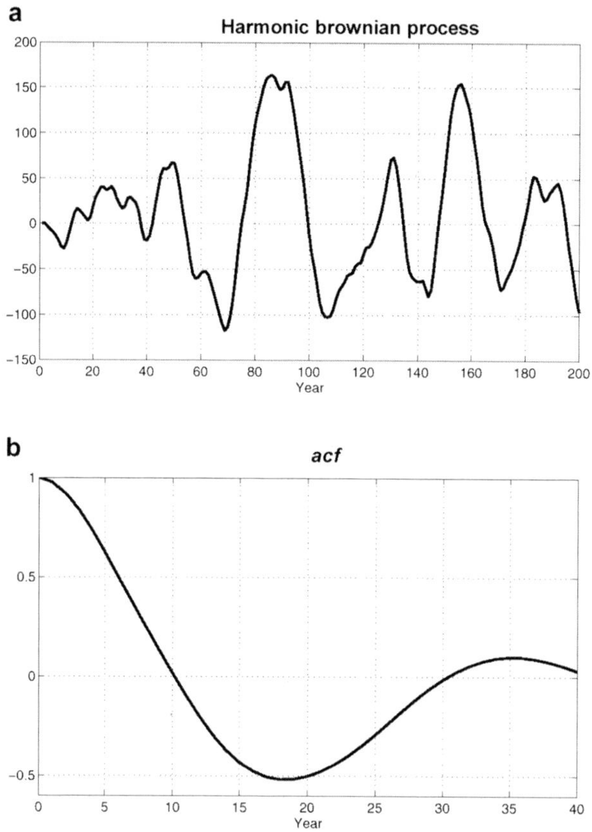

Fig. 6.4. a) A 200 year realization of the harmonic Brownian process (the continuous time process is described by Eq. 6.6) using an inverse damping time $\beta^{-1} = 10$ years, period $2\pi / \omega_0 = 33$ years. b) The *acf* for 400 years of the harmonic Brownian process (solid line), including the 200 years of (a).

6.4.2 Dominant patterns of variability

A useful means for quantifying the amount of spatio-temporal variability exhibited by a field is to compute its covariance matrix. This is an $N \times N$ symmetric matrix, where N represents the number of discrete grid points from which the field is sampled. As the covariance matrix is symmetric, it possess a complete set of mutually orthogonal eigenvectors with non-negative eigenvalues. These eigenvectors are the empirical orthogonal functions mentioned earlier. The EOF with the largest eigenvalue represents the most dominant pattern of variability, and the EOF with the next largest eigenvalue represents the next most dominant pattern, and so on. As the EOFs form a complete set, they can be used as a basis for representing the field. The weighting factors in this expansion change in time, since the field changes in time, and they form what are known as principle components (PCs). For many geophysical phenomena, the dominant EOF patterns have both the largest space scales and their corresponding PCs have the longest time scales.

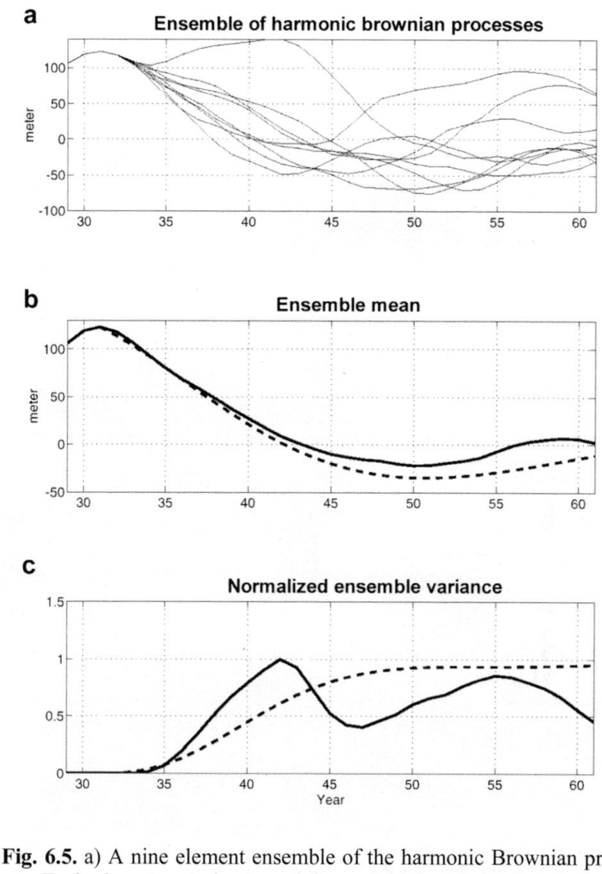

Fig. 6.5. a) A nine element ensemble of the harmonic Brownian process. Each element was integrated for an initial period using the same white noise forcing. Afterwards, the noise is unique to each element. b) The ensemble mean (solid) and the damped harmonic persistence forecast (Eq. 6.7; dashed line). c) The ensemble variance (solid) and the error function of the damped harmonic persistence forecast (Eq. 6.8); dashed line). The parameters for these forecasts were estimated from the *acf* of Fig. 6.4. Note the approximately zero initial slope in the variance representing the slower initial de-correlation of the ensemble elements relative to the Brownian process ensemble shown in Fig. 6.3.

Griffies and Bryan (1997a,b) analyzed predictability of the dominant EOFs for a set of fields within the North Atlantic region of their simulation. In particular, the model's North Atlantic variability is seen quite clearly in the dynamic topography, which is the ocean's analog of the pressure surfaces seen on atmospheric weather maps. Dynamic topography is basically the sea surface height taken with respect to a reference level at some depth in the ocean (1100 meters is used here), and it represents a summary of the ocean's large-scale circulation over the middle to upper ocean.

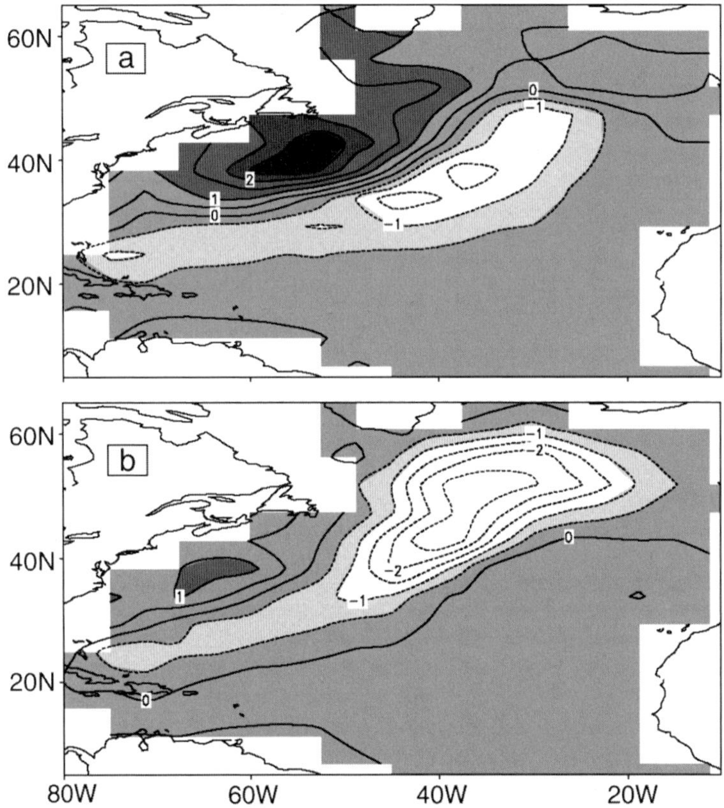

Fig. 6.6. a) EOF-1 (21% of the variance) and b) EOF-2 (17% of the variance) patterns for the climate model's North Atlantic dynamic topography defined from annual mean model data over years 1-200. The patterns are scaled by the standard deviations of their associated principal components. Note that an outline of the realistic geography, with Labrador to the left, is provided for orientation. The model grid cells are visible on the boundaries between land and sea. This figure is taken from Schneider and Griffies (1999).

Fig. 6.6a and 6.6b show the first and second EOF patterns for the climate model's North Atlantic annual mean dynamic topography. These patterns account for 21% and 17%, respectively, of the variability in model years 1-200. Note that these percentages are determined by the relative values of the covariance matrix eigenvalues. Fig. 6.7a and 6.7b show the corresponding coherent oscillations exhibited by the first principle component (statistics for the second PC are similar to the first). This form of variability is reminiscent of the noise-driven damped linear oscillator process considered in Sect. 6.3.2. Indeed, as shown by Griffies and Tziperman (1995) and Griffies and Bryan (1997a,b), this simple model does a very good job at capturing the statistical behavior of the large-scale North Atlantic variability seen in the coupled climate model.

In order to investigate the predictability of the climate model's behavior, Griffies and Bryan ran many realizations of the climate model, in which the initial conditions of the ocean were identical yet the atmospheric initial conditions differed. This approach is based on the ideas mentioned earlier, in which the Atlantic's THC is thought to be stochastically forced by the atmosphere. To characterize predictability, they found it useful to project the EOFs onto the variability seen for each ensemble member. The trajectories for the principal components are initially coincident in phase space. In time, however, the trajectories for the different ensemble members spread. The rate at which they spread quantifies the pattern's sensitivity to the randomly chosen atmospheric initial conditions. The approach is directly analogous to that taken in Sect. 6.3 when discussing predictability of the red noise and damped oscillator.

Fig. 6.7c and 6.7d show trajectories from the climate model, and the deviation from the ensemble mean, for PC-1 from the first of the four ensembles generated by Griffies and Bryan. Also shown is the result expected if the climate model variability was acting just as the noise driven damped linear oscillator (i.e., Eq. 6.7). The agreement between the climate model and the damped oscillator is quite good for the first 15-20 years, further indicating the relevance of the linear oscillator model for summarizing the climate model's variability. Note that the spread in the principle components reaches one-half the climatological variance after some 10-15 years, which provides a rough measure of the predictability time scale for the field.

6.4.3 Dominant patterns of predictability

Griffies and Bryan (1997a,b) considered predictability of EOF patterns. Again, these patterns provide an optimal partitioning of a field into its orthogonal modes of variability. The Schneider and Griffies (1999) study considered the analogous question: What are the most predictable patterns, or how can the field be partitioned into patterns according to their predictability? Schneider and Griffies (1999) provided an algorithm for doing so, a so-called predictable component analysis, and they then applied the techniques to the Griffies and Bryan model dataset. The results largely confirmed the original findings based on EOF predictability, yet the new analysis put the earlier results on a firmer theoretical foundation. The interested reader should refer to the Schneider and Griffies study for complete details and elaboration. Additional discussion, from an alternative perspective, can be found in the study by Yang et al. (1998).

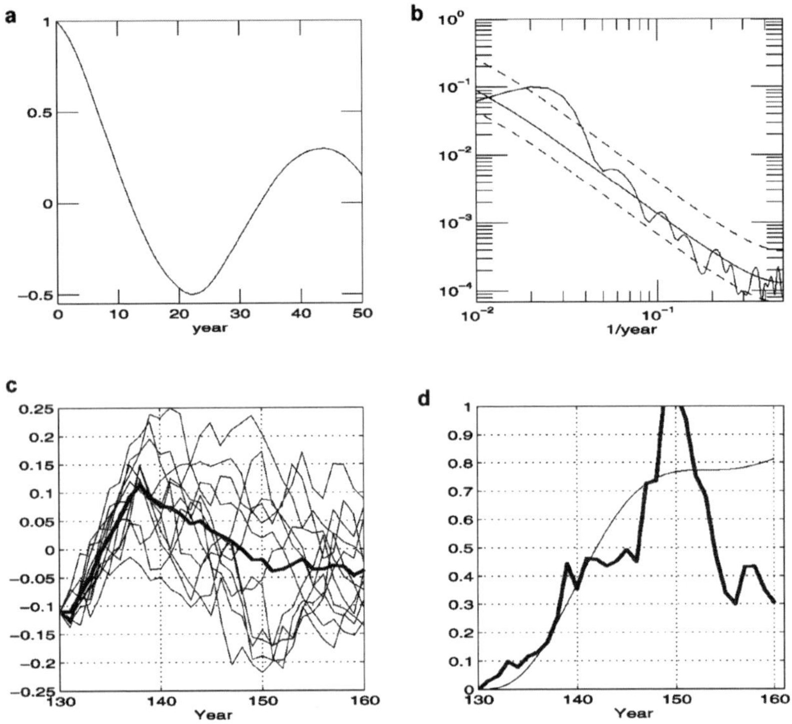

Fig. 6.7. a) Auto-correlation function for PC-1 for the North Atlantic dynamic topography defined from yearly mean fields over model years 1-200. The damped sine-wave behavior is characteristic of oscillatory variability (e.g., see Fig. 6.4). This sample autocorrelation function can be fit to that from a damped harmonic oscillator driven by white noise (Eq. 6.17), where the oscillator period and damping time are roughly 40 years each. b) Power spectrum (ordinate) corresponding to the autocorrelation function. The power rises above the 95% red noise null hypothesis bounds (the dashed lines) in the 40-60 year time scale. c) The PC-1 ensemble member trajectories for dynamic topography. The dark solid line is the ensemble mean. d) The normalized ensemble variance (ordinate), with the thin smooth line being the variance predicted from an infinite sized ensemble of the noise-driven damped harmonic oscillator (i.e., Eq. 6.8). This figure is taken from Griffies and Bryan (1997a). It is reprinted with permission from Science, volume 275, pages 181-184: copyright 1997 by The American Association for the Advancement of Science.

6.5 Closing remarks.

Decadal to centennial climate predictability is a rather young area of research when compared to the more mature area of weather and ENSO predictability. It is anticipated that with increased understanding and observations of long-term climate phenomena, there will be an increased interest in understanding and quantifying its predictability.

There are many items missing from this introduction. In particular, absent is the very important issue of estimating parameters for linear stochastic models. In general, neither the optimal model nor its parameters are known. This information must be estimated based on the given observed or simulated datasets. Such data is often sparse and under-sampled, even when taken from climate models due to the expense of running them. The reader is referred to the paper by Neumaier and Schneider (1997) for a thorough, and readable, discussion of these issues. Additionally, the *Matlab®* program documented by Schneider and Neumaier (1997) is very handy for performing the necessary computations. The Schneider and Griffies (1999) study provides some discussion with application to the Griffies and Bryan (1997a,b) climate model dataset.

Also missing is mention of nonlinear methods of analysis, such as those useful in studies of chaotic dynamical systems. These methods have found some use in idealized studies of ENSO (e.g., Tziperman et al. 1994 and Jin et al. 1994). Yet for decadal to centennial modes of variability, these methods have not found much use. One reason is that the nonlinear methods generally require an extremely long time series, which is typically not available for multi-decadal variability. Another is the empirical fact, mentioned in the introduction, that decadal to centennial modes of variability are often well modeled by linear noise-driven systems. The studies of Griffies and Bryan (1997a,b) and Schneider and Griffies (1999) provide examples.

Acknowledgements

This chapter is based on the second of two talks given during a conference on "Global Climate" held March 1999 at the Museu de la Ciència de la Fundació "La Caixa" in Barcelona, Catalunya, Spain. Many warm thanks go to the organizers of this conference for holding a wonderful meeting in this splendid city. In particular, thanks go to Francisco Comín, Paquita Ciller, Xavier Rodó, and Jorge Wagensberg. Thanks also go to the participants, and to the translators who brilliantly kept up with the lectures. Further thanks go to my predictability mentors and collaborators, including Jeff Anderson, Kirk Bryan, Tapio Schneider, Eli Tziperman, and many others.

Appendix

The stationary statistics of the red noise and harmonic Brownian processes are presented in this appendix. More elaboration of these processes useful for the derivations here can be found in the books by Gardiner (1985) and Kubo et al. (1985).

The red noise process

A particular solution to the red noise process (Eq. 6.1) is given by

$$v_p(t) = \int_{-\infty}^{t} e^{-\alpha(t-u)} \xi(u)du \tag{6.10}$$

For the subtleties of interpreting the integral of white noise, refer to Gardiner (1985). Choosing the lower limit at $-\infty$ for the particular solution allows the stationary statistics of the process to be found straightforwardly from this expression. The stationary mean $\langle v \rangle$ of the process vanishes since $\langle \xi \rangle = 0$. The stationary auto-covariance function (*acvf*) can be found by multiplying $v_p(t)$ by v_p (s) and taking an expectation

$$\langle v(t)v(s) \rangle = \langle v^2 \rangle e^{-\alpha|t-s|} \tag{6.11}$$

Where

$$\langle v^2 \rangle = (\sigma_\xi^2 / 2\alpha) \tag{6.12}$$

is the zero lag variance. The covariance between adjacent points in time falls off exponentially with e-folding time $1/\alpha$. Note the dependence on the time difference $(t - s)$; a property characteristic of statistically stationary processes. As expected, the larger the damping coefficient α, the smaller the variance; conversely, the larger σ_ξ^2, representing the power of the noise forcing, the larger the variance. The normalized auto-covariance

$$\langle v(t)v(s) \rangle \langle v^2 \rangle^{-1} = e^{-\alpha|t-s|} \tag{6.13}$$

is called the auto-correlation function (*acf*). The Fourier transform of Eq. 6.1 yields the frequency space solution

$$\tilde{v}(\omega) = \tilde{\xi}(\omega) / (\alpha + i\omega) \tag{6.14}$$

Since the noise is white, $\tilde{\xi}(\omega) = \sigma_\xi$, the absolute square $|\tilde{v}(\omega)|^2$ gives

$$S(\omega) = \frac{2\alpha \langle v^2 \rangle}{\alpha^2 + \omega^2} \tag{6.15}$$

The form of this spectrum motivates the name red noise since the power is concentrated in the low or red frequency end of the spectrum. For correlation times longer than the e-folding time α^{-1} of the *acf*, there is basically no correlation or memory remaining in the process. Correspondingly, for frequencies longer than the angular frequency $\omega = \alpha$, the spectrum flattens out to approximate that of a white noise process with power $2\langle v^2\rangle \alpha^{-1} = \sigma^2_\xi / \alpha^2$.

The harmonic Brownian process

The harmonic Brownian process, governed by Eq. 6.6, has a particular solution given by

$$x_p(t) = \int_{-\infty}^{t} \frac{e^{-\beta(t-u)}}{\Omega} \sin[\Omega(t-u)]\xi(u)du \tag{6.16}$$

and *acvf*

$$\langle x(t)x(s)\rangle = \langle x^2\rangle e^{-\beta|t-s|}\left(\cos[\Omega|t-s|] + \frac{\beta}{\Omega}\sin[\Omega|t-s|]\right) \tag{6.17}$$

Setting $t = s$ yields the zero lag variance of the process $\langle x^2\rangle = \sigma^2_\xi / (4\beta\omega^2_0)$. The *acf* $\langle x(t)\, x(s)\rangle \langle x^2\rangle^{-1}$ is that function which is fit to the sample *acf*s throughout this section. The Fourier transform of the *acvf* gives the spectrum

$$S_{ho}(\omega) = \frac{4\beta\omega_0^2\langle x^2\rangle}{4\beta^2\omega^2 + (\omega_0^2 - \omega^2)^2} \tag{6.18}$$

References

Anderson, J. L., and W. F. Stern (1996) Evaluating the potential predictive utility of ensemble forecasts. Journal of Climate, 9, 260-269.

Bryan, K. and F.C. Hansen (1995) A stochastic model of North Atlantic climate variability on a decade to century time-scale. Proceedings of the Workshop on Decade-to-Century Time Scales of Climate Variability, National Research Council, Board on Atmospheric Sciences and Climate, National Academy of Sciences, Irvine, CA.

Chandrasekhar, S. (1943) Stochastic problems in physics and astronomy. Reviews of Modern Physics, 15, 1-89.

Chatfield, C. (1989) The Analysis of Time Series: An Introduction. Chapman and Hall. 241 pages.

Delworth, T., S. Manabe, R.J. Stouffer (1993) Interdecadal variations of the thermohaline circulation in a coupled ocean-atmosphere model. Journal of Climate, 12, 1993-2011.

Dickson, R., J. Lazier, J. Meincke, and P. Rhines (1996) Long-term coordinated changes in the convective activity of the North Atlantic. In NATO Advanced Study Institute: Decadal Climate Variability: Dynamics and Predictability, edited by D.L.T. Anderson and J. Willebrand. Springer, Berlin.

Folland, C. K., T. N. Palmer, and D. E. Parker (1986) Sahel rainfall and worldwide sea temperature. Nature 320, 602-606.

Gardiner, C.W. (1985) Handbook of Stochastic Methods for Physics, Chemistry, and the Natural Sciences. Springer-Verlag. 442 pages.

Ghil, M., and S. Childress (1987) Topics in Geophysical Fluid Dynamics: Atmospheric Dynamics, Dynamo Theory, and Climate Dynamics. Springer-Verlag Publishers. 485 pages.

Gnanadesikan, A. (1999). A simple predictive model for the structure of the oceanic pycnocline. Science, 283, 2077-2079.

Griffies, S. M and K. Bryan (1997a) Predictability of North Atlantic multidecadal climate variability. Science, 275, 181-184.

Griffies, S. M. and K. Bryan (1997b). A Predictability Study of Simulated North Atlantic Multidecadal Variability. Climate Dynamics, 13, 459-488.

Gri_ies, S. M. and E. Tziperman (1995). A linear thermohaline oscillator driven by stochastic atmospheric forcing. Journal of Climate, 8, 2440-2453.

Hansen, D. V. and H. F. Bezdek (1996). On the nature of anomalies in North Atlantic sea surface temperature. Journal of Geophysical Research, 101(CA), 8749-8758.

Hasselmann K (1976). Stochastic climate models, part 1- theory. Tellus, 18, 473-484.

Hasselmann, K. (1988). PIPs and POPs: the reduction of complex dynamical systems using principal interaction and oscillation patterns. J. of Geoph. Res., 93, 11015-11021.

Honerkamp, J. (1994). Stochastic Dynamical Systems: Concepts, Numerical Methods, Data Analysis. 535 pages.

Huang, R.X., J.R. Luyten, and H.M. Stommel (1992). Multiple equilibria states in combined thermal and saline circulation. Journal of Physical Oceanography, 22, 231-246.

Jenkins, G.M. and D.G. Watts (1968). Spectral Analysis and its Applications, Holden-Day.

Jin, F.-F., J.D. Neelin, and M. Ghil (1994) Science, 94, 70-72.

Jolliffe, I.T. (1986) Principal Component Analysis, Springer Series in Statistics. New York. 271 pages.

Kloeden, and Platen (1992). Numerical Solution of Stochastic Differential Equations, Springer-Verlag.

Kubo, R., M. Toda, and N. Hashitsume (1991). Statistical Physics II. Springer-Verlag. 279 pages.

Kushnir, Y. (1994). Interdecadal variations in North Atlantic sea surface temperature and associated atmospheric conditions. Journal of Climate, 7, 141-157.

Lorenz, E. (1969). Three approaches to atmospheric predictability. Bulletin of the American Meteorological Society, 50, 345-349.

Lorenz, E. (1973). On the existence of extended range predictability. Journal of Applied Meteorology, 12, 543-546.

Lorenz, E. (1975). Climatic Predictability. In The Physical Basis of Climate and Climate Modelling, GARP Publications Series No. 16, Geneva.

Lütkepohl, H. (1993). Introduction to Multiple Time Series Analysis, Springer-Verlag, Berlin. 545 pages.

MacDonald, D.K.C. (1962). Noise and Fluctuations: an Introduction, John Wiley & Sons.

Manabe, S., and R.J. Stouffer (1988). Two stable equilibria of a coupled ocean-atmosphere model. Journal of Climate, 1, 841-866.

Manabe, S., R.J. Stouffer, M.J. Spelman, and K. Bryan (1991). Transient response of a coupled ocean-atmosphere model to gradual changes of atmospheric CO2 : Part I: Annual mean response. Journal of Climate, 4, 785-818.

Mann, M. E., and J. Park (1994). Global-scale modes of surface temperature variability on multiannual to century timescales. Journal of Geophysical Research, 99, 25819-25833.

Marshall, J., and R. Schott (1999). Open ocean convection: observations, theory, and models. Reviews of Geophysics, 37, 1-64.

Neumaier, A., and T. Schneider (1997). Multivariate autoregressive and Ornstein-Uhlenbeck processes: Estimates for order, parameters, spectral information, and confidence regions. Submitted to ACM Transactions of Mathematical Software.

Available at http://www.aos.princeton.edu/WWWPUBLIC/tapio/ar_t/index.html

Palmer, T. (1996). Predictability of the atmosphere and oceans: from days to decades. In NATO Advanced Study Institute: Decadal Climate Variability: Dynamics and Predictability, edited by D.L.T. Anderson and J. Willebrand. Springer, Berlin.

Penland, C. (1989). Random forcing and forecasting using principal oscillation pattern analysis. Monthly Weather Review, 117, 2165-2185.
Penland, C. (1996). A stochastic model of IndoPacific sea surface temperature anomalies. Physica D, 98, 534-558.
Reif, F. (1965). Fundamentals of statistical and thermal physics, McGraw-Hill. 651 pages.
Robinson, A. R., and M. Golnaraghi (1994). The Physical and Dynamical Oceanography of the Mediterranean Sea, in Ocean Processes in Climate Dynamics, P.Malanotte-Rizzoli and A. R. Robinson (editors), pp. 255-306, Kluwer Academic.
Sarachik E.S., M. Winton, and F.L. Yin (1996). Mechanisms for decadal-to-centennial climate variability. In NATO Advanced Study Institute: Decadal Climate Variability: Dynamics and Predictability, edited by D.L.T. Anderson and J. Willebrand. Springer, Berlin.
Schneider, T., and S. M. Griffies (1999). A conceptual framework for predictability studies. Journal of Climate, 12, 3133-3155.
Schneider, T., and A. Neumaier (1997). Algorithm: ARfit - A Matlab® package for the estimation and spectral decomposition of multivariate autoregressive models. Submitted to ACM Transactions of Mathematical Software.
Available at http://www.aos.princeton.edu/WWWPUBLIC/tapio/ar_t/index.html
Stommel, H. (1961). Thermohaline convection with two stable regimes of ow. Tellus, 13, 224-230.
Toggweiler, J. R., and B. Samuels (1995). Effect of Drake Passage on the global thermohaline circulation. Deep-Sea Research I, 42(4), 477-500.
Tziperman, E., L. Stone, M.A. Cane, and H. Jarosh (1994). El Niño chaos: overlapping of resonances between the seasonal cycle and the Pacific Ocean-Atmosphere oscillator. Science, 264, 72-74.
Vallis, G. (1999). Large-scale circulation and production of stratification: effects of winds, geometry, and diffusion. Journal of Physical Oceanography, in press.
von Storch, H., and A. Navarra (1995). Analysis of climate variability : applications of statistical techniques., Proceedings of an Autumn School organized by the Commission of the European Community on Elba from October 30 to November 6, 1993. Springer, Berlin. 334 pages.
Sutton, R.T., and M.R. Allen (1998). Decadal predictability of North Atlantic sea surface temperature and climate. Nature, 388, 563-567.
Wilks, D. S. (1995). Statistical methods in the atmospheric sciences : an introduction, Academic Press, San Diego. 467 pages.
Wunsch, C. (1992). Decade-to-century changes in the ocean circulation. Oceanography 5, 99-106.
Yang, X-Q., J. L. Anderson, and W. F. Stern (1998). Reproducible forced modes in AGCM ensemble integrations and potential predictability of atmospheric seasonal variations in the extratropics. Journal of Climate, 11, 2942-2959.

7 Carbon Cycling over Lands and Oceans

Philippe Ciais

Laboratoire des Sciences du Climate et l'Environement, LSCE (CNRS). Orme des Merisiers, Bât. 709, 91191 Gif-sur-Yvette, Cedex, France
ciais@lsce.saclay.cea.fr

7.1 Introduction

Increasing carbon dioxide (CO_2) is the main driving force of climate change. We learned from ice core measurements that prior to the Industrial Revolution, 200 years ago, the CO_2 concentration in the atmosphere was fairly stable, indicating some quasi-equilibrium state for the sum of all processes which controlled the atmospheric carbon content. Since then, growing industries have put a strong demand on energy, which was met by extracting fossil fuels from the earth's crust and burning them in engines. Furthermore, the population increase from 0.6 to 6 billion has fostered the need for new lands for agriculture, and the extension of cropland and pastures has led to deforestation. Many forests have been cut or degraded, and today, about 80% of the terrestrial ecosystems are directly influenced by human activities. Land use and land use changes over the past 200 years have caused land ecosystems to lose carbon. Overall, in response to both land use changes and fossil fuel burning, the amount of CO_2 in the air has risen from 280 ppm before the Industrial Revolution to the present 365 ppm. The magnitude of the ongoing rise in CO_2 is comparable to changes which occurred in the distant past. More than 10 000 years ago, our planet warmed up by 5°C as extensive ice sheets covering North America and Scandinavia melted away. During that period, CO_2 increased from 200 to 280 ppm. Yet this 80 ppm increase took place within 5000 years whereas human activities have fostered a comparable rise within only 200 years.

The concentration of CO_2 in the atmosphere is regulated by natural processes that exchange carbon between the atmosphere, the ocean and the land biota. Carbon cycle studies aim at better understanding and quantifying the mechanisms which transfer carbon between and within these reservoirs. A good understanding of the carbon cycle is necessary to predict future CO_2 levels, and therefore future climate change. Climate models predict an important increase of the surface temperature (1.5 to 4.5°C globally) under a doubling of CO_2. At the present rate, CO_2 will have doubled by the next century. The only real long term player in mitigating the atmospheric CO_2 increase is the ocean. The ocean will eventually absorb most of the CO_2 in excess in the atmosphere, and atmospheric CO_2 will slowly stabilize towards an asymptotic concentration above its pre-industrial value. Yet, reaching this new equilibrium will take centuries after fossil fuel emissions have stopped. The coming decades, however, will see both plants and soils play a significant role in helping the ocean to control CO_2. If trees for instance, take advantage of rising CO_2 to become larger, the land biosphere would gain carbon and mitigate the atmospheric CO_2 increase. A major research challenge is to evaluate the transient response of both the oceans and of land ecosystems during the time period when anthropogenic emissions will continue. Better estimating the future trajectory of atmospheric CO_2 in the atmosphere is a step forward to better forecasting climate change. In the following, I will first present the "natural" carbon cycle as it stood during pre-industrial times, and later analyze the "perturbed" carbon cycle, as it is presently driven out of equilibrium by sustained human-induced CO_2 emissions.

7.2 The carbon cycle during pre-industrial time

7.2.1 Carbon pools

Ocean

The ocean (Fig. 7.1) is by far the largest carbon pool with 40 000 GtC. Because of the weak acid character of CO_2, bicarbonate ions (HCO_3^-) make up roughly 90 per cent of carbon dissolved in the sea, whereas carbonate ions (CO_3^{2-}) contribute 10 per cent, and dissolved CO_2 only 1 percent. A given relative variation of dissolved CO_2 in the ocean produces a 10-times smaller variation in the ocean's total carbon content. This is called the "buffer effect" of CO_2. The ocean is physically stratified, so that only carbon dissolved in the surface layer is exchangeable with the atmosphere. The mixing between surface waters and the deep sea is rather slow and requires from decades to centuries to occur. Yet, the deep ocean carbon reservoir, by its enormous size, exerts a slow but powerful control on the atmospheric CO_2 levels. For this reason, it is essential to understand how carbon is exchanged between the surface and the ocean interior, when deep waters outcrop to the surface, or when surface waters sink to the abyss. Ocean algae and animals are a carbon reservoir of negligible size (Fig. 7.1) compared to the dissolved carbon pools. Yet, the role of ocean biology is very important because it transfers large quantities of carbon from the sunlit layers near the surface, to below the mixed layer.

Land biosphere

Plants and soils on the Earth contain carbon both in living biomass (leaves, branches, trunks, roots, etc.,) and in soil organic matter (litter, soil compounds, etc.). The overall size of the terrestrial biosphere is about 2500 GtC. Carbon is distributed among various pools, each associated with a residence time that ranges from between a few months (as with leaves on deciduous trees) to few hundred years (as with organic matter in peat). The amount of carbon stored above and below ground (Table 7.1) is a function of the ecosystem type. The contrast is striking between a tundra where 90% of the carbon lies in the soil, and a tropical forest where large trees store 50% of the carbon above ground. The amount of carbon stored in soils depends on climatic and edaphic conditions (Dixon et al. 1994). The oxidation of soil organic compounds is generally faster at warmer temperatures. Soil carbon density is maximum in tundra soils, large in boreal forests, and minimum in agricultural lands, where plants are harvested each year with no delivery of litter to the soil. Likewise for the ocean, the question arises as to which terrestrial carbon pool can quickly exchange with the atmosphere. Over the course of approximately one year, a tropical forest exchanges 20% of its soil carbon with the atmosphere, in contrast with against 5-10% for a cultivated land and only 1% for a tundra (Jenkinson et al. 1991).

Table 7.1. Global carbon stocks in vegetation and the top 1m of soils (assembled by Pr. W. Cramer based on WBGU 1998)

Bioma	Area	Carbon Stocks (Gt C)		
	(10^6 km^2)	Vegetation	Soils	Total
Tropical forests	17.6	212	216	428
Temperate forests	10.4	59	100	159
Boreal forests	13.7	88	471	559
Tropical savannas	22.5	66	264	330
Temperate grasslands	12.5	9	295	304
Deserts & semideserts	45.5	8	191	199
Tundra	9.5	6	121	127
Wetlands	3.5	15	225	240
Croplands	16.0	3	128	131
Total	151.2	466	2011	2477

Atmosphere

The atmosphere contains gaseous CO_2. Unlike the ocean and the land biosphere, the atmosphere is a very homogeneous reservoir. The atmospheric circulation mixes up CO_2 within the entire troposphere within 1-2 years, and consequently, only small gradients of concentration can be detected between stations around the world. The atmosphere is connected to the ocean surface and to the land biosphere by large fluxes (Fig. 7.1). The net sea-to-air flux of CO_2 decomposes into two large fluxes of opposite direction, which compensate each other globally. The net flux between land ecosystems and the atmosphere, results from uptake by photosynthesis and release by respiratory processes. Photosynthesis is a flux which removes about 15% of the atmospheric CO_2 content every year. Respiration balances photosynthesis on a yearly basis. Measurements of air bubbles occluded in ice sheets, showed that over the past millennium preceding the Industrial Era, the CO_2 concentration in the atmosphere was fairly stable, indicating some quasi-equilibrium state at that time for the global carbon cycle. In the pre-industrial carbon cycle, photosynthesis and respiration balanced each other, as did ocean uptake and ocean out-gassing of CO_2.

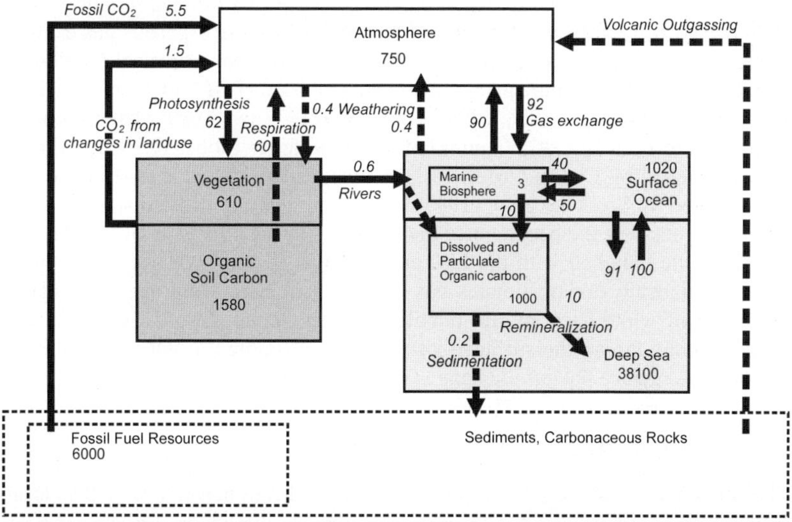

Fig. 7.1. A simplified scheme of the global carbon cycle and its main fluxes between the atmosphere, the oceans, and the terrestrial biosphere. Solid boxes and lines represent fluxes of the fast exchange carbon cycle. Dashed arrows indicate carbon exchange fluxes with the slow lithospheric carbon pools (dashed boxes). Numbers denote estimated carbon content (GtC) and those in cursive, carbon fluxes (GtC yr^{-1}).

7.2.2 Carbon within the ocean

Gas exchange

CO_2 is exchanged at the air sea interface when its concentration dissolved in seawater (C_o) differs from the one corresponding to thermodynamic equilibrium with the atmosphere as predicted from Henry's Law (C_a). The air-sea flux (F_o) is proportional to the concentration gradient across the interface, as expressed by $F_o = v\,(C_o - C_a)$, where fluxes are conventionally positive when the atmosphere gains carbon. The out-gassing flux (F_{oa}) and dissolution flux (F_{ao}), are respectively equal to $v\,C_o$ and $-v\,C_a$. The quantity v, dimensioned to a velocity, is called "piston velocity". Furthermore, Henry's Law states that $C_a = S\,P_a$ where S is the solubility of CO_2 in sea water and P_a its partial pressure in the air. Introducing the partial pressure of CO_2 in sea-water, defined by $P_o = S^{-1}\,C_o$, and the gas exchange coefficient, $K_{ex} = v S$, we have $F_o = v S (P_o - P_a) = K_{ex} (P_o - P_a)$. Unlike v and S, K_{ex} is rather insensitive to temperature.

There is no "unified" theory to predict the variations of K_{ex} over the globe. Generally, its is observed that K_{ex} increases with the wind speed above the ocean surface, as observed in many experimental studies, either in wind tunnels or in the open ocean. K_{ex} varies as a function of latitude, with more vigorous gas exchange under stronger wind conditions at high latitudes. Different empirical relationships between K_{ex} and the wind speed have been proposed (Liss and Merflivat, 1986 ; Wanninkhof and Mc Gillis, 1999; Frew, 1997). When extrapolated from the globe, the relationship of Wanninkhof yields a larger global K_{ex} value than the one established by Liss and Merlivat. An independent global average estimate of K_{ex} obtained by Broecker et al. (1985) using the inventory of Radiocarbon (14 °C) that has penetrated the ocean since nuclear tests in the 1950's, is 50% larger than the K_{ex} value obtained with the Liss and Merlivat. It has been suggested that the latter relationship underestimates gas exchange at high wind speeds, when the formation of white caps entrain bubbles that accelerate the gas exchange. Also, certain conditions on the surface that were not accounted in wind tunnel experiments may enhance the gas exchange (surfactants, algae, etc.) in the ocean.

Ocean dissolved CO_2

The air-sea flux of CO_2 is proportional to the difference $(P_o - P_a)$ that is also called ΔpCO_2. Numerous ΔpCO_2 measurements during ocean surveys yet do not allow global coverage, neither in space nor in time. These data indicate that upwelling areas at the equator in the Eastern Pacific and Atlantic oceans are net CO_2 sources to the atmosphere (e.g. Andrié et al. 1986 ; Feely et al. 1999). Conversely, the North Atlantic and the sub-Antarctic waters are important CO_2 sinks (Takahashi et al. 1997 ; Lefèvre et al. 1999), as observed in Fig. 7.2. The value of ΔpCO_2 is controlled by temperature, biological activity and ocean circulation. The impact of temperature on ΔpCO_2 involves the solubility which increases by 4% per °C of temperature decrease. Waters that are cooled absorb CO_2, whereas warmed waters loose CO_2 to the atmosphere. The role of ocean biology is more complex. On the one hand, marine organisms fix dissolved carbon and cause ΔpCO_2 to decrease in the sunlit layer of the surface ocean. This defines the "biological pump". On the other hand, the formation of carbonaceous shells utilizes carbonate ions, and causes dissolved CO_2 concentration, and therefore ΔpCO_2, to increase in the presence of algae. This latter "carbonate counter pump" does not completely counteract the effect of the "biological pump", and marine photosynthesis generally causes the ocean surface to be undersaturated in CO_2. As an example, observed values of ΔpCO_2 can be on the order of -80μatm during plankton "blooms" in spring and summer (Takahashi et al. 1994). Algae are formed in the sunlit layers, within the top 100 m of the oceans, and after they die and sink towards the bottom of the sea where they are mineralized. Intermediate and deep waters, where mineralization takes place, are thus enriched both in carbon and nutrients and are supersaturated in CO_2 when they outcrop to the atmosphere in upwelling areas. As an illustration, the ΔpCO_2 values observed in the eastern equatorial Pacific are on the order of 110 μatm (Feely et al., 1999). Fig. 7.2 shows a map of the annually averaged ΔpCO_2 field in February, obtained by extrapolating *in situ* measurements from the globe (Takahashi et al., 1997). The global biomass of marine organisms, on the order of 3 GtC, (Fig. 7.1) is negligible and has no storage capacity. Rather, the marine biology is analogous to a "pipe" transferring carbon from the surface to the deep sea, and inducing a positive gradient of carbon between the deep sea and the surface waters at a steady rate. An instant kill of all marine organisms would have the effect of cancelling out this vertical gradient, thus increasing the CO_2 partial pressure in the atmosphere by 150 ppm.

7.2.3
Terrestrial carbon exchange

Photosynthesis

Photosynthesis takes up atmospheric CO_2 and emits O_2 under a suite of reactions which reduce CO_2 into a monomer of cellulose:

$$CO_2 + H_2O \xrightarrow{h\nu} CH_2O + O_2 \tag{7.1}$$

Photosynthesis occurs in chloroplasts within leaf cells and uses solar photons of wavelengths comprised of between 380 nm and 680 nm (Photosynthetic Active Radiation, PAR). Carbon fixed by photosynthesis is transformed into organic compounds, which are incorporated into plant tissues. At the ecosystem level, the assimilation of CO_2 by plant photosynthesis corresponds to a flux (A) called Gross Primary Production (GPP) at the global scale. One can write that $A = g_s\,(P_a - P_i)$ where g_s is a conductance which regulates the diffusion of CO_2 between canopy air and chloroplasts, and P_a and P_i are partial pressures of CO_2 respectively in the canopy air and within the chloroplast. Leaves have very small holes at their surface, called stomates, through which CO_2 diffuses in (photosynthesis) and H_2O diffuses out (transpiration), as represented in Fig. 7.3.

It is believed that plant growth does regulate carbon inputs and water losses in an optimal manner, and that plants control the value of g_s by closing or opening their stomates. They adjust the value of g_s to maximize photosynthesis and to minimize transpiration. The stomatal control responds to external variables such as canopy humidity, and g_s varies over the course of the day, and over the growing season. As for the ocean gas exchange, there is no single theory that predicts g_s. Ball (1988) has expressed g_s for C3 plants, based on experiments under controlled conditions. The term C3 is a photosynthesis pathway common to all trees and temperate grasses, the alternate C4 photosynthesis pathway pertains to tropical grasses. The formulation of Ball (1988) is given by $g_s = g_l\,(Ah / P_a)$ where h is the relative humidity at the leaf surface, A the rate of CO_2 assimilation by photosynthesis and P_a the partial pressure of CO_2 in canopy air. The slope g_l is empirically determined to be relatively independent of the plant species considered.

Other expressions of g_s have been proposed (see for instance Leunning et al. 1995).

There is a third physiological relationship which describes photosynthesis (A) as a function of P_i, that was established by Farquhar et al., (1980) for C3 photosynthesis and by Collatz et al. (1991) for C4 photosynthesis. The set of equations above can be solved for A and P_i. It makes it possible to calculate A using as an input temperature, soil moisture, and solar radiation. Scaling up photosynthesis from the leaf to the globe requires additional knowledge of the spatial distribution of plant functional types and of the seasonal development of leaf growth (e.g. Sellers et al. 1992 ; Friend *et al.* 1997; Woodward *et al.* 1995). Remote sensing of the continental surfaces provides global independent information on the "greenness" of the Earth, which is linked to photosynthesis. Fig. 7.4 shows the global geographic distribution of Gross Primary Production, as derived from remotely sensed information. It shows that the most productive ecosystems are tropical forests, temperate forests and agricultural lands.

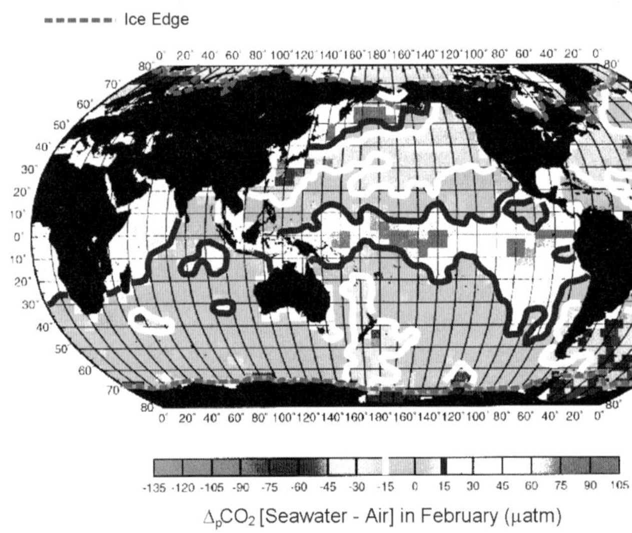

Fig. 7.2. Sea-to-air difference of CO_2 partial pressure (ΔpCO_2) in February averaged over several years. Measurements from different campaigns have been gridded using surface fields of the ocean circulation (Takahashi et al. 1997). Positive values indicate that the ocean is a source of CO2 to the atmosphere while the negatives indicate a sink. The solid blank line denotes positive values larger than 15 µatm, the solid white line denotes negative values below –15 µatm and the dashed gray line the extent of the marine ice (from Takahashi et al. 1997)

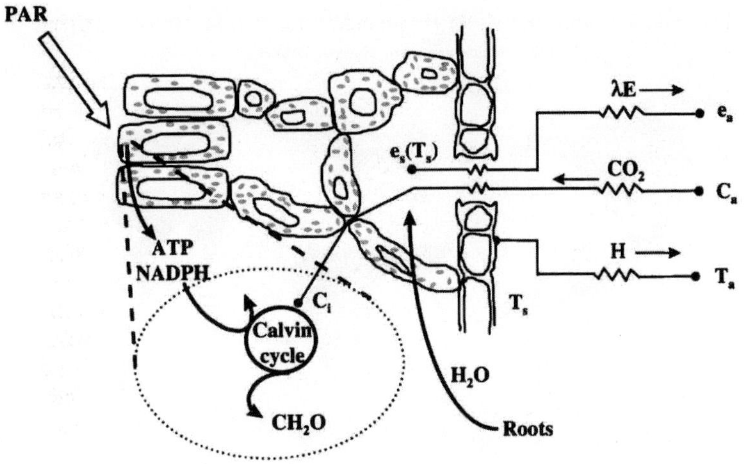

Fig. 7.3. Simplified diagram of physiological controls on heat, water vapor and CO_2 transfer between a leaf and the atmosphere. CO_2 fixation and assimilation, by way of the Calvin cycle, depend on the leaf interior CO_2 concentration (mainly controlled by stomatal conductance) and on the enzime metabolism (controlled by light supply).

Respiration

Respiration represents a variety of processes which oxidize organic carbon and release CO_2 into the atmosphere. Respiratory processes separate into plant respiration (PR) and heterotrophic respiration (HR). Photosynthesis (GPP) diminished by plant respiration (PR) is called Net Primary Production (NPP). The annual value of $NPP = GPP - PR$ represents the accumulated amount of carbon fixed into plant tissues over one growing season. Plant respiration represents the oxidation of organic compounds by vegetation (leaves, roots, stems) to produce energy for maintenance and growth. Heterotrophic respiration (HR) is the oxidation of dead organic matter by animals, soil microbes, and chemical reactions. Generally, heterotrophic respiration is proportional to the amount of biomass available for decay (M), as given by $HR(T) = M\,k(T)$. The rate of decay, $k(T)$ increases with temperature T, and is often represented by an exponential relationship:

$$k(T) = Q_{10}^{\,(T-T_0)/10} \tag{7.2}$$

where a 10 °C warming causes the decay rate to be fastened by Q_{10}. Incubation experiments suggest that Q_{10} ranges between 1.5 to 2.5 over most ecosystems (Raich and Schlesinger, 1992). An alternate formulation based on the Arhennius Law (Lloyd and Taylor, 1994) gives:

$$k(T) = R_{10}^{\,A/T} \tag{7.3}$$

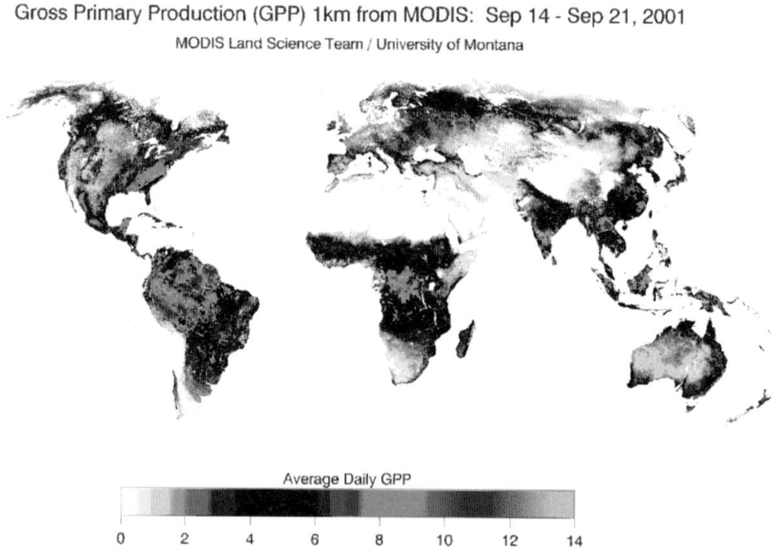

Fig. 7.4. Average daily estimation of land Gross Primary Production (GPP, gCm^{-2}) during the third week of September 2001. Map was produced using the Biome-BGC model by the NTSG (Numerical Terradynamic Simulation Group, School of Forestry, University of Montana). The model was fed with remote sensing surface radiance data from the MODIS spectroradiometer in Terra Satellite Platform. Spatial resolution is 1 km. The map is available at http://www.forestry.umt.edu/ntsg

Little is known about heterotrophic respiration at the ecosystem level and its variation with soil moisture or soil fauna. The oxidation of bound organic nitrogen by respiration further releases nitrogen in a mineral form which becomes available for vegetation growth. The coupling between the carbon and nitrogen cycles in soils has potentially large implications for plant growth, since nitrogen is a limiting nutriment in many forest ecosystems, over temperate and boreal regions.

Net terrestrial CO_2 exchange

The net terrestrial CO_2 exchange flux is the difference between net primary production (NPP) and heterotrophic respiration (HR). The seasonal cycle of production is different from the one of respiration (e.g. Fung et al. 1987). Tropical evergreen rainforests grow under rather constant moisture and temperature conditions throughout the year. Photosynthesis is nearly the same every month and always compensates for respiration, yielding a net CO_2 exchange close to zero (Fig. 7.5a). Temperate and boreal forests, have a well defined growing season, between leaf onset and senescence. Where water for transpiration is available, photosynthesis peaks during the summer, whereas respiration is more evenly distributed during the year, as long as soils are not frozen at depth. Temperate and boreal forests act as a net sink of atmospheric CO_2 in summer when photosynthesis exceeds respiration, and as a net source from the fall to the early spring when the opposite is true (Figs. 7.5b and 7.5c). In the Northern hemisphere, the timing of the growing season, together with a large continental surface with vegetation, causes a pronounced seasonal cycle in atmospheric CO_2, with a minimum in July-August and a maximum in winter. The peak to peak amplitude in the atmospheric CO_2 signal is maximum at around 65°N. At the station of Point Barrow for instance at 71°N (Fig. 7.6), the concentration in CO_2 is 15 ppm lower in August than in January. The amplitude of the atmospheric seasonal cycle decreases towards the Tropics. At Mauna Loa (20°N), it is half the one measured at Point Barrow. In the southern hemisphere, there is only a very small seasonality, with a peak to peak amplitude minimum of 0.5 ppm at around 40°S. The CO_2 seasonal cycle in the southern hemisphere is opposite in phase with the seasonal cycle in the northern hemisphere. Carbon exchange over the southern oceans, biomass burning, and the transport of northern hemisphere air to the southern hemisphere altogether contribute to the observed seasonal changes of CO_2 south of 40°S.

7.3 The carbon cycle over the industrial period

7.3.1 Sources

Fossil fuel emissions

Since approximately 1800, the increasing demand for energy (electricity production, transports, heating...) is largely met by the combustion of fossil fuel reserves. The use of coal, oil, and natural gas in engines and heaters results in the injecting of fossil CO_2 into the atmosphere. Over the period 1860-1998, the amount of fossil CO_2 that was released to the atmosphere is estimated to be approximately 265 GtC. Fossil CO_2 emissions are currently inventoried in each country from statistics with an accuracy on the order of 10% in industrialized countries (Marland et al. 1985). Today, the fossil CO_2 emission amounts to 6 GtC per year and occurs over Europe, the United States, Japan, and countries with fast growing economies (China, India, South East Asia). Fossil CO_2 emissions are small in magnitude compared to the natural fluxes, but load into the atmosphere carbon that was immobilized into the Earth's crust for millions of years. By burning fossil fuels which remaining reserves are estimated to be on the of 10 000 GtC, humans increase the total amount of carbon that is distributed between the atmosphere, the ocean and the land biosphere. Once it has entered the global carbon cycle, fossil CO_2 is not likely to be returned to the Earth's crust any time soon, because the only process which stores carbon durably away from the ocean is the formation of carbonated sediments, and it is a very small sink compared to the fossil source (Fig. 7.1).

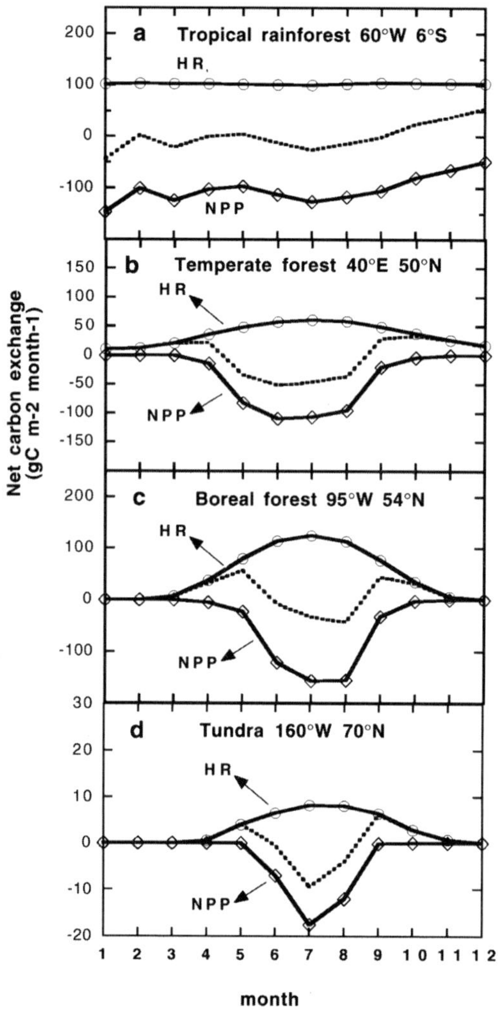

Fig. 7.5. Seasonal cycle of Net Primary Production (NPP), heterotrophic respiration (HR), and net exchange of CO_2 between the atmosphere and different of ecosystems (dotted line) (a) Tropical rainforest in the Amazon, (b) Deciduous temperate forest, (c) Evergreen boreal forest, (d) Arctic tundra. Negative fluxes represent a gain of carbon by the terrestrial biosphere, positive ones represent a release of CO_2 from the biosphere to the atmosphere. After Fung et al. 1987.

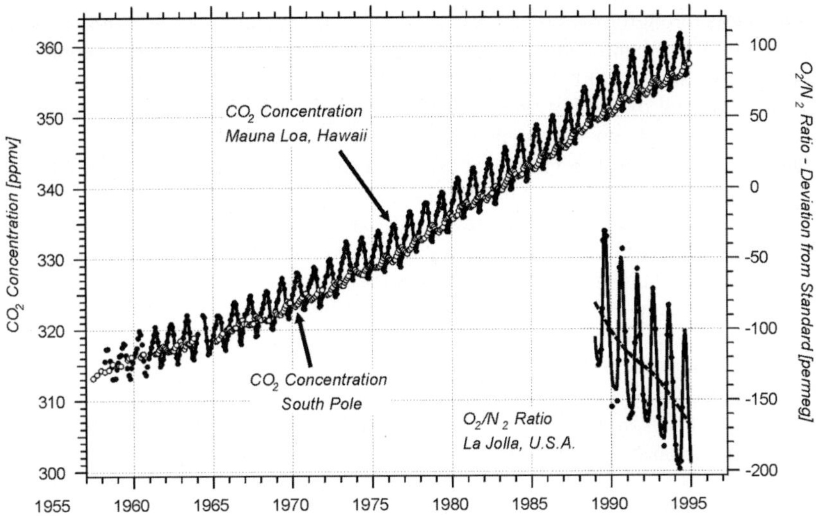

Fig. 7.6. Atmospheric CO_2 measurements at Mauna-Loa, 20°N, the South Pole and atmospheric measurements of the O_2/N_2 ratio at La Jolla, 40°N. A seasonal cycle is observed at Mauna-Loa, as driven by seasonal changes in Net Primary Production and Heterotrophic Respiration over northern hemisphere ecosystems. A small seasonal cycle is apparent at the South Pole, as driven by the transporing of air from the northern hemisphere, and seasonal changes in the southern hemisphere terrestrial and oceanic carbon fluxes. The long term increase in CO_2 because of fossil fuels and land-use related emissions in the atmosphere implies a decrease in atmospheric O_2/N_2. The magnitude of the O_2/N_2 trend is used to infer the ocean and land net carbon fluxes.

Land use and land use change

Changes in land use and in land cover have taken place since remote times, but have dramatically accelerated over the last century in response to the growing population. Changes in land use include the conversion of forests into pasture or cultivated land, the planting of new trees, the irrigation of arid zones, etc. Generally, the suppression of forests results in carbon emissions to the atmosphere. Overall, for the period 1860-1980, changes in land use are estimated to have released approximately 120 GtC to the atmosphere, however, this number is more uncertain (± 40%) than the fossil source. Deforestation has now stopped in Europe and in North America, where it reached maximum rates during the first half of the 20th century. After the 1950's however, the fast increasing population in the Tropics yielded an important pressure on forests there, with intensified forest exploitation of and forest conversion to agriculture and pasture. Deforestation still occurs today in Brazil, Equatorial Africa and South East Asia where forests are destroyed by fire or logging, and new lands are used for rearing cattle or for growing crops. Forests contain large carbon stocks, one fraction being released to the atmosphere immediately after deforestation, and another fraction subsequently afterwards through losses of soil carbon. On the other hand, reforestation or forest regrowth after abandonment of agricultural lands can increase carbon storage in ecosystems after disturbance. The carbon trajectory of a forest stand after it has been converted to another type of land depends on which type of vegetation replaced the forest, on soil management practices, on nutrients availability, etc. Overall, land use in the Tropics is estimated to contribute a net source to the atmosphere of 1.6 GtC yr^{-1} for the 1980's (Houghton et al. 1987; 1991a,b). Uncertainties on these numbers are large ($\pm$ 1GtC yr^{-1}) because of the difficulties in estimating changes in carbon stocks when ecosystems are affected by changes in land use. The land use induced CO_2 source includes any carbon flux (e.g. regrowth) associated to the disturbance of pristine ecosystems by humans. Most ecosystems are however, affected sporadically by natural disturbances (wind storms, fires, pests outbreaks, severe droughts...) in absence of land use changes. For instance, many savannas are destroyed by fires every year, but regrow during the next wet season and can have a neutral carbon balance on an annual basis. In this case, the burning of savannas is not a source of CO_2 to the atmosphere on an annual basis.

7.3.2
Atmospheric CO_2 increase

The rate of increase of atmospheric CO_2 is currently determined with high precision at over 60 stations around the world. Because the atmosphere is fast mixed by winds within 1-2 years, it is a superb integrator of the surface fluxes over the Globe. The average CO_2 rate of increase over the 1980's is 3.3 $\pm$ 0.2 GtC per year (Conway et al. 1994). Therefore, CO_2 accumulates in the atmosphere at a smaller rate than would be expected if all fossil and land use induced emissions were accumulating into the atmosphere. An amount of CO_2 of only about 50% the anthropogenic emissions is retained in the atmosphere each year, the other half being absorbed by the oceans and by the terrestrial biosphere.

The average global uptake of CO_2 by oceans and lands amounts to 2.2 GtC per year in the 1980's and to 3 GtC per year over the 1990's. The apportionment of that sink between land and ocean reservoirs is uncertain, as shown by errors in Table 7.2. In Table 7.2, the ocean uptake has been estimated from global ocean carbon models results and the land net carbon flux is inferred as a residual term. The net land flux is therefore, the most poorly known number of the perturbed carbon cycle. Recently however, measurements of atmospheric oxygen-nitrogen ratio (O_2/N_2) have independently confirmed the estimates of Table 7.2 for the carbon balance of terrestrial ecosystems (Battle et al., 1996; Keeling et al. 1996). It is yet crucial to better determine which reservoir is the most active sink of carbon, in order to be able to make better predictions of the future CO_2 concentration in the atmosphere. An ocean sink will last longer (a few centuries) than a terrestrial sink (a few decades). Carbon storage on land reflects a slight excess of photosynthesis over respiration. It is a function of both changes in photosynthesis and in respiration, as well as of the mean residence time of carbon in ecosystems. The mean residence time of carbon in the terrestrial biota represents the period during which carbon is immobilized into terrestrial ecosystems before it is respired back to the atmosphere. Its value depends on which carbon pools store carbon. The mean residence time of carbon in ecosystems ranges from a few years in grasslands to a few decades in forests. In general, carbon storage in plants and soils is vulnerable to climate change and to land use and land management changes. Table 7.2 shows that the global land biosphere is approximately neutral for the 1980s and that it is a slight sink over the 1990's. Subtracting to this net flux the estimated land use source, it becomes apparent that a part of the global biosphere is currently storing carbon, or in other words that the Earth is "greening" (Table 7.2). We know little about which regions, or which type of ecosystems are storing carbon, and through which processes "greening" occurs.

Table 7.2. Estimate of the global carbon budget for the 1980's and the 1990's. The global net terrestrial carbon sink (4) is decomposed into a land use induced flux (5), source to the atmosphere, and a residual carbon storage (6), sink of atmospheric CO_2. The fluxes are given in GtC yr^{-1}; error limits correspond to an estimated 90% confidence interval. After IPCC, 1995.

Sources and sinks	1980 to 1989	1989 to 1998
(1) Fossil emissions	5.5 ± 0.5	6.3 ± 0.6
(2) Storage in the atmosphere	3.3 ± 0.2	3.3 ± 0.2
(3) Storage in the ocean	2.0 ± 0.8	2.3 ± 0.8
(4) Storage in land ecosystems = (1)-[(2)+(3)]	0.2 ± 1.0	0.7 ± 1.0
(5) Tropical land use induced source	1.6 ± 1.0	1.2 ± 1.0
(6) Residual land sink = "greening" = (4)-(5)	1.8 ± 1.4	1.9 ± 1.4

7.3.3 Ocean uptake

Oceanic uptake can be estimated using 1) global models of the ocean carbon cycle as shown by Table 7.2, 2) direct sea-to-air flux measurements (ΔpCO_2) extrapolated from the globe (Fig. 7.2), 3) direct measurement of the increase in the ocean inventory of dissolved inorganic carbon, and 4) indirect inferences by atmospheric tracer measurements (CO_2, $^{13}CO_2$ and O_2/N_2). Ocean carbon models of increasing complexity have been developed since the first prototype of Oeschger et al. (1975). Such models compute the amount of anthropogenic CO_2 which dissolves into the ocean based on the air-sea exchange and on the transport of carbon within the ocean. The first ocean carbon cycle models were rudimentary, and contained a well mixed surface layer rapidly balanced with the atmosphere, capping a deep ocean reservoir where the transport of carbon is much slower. The transport of an excess of carbon invading the ocean from the atmosphere was calibrated using radiocarbon, a long-lived tracer present in the atmosphere over the pre-industrial period, and emitted in large amounts by nuclear tests during the 1950s.

With three-dimensional models representing the ocean circulation in a more realistic way, now in use, a geographic and seasonal distribution of the ocean carbon uptake can be estimated (see for instance Maier-Raimer 1983, IPCC 1995). A current consensus amongst ocean modelers estimates the value of the ocean sink at 2 ± 0.8 GtC per year (Table 7.2 after IPCC, 1995), roughly 40% of current anthropogenic emissions.

7.3.4 Land uptake

The land uptake of carbon is difficult to measure directly because the biospheric reservoir is the most heterogeneous, and the biospheric fluxes are variable in space and time. In addition, processes that control the carbon storage in the terrestrial biota are not well quantified. The land uptake of carbon can be estimated using 1) global models of the terrestrial biogeochemical carbon cycle, and 2) indirect inferences from atmospheric tracers such as CO_2, O_2/N_2 and CO_2 isotopes. Recently, the eddy-correlation technique makes it possible to routinely monitor the net ecosystem exchange fluxes over a small area (1-5 km). As more flux measurements become progressively implemented over different types of ecosystems, it will be possible to use this information to scale up the fluxes over large continental areas (Baldocchi et al. 1996; Christenssen et al. 1996). One difficulty that models of the terrestrial carbon fluxes have to treat is the existence of a large spectrum of time scales for processes involved in vegetation and soils carbon cycling, ranging from a few minutes for leaf photosynthesis to a few decades for soil carbon oxidation. Simplified hypothesis are generally made in global models. For instance, whereas real ecosystems are composed of numerous species of different age, types, etc., models assume that they can be represented by a limited number of plant functional types. Similarly, the interception of solar radiation, and the photosynthesis fluxes at different heights in the canopy are scaled from top leaves in a simplified manner. A major difficulty is that terrestrial biogeochemical carbon models simulate the net exchange of carbon from the calculated value of photosynthesis and respiration separates, which are much larger than the net carbon flux, and both change with simultaneous changes in soil moisture and in temperature.

Several factors are expected to control carbon storage by land ecosystems, 1) the fertilization effect of CO_2, 2) the effects of nitrogen deposition, and 3) the effects of climate variability and of climate change, and 4) changes in natural (fires, storms, pests…) and man-made (harvest and other forestry activities) disturbance regimes. In addition to biogeochemical processes acting at a large scale, forest regrowth is presently occurring in temperate and boreal regions, and tree biomass increases due to forest management practices (Dixon et al. 1994). None of the above factors are independent one from the other, and carbon storage due to forest regrowth can be controlled by nitrogen deposition and CO_2 fertilization, as well as by climate conditions. Under current conditions, C3 photosynthesis is limited by carbon dioxide availability, and doubling CO_2 concentration has been shown in laboratory experiments to enhance net plant productivity by 20 to 50%. At the same time, elevated CO_2 levels tend to decrease the stimulate conductance, and for a given assimilation rate to reduce transpiration, with an indirect effect on plant growth. The production of ecosystems limited by water would then be expected to increase if plant transpiration is reduced when atmospheric CO_2 augments. Long term carbon storage in response to CO_2 fertilization will occur however, if the increase in photosynthesis is not immediately offset by an increase in plant respiration. The long term impact of rising CO_2 on the carbon storage on land is yet difficult to assess, as it is dependent among other factors, on temperature, water, and nitrogen availability.

Another mechanism proposed to account for carbon storage on land, is nitrogen deposition. Nitrogen oxides produced by the combustion of fossil fuels in engines are dispersed into the atmosphere and shortly deposited on plants. Additionally, the use of fertilizers in agriculture may also deliver an extra input of nitrogen to ecosystems. Nitrogen deposition occurs at mid-northern latitudes in the vicinity of populated areas. Forest ecosystems limited by nitrogen, such as temperate and boreal forests, can take advantage of additional nitrogen inputs, to fix more carbon (Townsend et al. 1996). Nitrogen fertilization is however effective as long as nitrate concentrations remain below a given level (Aber et al. 1989). As for CO_2 fertilization, the role of nitrogen deposition is difficult to quantify, and most biogeochemical global models lack a realistic description of the nitrogen cycle. The impact of nitrogen deposition on net carbon storage depends on the fate of nitrogen, whether nitrogen is leaked to rivers or retained in ecosystems, and whether it is available or fixed by plants that foster carbon sequestration in woody tissues, or in soil organic matter.

Finally, climate variability likely influences the exchange of carbon by terrestrial ecosystems. Beyond year to year changes in climate which have a strong impact on the terrestrial sources and sinks, decadal variability is can induce storage or release of carbon by land ecosystems, as it can promote a slight imbalance between photosynthesis and respiration. Dai and Fung (1993), using a very simple global biogeochemical model, driven by climate fields, have inferred that warmer temperatures and moister conditions over the last 40 years created about 0.1 GtC yr^{-1} of extra carbon uptake on land. In general, terrestrial sinks have been measured at the stand scale by direct flux measurements over middle-aged forests in North America, Europe and Brazil during the past few years, but there is no direct observational evidence for a global scale increase in the carbon content of ecosystems, either in soils or in above ground biomass. Yet, two global indicators suggest that some changes are presently going on in the terrestrial biota. Firstly, satellite measurements of vegetation indices, correlated with photosynthesis, exhibit a long-term positive trend over the past 15 years (Miyeni et al. 1997) with longer growing seasons and increased greenness in boreal regions, suggesting enhanced biospheric activity there. Secondly, the amplitude of the seasonal cycle in the northern hemisphere has increased by 20% over the past 30 years (Keeling et al. 1996), consistently with an increasing storage in the land biota at mid and high northern latitudes.

7.3.5
Regional carbon fluxes: inferences from atmospheric tracers

Atmospheric CO_2 measurements

Valuable information on the distribution of CO_2 fluxes in space and time over the globe is delivered by global atmospheric measurements. Small CO_2 concentration gradients among different stations reflect patterns in surface fluxes, despite vigorous atmospheric mixing. One can see in Fig. 7.6 that over the past 40 years, the CO_2 concentration difference between the Northern hemisphere (Hawaii) and the Southern Hemisphere (South Pole) has been increasing, which reflects the increase in fossil fuel emissions in the North, as well as long term changes in the spatial distribution of the ocean and terrestrial net fluxes. Atmospheric transport models can be used to simulate concentrations changes at a set of stations in response to regional sources and sinks. Such models solve the continuity equation over a 3D grid given a set of prescribed surface CO_2 fluxes. Global wind fields used to compute the atmospheric transport originate from climate models or from weather forecast models, the latter offering the advantage to simulate synoptic scale atmospheric CO_2 transport with the same timing as those observed in the real world. In addition to the varying winds, convection and vertical diffusion also transport CO_2 in the vertical. Transport models are far from perfect in the way they describe the large scale atmosphere dynamics especially in the lower troposphere, which is critical for matching their predictions with surface CO_2 concentration measurements. An additional source of uncertainty is the parameterization of vertical mixing, which can differ strongly among models. Measurements of inert tracers emitted by sources that are relatively well known help to design more realistic transport models. For instance, Radon-222 with a radioactive period of 3.8 days is suitable to constrain vertical mixing, while inert tracers ^{85}Kr and SF_6 constrain the horizontal transport. An inter-comparison of different transport models used for CO_2 has revealed substantial differences among their predicted concentration fields.

Fossil fuel CO_2 emitted primarily over Europe, North America and North East Asia causes atmospheric transport models to infer CO_2 concentrations on average higher in the northern hemisphere by 6 ppm compared to the southern hemisphere. In reality, the observed annual CO^2 difference between both hemispheres is only 3 ppm. From the mismatch between the observed inter-hemispheric difference and model results, one can infer that there is an active sink of CO_2 in the North hemisphere, which compensates part of fossil fuel emissions there. Whether this northern hemispheric sink is oceanic or biospheric is more difficult to quantify. Surveys of ΔpCO_2 over the North Atlantic and Pacific oceans (Takahashi et al. 1986) suggest that the Northern oceans do not absorb more than 0.6 GtC per year, which would implies the existence of a land sink of 2.3 GtC per year in the Northern biosphere. This reasoning proposed by Tans et al. (1990) concludes that temperate and boreal forests absorb large quantities of CO_2. One major uncertainty in that approach comes from the sparse and uneven coverage of oceanic ΔpCO_2 measurements. As CO_2 atmospheric measurements generalized from 20 stations in the early 1980's to about 80 stations now, it has been attempted to infer the carbon fluxes around the globe at the scale of continents or oceans. Atmospheric CO_2 gradients in longitude and in latitude are used in atmospheric transport models to retrieve the fluxes from the atmospheric concentrations. Several inversions have been carried out using various mathematical techniques, and applied to different CO_2 dataset. The principle consists in determining an optimal set of surface fluxes that minimizes a distance between the modeled simulated CO_2 and the observed CO_2 fields. Carbon fluxes inferred from recent inversion studies are given in Table 7.3. Some common findings are observed, such as the inference of a land uptake north of the equator, but the magnitude of the inverted fluxes are very different over each region. Generally, the continents are poorly constrained by observations, which makes it more difficult to infer the carbon fluxes there. On the other hand, the oceans benefit from a denser coverage by the present ground based network, and the fluxes inverted over large ocean regions generally are in better agreement.

Table 7.3. Net carbon flux inferred by atmospheric inversions over continents (positive are carbon sources, negative are carbon sinks). SH is southern hemisphere.

Reference	Period	Total continents	Europe	Siberia	North America	Tropics and SH
Bousquet et al. 1999	85-95	-1.3 ± 1.6	-0.3 ± 0.8	-1.5 ± 0.7	-0.3 ± 0.5	0.8 ± 1.0
Peylin et al. 1999	90-95	-1.2 ± 2.8	-0.6 ± 1.5	-0.8 ± 1.2	-1.0 ± 1.2	1.2 ± 1.2
Rayner et al. 1997	80-95	-0.7	-0.2	-0.1	-0.5	0.1
Kaminski et al. 1999	81-87	-1 ± 0.5	-0.1 ± 0.2	-0.6 ± 0.3	-0.2 ± 0.3	-0.1
Fan et al. 1998	88-92	-1.6 ± 1.7	-0.1 ± 0.7		-1.7 ± 0.5	0.2 ± 0.9

Carbon cycle tracers to infer land and ocean CO_2 fluxes

Stable isotope ^{13}C of carbon is of great interest to infer the biospheric carbon flux. Plant photosynthesis fixes preferentially the lightest isotope ^{12}C in atmospheric CO_2, implying that the isotope ratio $^{13}C/^{12}C$ of plant biomass is lower than that the one of atmospheric CO_2. Carbon uptake by the vegetation thus increases the $^{13}C/^{12}C$ ratio in the atmosphere (Fig. 7.7), whereas on the other hand, dissolution into the oceans barely affects the isotope ratio of CO_2. Isotopic ratios of reservoirs are expressed in per mil in the δ-scale as a deviation from a given reference material, $\delta\,^{13}C_{reservoir} = 1000\,[(^{13}C/^{12}C)_{reservoir} / (^{13}C/^{12}C)_{reference} - 1]$; the more negative the $\delta\,^{13}C$, the more depleted the reservoir is in heavy isotope. Global atmospheric $\delta\,^{13}C$ observations make it possible to separate ocean and land in the global carbon budget. Further, the spatial distribution of terrestrial carbon fluxes can be estimated from combined CO_2 and $\delta\,^{13}C$ measurements. Keeling et al. (1989b) constrained a global transport model by the inter-hemispheric difference in CO_2 and $\delta\,^{13}C$ to estimate the ocean and terrestrial carbon fluxes as a function of latitude. Although their total fluxes within each large latitude band are similar to those of

Tans et al. (1990), they locate the northern hemisphere sink predominantly in the ocean. Over the 1980's, there was therefore, a disagreement between the approach combining CO_2 and ΔpCO_2 measurements on the one hand (Tans et al.) and the approach using CO_2 and isotopes on the other hand (Keeling et al. 1989b). During the early 1990's, a larger set of both CO_2 and $\delta^{13}C$ atmospheric data has been improving the apportionment of the ocean vs. land carbon fluxes.

Apart from $\delta^{13}C$, measuring atmospheric oxygen provides a simple and powerful way to infer the land uptake of carbon. Atmospheric O_2 proves very difficult to measure, a precision of 0.1 ppm being required, whereas the ambient concentration in the atmosphere is of 210 000 ppm. Plant photosynthesis fixes carbon and emits O_2 in a ratio very close to 1, whereas respiration does exactly the opposite. The burning of fossil fuel utilizes O_2 in a ratio close to 1.4 for carbon, which causes oxygen to decrease in the atmosphere as a result of fossil fuel emissions. In contrast, atmospheric oxygen is in equilibrium with the small ocean reservoir of dissolved O_2. Unlike for carbon, the ocean contains very little oxygen, because O_2 is poorly soluble into sea water. Consequently, any carbon uptake by the ocean has no effect on the atmospheric O_2 concentration. Atmospheric O_2 being expressed normalized to N_2 to remove the effect of pressure variations, O_2/N_2 measurements provide an immediate fingerprint of the net land exchange. It is observed in Fig. 7.6 that as a result of the sum of fossil CO_2 release and terrestrial net CO_2 exchange, atmospheric O_2 measured routinely since 1990 has been decreasing in the atmosphere.

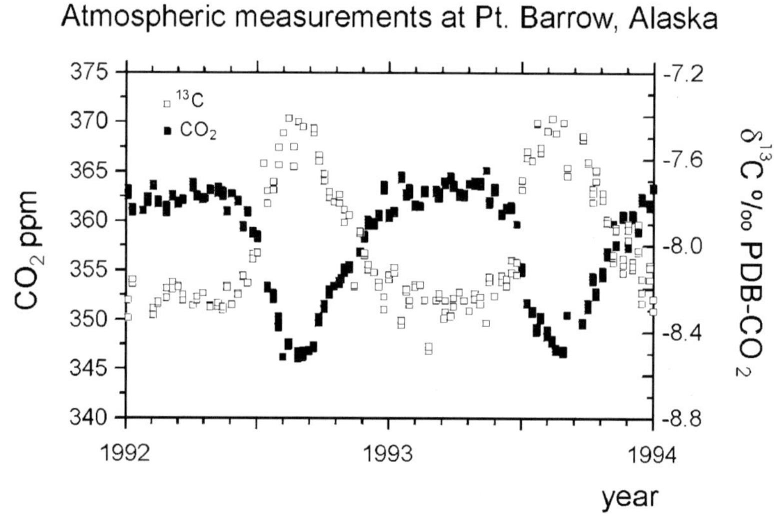

Fig. 7.7. Observed seasonal cycle of atmospheric CO_2 and its isotopic composition in ^{13}C at Point Barrow, Alaska, 74°N. When CO_2 decreases because of plant uptake in summer, ^{13}C increases as a result of the isotopic fractionation associated to photosynthesis.

Both routine monitoring in the atmosphere since 1990 (Keeling 1995), firn air measurements (Battle et al. 1996) and recent air archive measurements independently confirmed that the biosphere was only a slight sink over the 1980's (see Table 7.2).

Internal variations in the carbon cycle

The global growth rate of atmospheric CO_2 shows interannual changes of large amplitude, implying variations of approximately ± 2 GtC in the annual atmospheric carbon storage from one year to the next (Fig. 7.8). It is important to understand the causes of interannual variability in the carbon cycle, to better quantify processes by which the ocean and land carbon reservoirs respond to climate. Since fossil emissions are fairly stable, the observed CO_2 growth rate variations must reflect changes of ocean and land ecosystems fluxes. Interannual variations in the CO_2 growth rate statistically correlate with global climate anomalies. El Niño periods associated with warmer sea surface temperatures over the Eastern Pacific ocean, and warmer and dryer climate over the tropical continents are in phase with higher than average CO_2 growth rates in 1983, 1987 and 1998. During an El Niño event, Eastern equatorial pacific waters warm up, upwelling over that region and along the coast of Peru weaken, turning off the outcrop to the surface of deep waters rich in CO_2 and in nutrients, and reducing the upwelled ocean CO_2 source to the atmosphere. This change in ocean circulation and biological activity dominates the direct effect induced by the warming itself via the solubility, and the equatorial Pacific Ocean acts as a smaller source to the atmosphere during El Niño, opposite the observed growth rate of atmospheric CO_2.

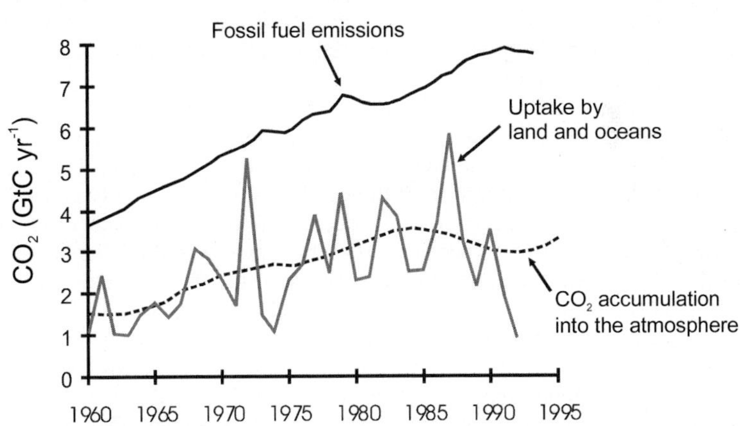

Fig. 7.8. Global fossil CO_2 release to the atmosphere (estimated from economic inventories) and observed accumulation of CO_2 in the atmosphere. The difference between these two terms changes from one year to the next and reflects variations in the ocean and terrestrial fluxes (after IPCC 1995 and Heimann et al. 1997).

In fact, El Niño causes a decrease in precipitation and a warming over the Amazon and over South East Asia. Changes in temperature impact plant respiration, as well as changes in soil moisture which is controlled both by temperature and precipitation. Biogeochemical models of the terrestrial biospheric carbon exchange simulate a decrease in the production (Tian et al. 1998, Kindermann et al. 1996) of rain forests and of savannas, which yield to decreased carbon storage over tropical ecosystems. All models however, do not attribute these changes in the net carbon fluxes to the same processes. The period 1991-95 was a weak El Niño, but during 1992-93, surprisingly low growth rates have been recorded. The climate anomaly induced by El Niño in the tropics was concomitant with cooler temperatures in the northern hemisphere, caused by volcanic aerosols emitted by the Pinatubo eruption in June 1991. Analysis of both $\delta^{13}C$ and O_2/N_2 atmospheric data have attributed a large northern hemisphere terrestrial carbon sink in 1992-93 (Ciais et al. 1995b, Keeling et al. 1996). Processes that fostered such an abnormally large carbon uptake can be an increased plant growth (NPP) in temperate and boreal regions, a reduction of the soil respired CO_2 flux, or a conjunction of both.

7.3.6 Impacts of future climate change on the carbon cycle

The ocean

The long term response of the ocean carbon cycle to changing climate involves changes in sea surface temperatures, in ocean biology, and in the ocean circulation of dissolved carbon and dissolved nutrients. These processes are not independent of each other. For instance, decreased upwelling of deeper waters, rich in dissolved nutrients may have two opposite consequences. Firstly, it will reduce the partial pressure of CO_2 by decreased upwelling of supersaturated waters to the surface, but it may as well weaken the biological pump, which helps maintain the surface waters unsaturated. The behavior of marine ecosystems and their adaptation to their changing environment in the ocean is still rather poorly understood. Yet another additional effect is how the ocean will absorb the excess CO_2 from the atmosphere. If the oceans become, on average more stratified, the ocean volume which will absorb anthropogenic CO_2 on a 20 to 100 years time scale will diminish, and atmospheric CO_2 will increase faster than it does now. A few studies have addressed this climate-CO_2 feedback by using global models of the ocean carbon cycle where the physical circulation changes are predicted from coupled ocean atmosphere models (e.g. Sarmiento et al. 1998; Joos et al. 1999). In most of these studies, in response to warming, the ocean thermohaline overturning reduces having the overall effect of slightly diminishing CO_2 absorption by the oceans. These studies somewhat disagree on the magnitude of the impact of climate change on the sea-to-air global fluxes, but the climate CO_2 feedback appears to be rather small on a 20 to 100 years timescale. On the one hand, a more stratified ocean with reduced thermohaline overturning might see a more efficient utilization of surface nutrients, but the reduced advection of nutrients to the surface may also limit the activity of marine biology. Similarly, reduced ocean overturning is expected to reduce the outcrop of CO_2 rich waters to the surface, and therefore to globally diminish CO_2 outgassing by the ocean.

Land ecosystems

The distribution and the production of vegetation on the surface of the Earth, as well as the storage of carbon in land ecosystems, are strongly driven by climate. The carbon balance of an ecosystem depends primarily on temperature, moisture and radiation. Increased CO_2 concentration has the primary effect of decreasing stomatal conductance, and in this way, diminishing plant transpiration. Its was simulated in biogeochemical models forced by the output of climate models predicting future climate change, that the reduction in transpiration may increase the amount of sensible heat exchanged by vegetated areas, which in turn, may warm up the surface (e.g. Sellers et al. 1996). This positive feedback loop of increasing CO_2 on climate change could be at least partially counteracted by changes in the vegetation structure (Betts et al, 1997). For instance, augmenting the surface of leaves available for transpiration could simply offset the reduction in transpiration from individual leaves. Increasing CO_2 concentration on a 20 to 100 years timescale will possibly have indirect impact on plants, through climate change that will amplify or counteract the direct CO_2 fertilization effect (e.g. Cao and Woodward, 1998). In turn, changes in ecosystems carbon stocks lead to changes in atmospheric CO_2 concentration, inducing CO_2-climate feedback. The magnitude of such feedback is rather uncertain, but few model studies have suggested that this feedback is positive. The land biosphere was simulated to lose significant amounts of carbon to the atmosphere in response to temperature and precipitation changes, especially in the tropics, which will contribute more radiative forcing and may in turn amplify climate changes. The global carbon cycle has already experienced such feedback in the past, as witnessed by large variations observed in the vegetation distribution and in the global distribution of carbon stocks. Between the last Ice Age 21 000 years ago and the Holocene warm period 11 000 years ago, it is estimated that the biospheric carbon stocks have augmented from 500 GtC to 2500 GtC today. Yet atmospheric CO_2 increased by 100 ppm over the deglaciation which lasted approximately 7000 years. Future climate changes on the 20 to 100 years time frame that are expected in response to increasing radiative forcing since the last century, will occur simultaneously with a very large rise in atmospheric CO_2, on the order of 200-400 ppm. The impact of both rapidly rising CO_2 and of climate change on terrestrial ecosystems must therefore be accounted for in models.

Today's global vegetation has adapted to seasonal and interannual variations of temperature, moisture and radiation (infrequent droughts and frost for instance). Climate variability, especially the frequency of extreme events influences the carbon balance of ecosystems. However, as far as climate conditions vary around a more or less stable long term mean, ecosystems recover from perturbations, and the distribution of vegetation types is not expected to change, excepted by direct human induced changes in land use. When changes in climate, and in atmospheric CO_2 persist for several decades however, some vegetation types will be replaced by others, which are better adapted to the new environment. For instance, C3 grasses may better compete with C4 grasses in dry environments under elevated CO_2. Vegetation changes are more likely to happen in climate transition zones, where one vegetation type could shift for another under a slight disequilibrium in climate. For instance, the establishment of drier conditions may yield grasslands to replace forests, leading to the release of the carbon stored in forest biomass. The establishment of longer growing seasons in the Arctic with warmer temperatures earlier on during the year may yield trees to establish in tundra areas. Warmer and wetter climate might favor vegetation growth and carbon accumulation in living vegetation, but this may have also undesirable effects on soil organic matter. Increased temperature and moisture may enhance the loss of soil carbon by respiration, a positive feedback on climate change. Finally, nitrogen might be mineralized faster, being more readily available for plants, yielding an increase in productivity and to a gain of carbon by soils through litter fall, a negative feedback on climate change.

7.4 Conclusions

Our knowledge of the contemporary carbon cycle has notably progressed during the past 20 years. The atmosphere is the reservoir we know best. On the one hand, ice core data were used to trace back the increase in atmospheric CO_2, which lasted for over two centuries and continues to amplify. On the other hand, atmospheric CO_2 is measured at several dozens of stations around the world, now part of an international effort. From the monitoring of CO_2 and its isotopic composition, as well as of other trace gases related to the carbon cycle, such as oxygen, modeling of the atmospheric concentrations establishes a rough diagnosis of the geographical and seasonal distribution of the sources and sinks of CO_2. Such atmospheric models generally infer a strong sink of CO_2 at mid-latitude of the northern hemisphere. The ocean sink only accounts for the uptake of a 30% fraction of anthropogenic CO_2: thus part of fossil fuel emissions must be absorbed by the vegetation.

However, more studies are underway as there is a need to predict the evolution of the carbon cycle given a priori scenarios of anthropogenic emissions for the next century. To pursue this goal, ocean carbon-cycle models are currently available. Although such models require large amounts of computational time, they provide a fine description of ocean dynamics and integrate some of the fundamental biological processes. The biospheric reservoir is not so well-known. However, the vegetation activity can be remote-sensed using satellite global observations, which then helps the development of global diagnosis for the primary production of main ecosystems. Realistic estimates of photosynthesis and respiration, and of their imbalance which yields carbon storage or loss on the continents, based on biological processes can be inferred by global models of the land biosphere. The ultimate goal remains to integrate atmospheric, oceanic, and biospheric models in order to simulate realistic transfer of CO_2 between these three reservoirs on time-scales going from years to centuries. Such models can be tested first against the intra and interannual fluctuations of atmospheric CO_2, and then used to estimate future concentrations.

References

Aber, J. D., Nadelhoffer, K. J., Steuder, P., & Melillo, J. M. (1989). Nitrogen saturation in northern forest ecosystems. Bioscience, 39(6):378-386.

Andrie C., Oudot C., Genthon C., and Merlivat L. (1986) CO_2 Fluxes in the tropical Atlantic during FOCAL cruises. *Journal of Geophysical Research* 91(C10):11741-11755.

Baldocchi D, Valentini R, Running S, Oechel W, Dahlman R (1996) Strategies for measuring and modelling carbon dioxide and water vapour fluxes over terrestrial ecosystems. Global Change Biol 3:159-168

Ball, J. T., (1988). An analysis of stomatal conductance, PhD Thesis, Stanford University, Stanford, Calif. 89 pp.

Battle M., Bender M., Sowers T., Tans P. P., Butler J. H., Elkins J. W., Ellis J. T., Conway T., Zhang N., Lang P. M., and Clarke A. D. (1996) Atmospheric gas concentrations over the past century measured in air from firn at the South Pole. *Nature* 383: 231-235.

Barnola, J. M., Anklin, M., Porcheron, J., Raynaud, D., Schwander, J., & Stauffer, B. (1995). CO2 evolution during the last millennium as recorded by Antarctic and Greenland ice. Tellus, 47(B):264-272.

Betts R. A., Cox P. M., Lee S. E., and Woodward F. I. (1997) Contrasting physiological and structural vegetation feedbacks in climate change simulations. *Nature* 387, 796-799.

Broecker, W. S., Peng, T.-H., Ostlund, G., & Stuiver, M. (1985). The distribution of bomb radiocarbon in the ocean. Journal of Geophysical Research, 90(C4), 6953-6970.

Bousquet-P; Ciais-P; Peylin-P; Ramonet-M; Monfray-P (1999). Inverse modeling of annual atmospheric CO_2 sources and sinks 1. Method and control inversion. Journal of Geophysical Research 104 (D21):26161-26178

Cao M. and Woodward F. I. (1998) Dynalic responses of terrestrial ecosystem carbon cycling to global climate change. *Nature* 393, 249-252.

Christensen NL, Bartuska AM, Brown JH, Carpenter S, D'Antonio C, Francis R, Franklin JF, MacMahon JA, Noss RF, Parsons DJ, Peterson CH, Turner MG, Woodmansee RG (1996) The report of the Ecological Society of America committee on the scientific basis for ecosystem management. Ecol App 6:665-691

Ciais, P., Tans, P. P., Trolier, M., White, J. W. C., & Francey, R. J. (1995a). A large northern hemisphere terrestrial CO_2 sink indicated by the $^{13}C/^{12}C$ ratio of atmospheric CO_2. Science, 269, 1098-1102.

Ciais, P., Tans, P. P., White, J. W., Trolier, M., Francey, R., Berry, J. A., Randall, D., Sellers, P. J., Collatz, J. G., & Schimel, D. S. (1995b). Partitioning of ocean and land uptake of CO_2 as inferred by d13C measurements from the NOAA Climate Monitoring and Diagnostic Laboratory global air sampling network. Journal of Geophysical Research, 100(D3), 5051-5070.

Collatz, G. J., Ball, J. T., Grivet, C., & Berry, J. A. (1991). Physiological and environmental regulation of stomatal conductance, photosynthesis and transpiration: a model that includes a laminar boundary layer. Agricultural and Forest Meteorology, 54, 107-136.

Conway, T. J., Tans, P. P., Waterman, L. S., Thoning, K. W., Kitzis, D. R., Masarie, K. A., & Zhang, N. (1994). Evidence for interannual variability of the carbon cycle from the National

Oceanic and Atmospheric Administration/Climate Monitoring and Diagnostic Laboratory Global Air Sampling Network. Journal of Geophysical Research, 99(D11), 22831-22855.

Dai, A., & Fung, I. (1993). Can climate variability contribute to the "missing" CO_2 sink? Global Biogeochemical Cycles, 7(3), 599-609.

Dixon, R. K., Brown, S., Houghton, R. A., Solomon, A. M., Trexler, M. C., & Wisniewski, J. (1994). Carbon pools and flux of global forest ecosystems. Science, 263, 185-190.

Farquhar, G. D., Caemmerer, S. v., & Berry, J. A. (1980). A biochemical model of photosynthetic assimilation in leaves of C3 species. Planta, 149, 78-90.

Feely R. A., Wanninkhof R., Takahashi T., and Tans P. (1999) Influence of El Niño on the equatorial Pacific contribution to atmospheric CO_2 accumulation. *Nature* 398, 597-601.

Francey, P. P. Tans, R. J., Allison, C. E., Enting, I. G., White, J. W. C., and Trolier, M. (1994). Changes in the oceanic and terrestrial carbon uptake since 1982. Nature 373, 326-330.

Friend, A.D., A.K. Stevens, and M.G.R. Cannell, A process-based, terrestrial biosphere model of ecosystem dynamics (Hybrid v.3.0), *Ecological Modelling*, *95*, 249-288, 1997.

Frew, N. M. The role of organic films in air-sea gas exchange; in *The sea surface and Global change* (ed. Liss, P. S., and Duce, R. A.) 121-171 (Cambridge university Press, Cambridge, 1997)

Fung, I. Y., Tucker, C. J., & Prentice, K. C. (1987). Application of advanced very high resolution radiometer vegetation index to study atmosphere-biosphere exchange of CO_2. Journal of Geophysical Research, 92(D3), 2999-3013.

Hao, W. M., Liu, M. H., & Crutzen, P. J. (1990). Estimates of annual and regional release of CO_2 and other trace gases to the atmosphere from fires in the Tropics, based on the FAO statistics for the period 1975-1980. In G. J. Goldammer (Ed.), Fires in the tropical biota: ecosystem processes and global challenges (pp. 440-462). Berlin: Springer-Verlag.

Houghton, R. A. (1991). Releases of carbon to the atmosphere from degradation of forests in tropical Asia. Canadian Journal of Forest Research, 21, 132-142.

Houghton, R. A., Boone, R. D., Fruci, J. R., Hobbie, J. E., Mellilo, J. M., Palm, C. A., Peterson, B. J., Shaver, G. R., Woodwell, G. M., Moore, B., Skole, D. L., & Myers, N. (1987). The flux of carbon from terrestrial ecosystems to the atmosphere in 1980 due to changes in land use : geographic distribution of the global flux. Tellus, 39B, 122-139.

Houghton, S. (1991). Biomass burning from the perspective of the global carbon cycle. In J. S. Levine (Eds.), Global Biomass Burning: Atmospheric, Climatic and Biospheric implications Cambridge: MIT Press. 321-325.

IPCC (1995). Climate change 1994. Radiative forcing of climate change. An evaluation of the IPCC IS92 emissions scenarios. Cambridge university press. 329 pp.

Jenkinson D. S., D. E. Adams and A. Wild (1991) Model estimates of CO_2 emissions from soil in response to global warming. Nature 351: 304-306.

Joos F., Plattner G. K., Stocker T. F., Marchal O., and Schmittner A. (1999) Global warming and marine carbon cycle feedbacks on future atmospheric CO2. *Science* 284, 464-467.

Keeling, C. D. (1995). Interannual extremes in the rate of rise of atmospheric carcone dioxide since 1980. Nature, 375, 666-670.

Keeling R. F. (1995) The atmospheric oxygen cycle : the oxygen isotopes of atmospheic CO_2 and O_2 and the O_2/N_2 ratio. *Reviews of Geophysics*, 1253-1262.

Keeling, C. D., Bacastow, R. B., Carter, A. F., Piper, S. C., Whorf, T. P., Heimann, M., Mook, W. G., & Roeloffzen, H. A. (1989a). A Three-dimensional Model of Atmospheric CO_2 Transport Based on Observed Winds : 1. Analysis of Observational Data. In Aspects of Climate Variability in the Pacific and Western Americas, Geophysical monograph 55, AGU (Ed), Washington (USA) 164-236.

Keeling, C. D., Piper, S. C., & Heimann, M. (1989b). A Three-dimensional Model of Atmospheric CO_2 Transport Based on Observed Winds : 4. Mean annual gradients and interannual variations. In Aspects of Climate Variability in the Pacific and Western Americas, Geophysical monograph 55, AGU (Ed), Washington (USA), 304-363.

Keeling, C.D., J.F.S. Chin, and T.P. Whorf, Increased activity of northern vegetation inferred from atmospheric CO_2 measurements, Nature, 382, 146-149, 1996.

Kindermann J., Wurth G., Kohlmaier G. H., and Badeck F. W. (1996) Interannual variation of carbon exchange fluxes in terrestrial ecosystems. *Global Biogeochemical Cycles* 10(4), 737-755.

Lefèvre, N., A.J. Watson, D.J. Cooper, R.F. Weiss, T. Takahashi and S.C. Sutherland. Assessing the seasonality of the oceanic sink for CO_2 in the Northern Hemisphere. Global Biogeochemical cycles, 13, 273-286, 1999.

Leuning, R., A critical appraisal of a combined stomatal-photosynthesis model for C3 plants, *Plant, Cell and Environment 18*, 339-355, 1995.

Liss, P. S., & Merlivat, L. (1986). Air-sea gas exchange rates : introduction and synthesis. In P. Buat-Menard (Ed.), The role of air-sea exchange in geochemical cycling NATO-ASI Series Vol. 185, D. Reidel., Dordrecht, 113-127

Lloyd J. and Taylor J. A. (1994) On the temperature dependance of soil respiration. *Functional Ecology* 8, 315-323.

Maier-Reimer, E. (1993). Geochemical cycles in an ocean general circulation model. Preindustrial tracer distribution. Global Biogeochemical Cycles, 7, 645-677.

Marland, G., Rotty, R. M., & Treat, N. L. (1985). CO_2 from fossil fuel burning: global distribution of emissions. Tellus, 37(B), 243-258.

Myneni, R.B., C.D. Keeling, C.J. Tucker, G. Asrar, and R.R. Nemani, Increased plant growth in the northern high latitudes from 1981 to 1991, Nature, 386, 698-702, 1997.

Neftel, A., Oeschger, H., Schwander, J., Stauffer, B., & Zumbrunn, R. (1982). Ice core sample measurements give atmospheric CO_2 content during past 40 000 years. Nature, 295, 220-223.

Oeschger, H., Siegenthaler, U., Schotterer, U., & Gugelmann, A. (1975). A box diffusion model to study the carbon dioxide exchange in nature. Tellus, 27, 168-191.

Raich, J. W., & Schlesinger, W. H. (1992). The global carbon dioxide flux in soil respiration and its relationship to vegetation and climate. Tellus, 44(B), 81-89.

Randall, D. A., Sellers, P. J., Berry, J. A., Dazlich, D. A., Zhang, C., Collatz, J. A., Denning, A. S., Los, S. O., Field, C. B., Fung, I., Justuce, C. O., & Tucker, C. J. (1996). A revised land surface parameterization (SiB2) for GCMs. Part 3: the greening of the Colorado State University general circulation model. Journal of Climate, 9. 738-763.

Raynaud D., J. Jouzel, J. M. Barnola, J. Chappelaz, R. J. Delmas and C. Lorius (1993) The ice record of greenhouse gases. Science 259: 926-934.

Ruimy, A., Saugier, B., & Dedieu, G. (1994). Methodology for the estimation of terrestrial net primary production from remotely sensed data. Journal of Geophysical Research, 99, 5263-5283.

Sarmiento J. L., Hughes T. M. C., Stouffer R. J., and Manabe S. (1998) Simulated response of the ocean carbon cycle to anthropogenic climate warming. *Nature* 393, 245-249.

Sellers, P. J., Berry, J. A., Collatz, G. J., Field, C. B., & Hall, F. G. (1992). Canopy reflectance, photosynthesis, and transpiration. III. A reanalysis using improved leaf models and a new canopy integration scheme. Remote Sensing Environment, 42, 187-216.

Sellers P. J., Bounoua L., Collatz G. J., Randall D. A., Dazlich D. A., Los S. O., Berry J. A., Fung I., Tucker C. J., Field C. B., and Jensen T. G. (1996) Comparison of radiative and physiological effects of doubled atmospheric CO_2 on climate. *Science* 211, 1402-1406.

Takahashi, T., Goddard, J., Sutherland, S., Chipman, D. W., & Breeze, C. C. (1986). Seasonal and Geographic Variability of Carbon Dioxide Sink/Source in the Oceanic Areas. Final Technical Report for contract MRETTA 19X-89675C. New-York: Lamont-Doherty Geological Observatory.

Takahashi, T., Olafson, J., Goddard, J. G., Chipman, D. W., & Sutherland, S. C. (1994). Seasonal variation of CO_2 and nutrients in the high-latitude surface ocans: a comparative study. Global Biogeochemical Cycles, 7(4), 843-878.

Takahashi, T. Feely, R.A., Weiss, R.F., Wanninkhof, R. H., Chipman, D.W., Sutherland S.T. and T. T. Takahashi. Global air-sea flux of CO_2 : an estimate based on measurements of sea-air pCO_2 difference. (1997) Proc. Natl. Acad. Sci., USA. 94 8292-8299.

Tans, P. P., Fung, I. Y., & Takahashi, T. (1990). Observational Constraints on the Global Atmospheric CO_2 budget. Science, 247, 1431-1438.

Tian H., Mellilo J. M., Kicklighter D. W., Mc Guire A. D., Helfrich III J. V. K., Moore III B., and Vörösmarty C. J. (1998) Effect of interannual climate variability on carnon storage in Amazonian ecosystems. *Nature* 396, 664-667.

Townsend, A. R., Braswell, B. H., Holland, E. A., & Penner, J. E. (1996). Nitrogen deposition and terrestrial carbon storage: linking atmospheric chemistry and the global carbon budget. Ecological Applications, 6. 806-814.

Trolier M., White, J.W.C., Tans, P.P., Masarie, K.A. and Gemery, P.A. (1996). Monitoring the isotopic composition of atmospheric CO_2 : measurements from the NOAA global air sampling network. Journal of Geophysical Research,101, D20, 25897-25916, 1996.

Wanninkhof, R. and W. R. McGillis, A cubic relationship between air-sea CO_2 exchange and wind speed. Geophys. Res. Lett., 26, 1889-1892, 1999.

Woodward, F.I., T.M. Smith, and W.R. Emanuel, A global land primary productivity and phytogeography model, *Global Biogeochemical Cycles*, *9*, 471-490, 1995.

8 Changes in the Global Carbon Cycle and Ocean Circulation on the Millennial Time Scale

Thomas F. Stocker

Climate and Environmental Physics, Physics Institute, University of Bern. Sidlerstrasse 5, CH-3012, Bern, Switzerland
stocker@climate.unibe.ch

Abstract. Carbon dioxide is, after water vapor, the most important greenhouse gas. Naturally, its atmospheric concentration has varied between 190 and 290 ppmv over the last half million years. The man-made CO_2 increase of the last 250 years has already reached this amplitude with the potential of inducing significant global warming. Climate models suggest that the ocean circulation reacts in a sensitive way to global warming in that large circulation systems can break down. Here we discuss the question of potential feedback mechanisms between the global carbon cycle and the ocean circulation. This is done by first examining the paleoclimatic record of past CO_2 changes obtained from polar ice cores. Modeling experiments provide a quantitative interpretation of the changes found in these records.

8.1 Introduction

Greenhouse gases play a crucial role in the energy balance of the Earth. In order to illustrate the importance of these gases, the simplest climate model is considered. Were the Earth a black body which is in thermal equilibrium with the incoming short wave radiation from the sun, the following energy balance can be formulated:

$$\pi R^2 (1 - \alpha) S_0 = 4\pi R^2 \sigma T^4 \tag{8.1}$$

where R = 6730 km is the Earth's radius, $\alpha = 0.3$ is the global mean albedo (reflectivity), $S_0 = 1367\ Wm^{-2}$ is the solar constant, $\sigma = 5.67\ 10^{-8}\ Wm^{-2}\ {}^oK^{-4}$ is the Stefan-Boltzmann constant (Fig. 8.1). Solving for the global mean annual temperature, we obtain:

$$T = \left[\frac{(1-\alpha)\cdot S_0}{4\cdot\sigma}\right]^{0.25} = -18.3°C \tag{8.2}$$

This is in contrast with observations, and T is below the freezing point of water. Therefore life in its current form would not be possible. Instead, the mean annual temperature is about +15°C.

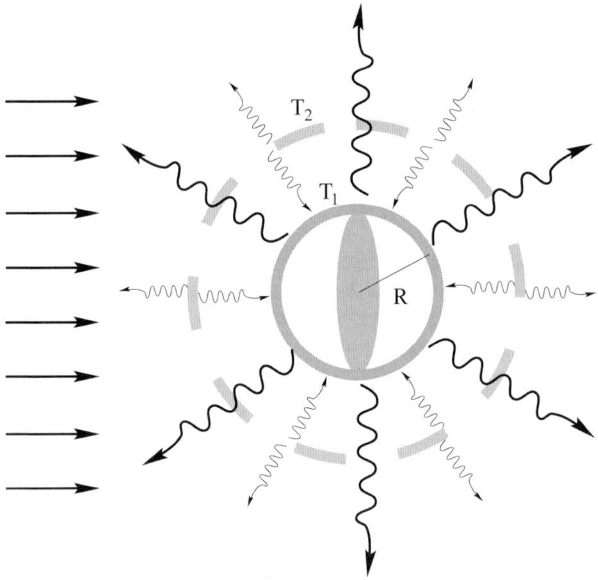

Fig. 8.1.: A minimum climate model in which the Earth is assumed to be a black body in radiative equilibrium with incoming solar short wave, outgoing longwave radiation at temperature T_1 and a layer at higher altitude ("clouds"), which emits radiation as a black body at temperature T_2.

An increase in surface temperature can be qualitatively obtained by introducing a second black-body shell at a higher altitude. This shell shall represent the combined effect of clouds, water vapor and other greenhouse gases and is assumed to be transparent to shortwave radiation and opaque to longwave radiation. We also assume that this shell is only partly present to mimic the effects of an incomplete "cloud" cover. This fraction is denoted by c. The energy balance for such a configuration then reads:

$$\pi R^2 (1-\alpha) S_0 + c\, 4\pi R^2 \sigma T_2^4 = 4\pi R^2 \sigma T_1^4 \qquad (8.3)$$

$$c\, 4\pi R^2 \sigma T_1^4 = 2 \cdot 4\pi R^2 \sigma T_2^4 \qquad (8.4)$$

where T_1 and T_2 are now the temperatures of the surface and the outer shell, respectively. Note that in the above equations, the "cloud" cover does not change albedo; this is not correct for clouds at mid and low altitudes. Assuming a fractional "cloud" cover of 77%, the equilibrium temperatures are now:

$$T_1 = +14.6\ ^\circ\mathrm{C} \text{ and } T_2 = -31.1\ ^\circ\mathrm{C} \qquad (8.5)$$

Equivalently, one could argue that the greenhouse effect is taken into account by simply assuming that the Earth is a grey body with emissivity ε. This planetary emissivity is estimated at about $\varepsilon = 0.6$ which yields an equilibrium temperature of:

$$T = \left[\frac{(1-\alpha)\cdot S_0}{4\cdot\varepsilon\cdot\sigma} \right]^{0.25} = +16.4°C \qquad (8.6)$$

What is parameterised here bluntly in the form of a 2-shell Earth or a simple "grey body" is in reality the cumulative effect of all radiatively active gases in the atmosphere. Their order of importance is water vapor, carbon dioxide (CO_2), methane (CH_4) and nitrous oxide (N_2O). It is thus of prime importance to document and understand past changes in these radiatively active agents in order to be able to estimate the effect of future changes in the radiative balance of the Earth.

Past changes in the content of the most important greenhouse gas, water vapor, cannot be directly reconstructed. Only indirect paleoclimatic evidence such as changes in snowline elevation is available which might be employed to estimate changes in the vertical lapse rate and hence in the water vapor content. Further, stable isotopes of the water molecule ($\delta\,^{18}O$ and $\delta\,^{2}H$) determined on ice cores from different altitudes may provide additional information. The concentrations of the other greenhouse gases, however, can be reconstructed for the past by measuring air entrapped in polar ice cores (Raynaud et al. 1993).

8.2 Carbon cycle

In order to understand changes in the global carbon cycle of the Earth due to anthropogenic activity during the last 250 years, the natural cycle is briefly revisited. For a comprehensive review the reader is referred to Siegenthaler & Sarmiento (1993). For time scales not exceeding a few thousand years, the most important reservoirs are the atmosphere, the ocean and the terrestrial biosphere (Fig. 8.2). Carbon appears in different forms in these reservoirs: in the atmosphere it is primarily CO_2, in the terrestrial biosphere in the form of fixed carbon in organic matter, whereas in the ocean the major part of the inventory is in the form of the bicarbonate ion (HCO_3^-). The ocean inventory is over 60 times larger than the atmospheric inventory and about 20 times larger than the terrestrial biosphere inventory and is thus the most important component in determining the level of atmospheric CO_2. Carbon is naturally cycled through these three reservoirs, and the atmosphere serves as a gateway with a relatively short residence time of about $600 / (100 + 74) \approx 3.5$ years. The exchange mechanisms between atmosphere and terrestrial biosphere are photosynthesis (uptake of CO_2 from the atmosphere) and respiration (release of CO_2 to the atmosphere), whereas gas exchange dominates the fluxes between the ocean and the atmosphere.

Carbonate chemistry in the surface waters of the ocean is central to the understanding of the influence of and dynamics of changes in the ocean carbon content upon atmospheric CO_2.

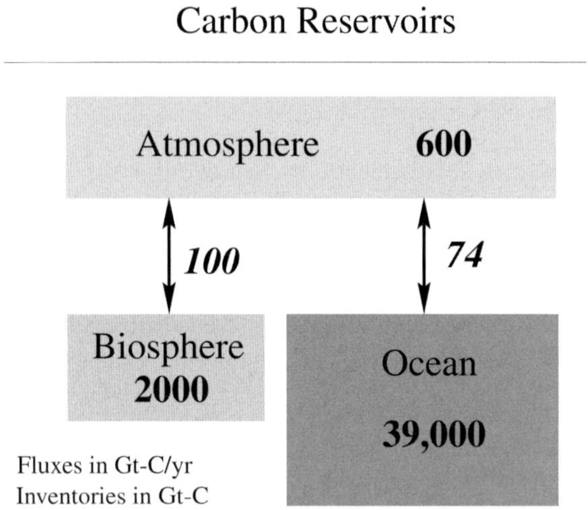

Fig. 8.2: The global carbon cycle involves three fast exchanging reservoirs: the atmosphere, the ocean and the terrestrial biosphere. Carbon inventories and exchange fluxes are indicated. This is a simplified version of the figure by Siegenthaler & Sarmiento (1993).

Here we follow Joos & Sarmiento (1995). In the ocean surface waters, dissolved inorganic carbon occurs in the three species

$$[CO_2]:[HCO_3^-]:[CO_3^{2-}] = 1:175:21 \tag{8.7}$$

which implies that only a very small fraction of dissolved inorganic carbon (about 1%) can exchange with the atmosphere. The three species are in chemical equilibrium:

$$CO_2 + CO_3^{2-} + H_2O \longleftrightarrow 2\ HCO_3^- \tag{8.8}$$

and thus

$$\frac{[HCO_3^-]^2}{[CO_2]\cdot[CO_3^{2-}]} = \text{constant} \tag{8.9}$$

Because of the dominance of the bicarbonate pool, (Eq. 8.9) can be approximated by

$$[CO_2]\ [CO_3^{2-}] \approx \text{constant} \tag{8.10}$$

An additional constraint is the conservation of charge. This is taken into account by considering alkalinity, which can be defined as the total excess positive charge of all the ions in solution. Again, considering only the most important ionic species, an approximation is given by

$$[HCO_3^-] + 2\,[CO_3^{2-}] \approx \text{constant} \tag{8.11}$$

With these approximations, we can calculate the ratio of the relative changes in CO_2 and total carbon in the ocean:

$$\zeta = \frac{\dfrac{d[CO_2]}{[CO_2]}}{\dfrac{d([HCO_3^-]+[CO_3^{2-}])}{[HCO_3^-]+[CO_3^{2-}]}} = \frac{[HCO_3^-]+[CO_3^{2-}]}{[CO_3^{2-}]} \approx \frac{175+21}{21} \approx 9.3 \tag{8.12}$$

where (Eq. 8.10) and (Eq. 8.11) have been used. ζ is the buffer factor or Revelle Factor which is central to the understanding of the effects of changes in the ocean carbon inventory on the atmospheric CO_2 concentration. ζ depends on the chemical composition of the sea water, its salinity and temperature and ranges from about 8 in equatorial to about 14 in highlatitude waters (Takahashi et al. 1993). Relative inventory changes in the atmosphere are about 10 times larger than the corresponding relative changes in the ocean. This simplified reasoning also explains how much of an initial carbon injection into the atmosphere will remain there (assuming no interaction with the carbonate sediments). Assume that the carbon inventory in atmosphere increases by a small fraction f. The mass of additional carbon in the atmosphere and ocean is then given, respectively, by $f \cdot I_A$ and $(f/\zeta) \cdot I_O$, where I_A and I_O are the inventories of atmosphere and ocean as given in Fig. 8.2. The fraction of carbon that remains in the atmosphere (airborne fraction) can be calculated according to:

$$r = \frac{f \cdot I_A}{f \cdot I_A + (f/\zeta) \cdot I_O} = \frac{1}{1 + \dfrac{1}{\zeta} \cdot \dfrac{I_O}{I_A}} \approx 13\% \tag{8.13}$$

So far, only the inorganic aspect of the marine carbon cycle was discussed. However, using the above relations one would obtain an atmospheric CO_2 concentration of about 670 ppmv (parts per million by volume) for the atmosphere in chemical equilibrium with an abiotic ocean. This is in strong contrast to current observations indicating about 370 ppmv, and reconstructions of pre-industrial CO_2 concentration of about 280 ppmv (Neftel et al. 1988; Indermühle et al. 1999).

Therefore, there must exist additional processes that keep the concentration of CO_2 in the surface ocean low. These are collectively referred to as oceanic carbon "pumps" (Volk & Hoffert 1985). There are three such pumps: (i) the solubility pump; (ii) the soft tissue, or organic matter pump; and (iii), the calcite pump, and they maintain a vertical gradient of total dissolved inorganic carbon (DIC) in the water column. The solubility pump is due to the fact that warmer waters are less soluble for CO_2, i.e. both cold deep and cold high-latitude surface waters are enriched in CO_2 relative to the main water masses of the "warmwater sphere". This leads to a reduction of DIC in the surface waters. The soft tissue pump is due to the DIC uptake during the formation of marine organisms. This organic matter is constantly exported to the deep ocean in particulate and dissolved form. This also tends to reduce the surface concentrations of DIC. The third pump operates in the opposite direction. When organic matter forms calcite shells ($CaCO_3$), CO_3^{2-} ions are extracted from the water column. Due to (Eq. 8.10) this results in an increase of atmospheric CO_2. In combination, these three pumps lead to a reduction of the equilibrium CO2 concentration in the atmosphere from 670 ppmv to 280 ppmv (Table 8.1).

Table 8.1. The effect of the three carbon pumps calculated using the carbon cycle model of Joos et al. (1999) [calculations courtesy G.-K. Plattner, alkalinity held constant].

solubility	organic matter	calcite	atmosph. CO_2
×	×	×	665 ppmv
✓	×	×	380 ppmv
✓	✓	×	250 ppmv
✓	✓	✓	278 ppmv

8.3 Ocean and Abrupt Climate Change

8.3.1 Climatically Relevant Ocean Circulation Types

The climate system consists of four major components. These are the atmosphere, the cryosphere, the terrestrial biosphere and the ocean. This subdivision is somewhat arbitrary and other approaches, e.g. according to the relevant cycles of heat, water and tracers, are equally valid. The terrestrial biosphere and the cryosphere (ice sheets) are important drivers of climate change. The former has a strong influence on the hydrological cycle and the albedo (Crowley and Baum 1997) whereas the ice sheets are mainly influencing atmospheric circulation and the surface radiative balance through the ice-albedo feedback (see Crowley and North 1991). For example, glacial-interglacial temperature reduction is not feasible without a strong contribution of land-surface changes. This is illustrated by taking the logarithmic derivative of (Eq. 8.6):

$$\frac{\Delta T}{T} = \frac{1}{4} \cdot \frac{\Delta S_0}{S_0} - \frac{1}{4} \cdot \frac{\Delta \alpha}{(1-\alpha)} - \frac{1}{4} \cdot \frac{\Delta \varepsilon}{\varepsilon} \tag{8.14}$$

Using estimates for the reduction of global mean annual temperature ($\Delta T \approx -4\ ^oK$), and solar radiation due to changes in orbital parameters ($\Delta S_O \approx 0.3\ Wm^{-2}$) during the last glacial maximum (Berger et al. 1993), we note from (Eq. 8.14), that changes in solar forcing cannot explain directly the colder climate.

In addition, a strong ice-albedo feedback is required. This can be estimated in the following way. Albedo at 60°N is *0.4* in summer and about *0.55* in winter due to snow and sea ice cover (Peixoto and Oort 1992). The regions polewards of 60° account for about 6% of the surface area, therefore $\Delta\alpha \approx 0.06 \cdot 0.15$. Thus,

$$\frac{-4}{298} = \frac{0.3}{4 \cdot 1367} - \frac{0.01}{4 \cdot 0.5} - \ldots \qquad (8.15)$$

demonstrates that albedo changes have the correct order of magnitude to change global mean temperature significantly. Such albedo changes are slow, because terrestrial ice sheets take several 10 000 years to build up. Rapid changes, which are abundantly found in the paleoclimatic record (e.g., Broecker 1997 for a review) cannot be explained by these processes alone.

Here, the ocean plays an important role. The ocean is a major important component of the climate system because it covers 70% of the Earth's surface. Considering ocean and atmosphere as the only components that are relevant for climate changes on time scales of less than 104 years, the ocean contains 95% of water and 99.9% of the heat content. However, it is the dynamics that is essential in providing a mechanism for abrupt change.

The present description of ocean circulation types is very basic and concerns only the most important, large-scale flows; here we follow Stocker (1999). The major processes that govern the dynamics are the action and regional distribution of momentum and buoyancy fluxes at the ocean's surface, the Earth's rotation and the presence of ocean basin boundaries. Four major circulation types characterize the flow in an ocean basin on large spatial scales (Fig. 8.3). The general circulation is forced by the input of momentum through surface wind stress τ and by the flux of buoyancy, indicated by the vertical arrow D and uniform upwelling Q (Fig. 8.3). The surface wind stress forces the wind-driven geostrophic circulation (WGC) which is intensified at the western boundary and forms the subtropical and the subpolar gyres. In the interior of the ocean basin, there is a balance between the pressure gradients and the Coriolis forces acting on the moving fluid. Western intensification, on the other hand, is a consequence of the spherical shape of the rotating Earth and frictional effects in the fluid and at the basin boundaries (Stommel 1948; Pedlosky 1996).

Winds blowing over the surface of the ocean lead to divergent (Ekman upwelling) or convergent (Ekman downwelling), frictionally driven flows and change the local depth of the near-surface isopycnals σ (lines of constant density) which set up horizontal pressure gradients with associated geostrophic flows. These flows are responsible for the fact that the wind driven gyres do not extend all the way to the bottom but are compensated by sloping isopycnals in the top few hundred meters. In other words, the geostrophic velocities exhibit a vertical structure and the wind-driven circulation remains confined to the top few hundred meters of the water column (Pedlosky 1996).

Turning now to the deep circulation, the source D feeds the deep western boundary current (DWBC) which flows southward and leaks into the deep interior where the geostrophic flow (DGF) is directed polewards at all latitudes. The DGF recirculates into the source area of the DWBC. There is a cross-interface mass flux Q upwelling into the upper 1000 m which supplies the mass lost due to D in the upper layer.

There are only a few locations in the ocean where new deep water is being formed. These are the Greenland-Iceland-Norwegian Seas in the north and the Weddell Sea in the south and a few other, minor sites (Killworth 1983; Marshall & Schott 1999). The dynamics of a fluid moving on a rotating sphere dictates that also the deep flow is confined to western boundary currents (Stommel 1958; Stommel & Arons 1960).

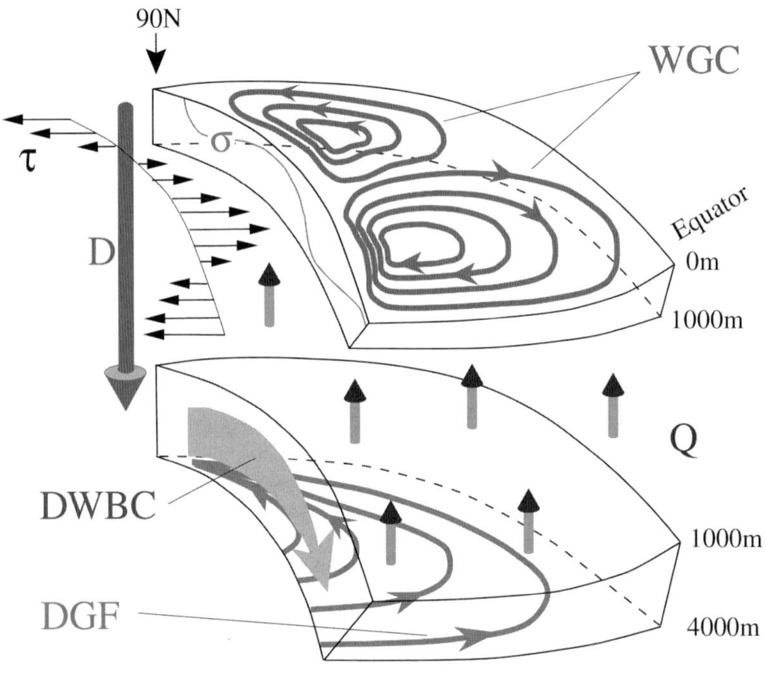

Fig. 8.3: Schematic view of the different types of steady-state circulations in a sectorial ocean basin extending from the equator to the pole with a longitudinal extent of roughly 60°. Wind stress τ drives a wind driven gyre circulation (WGC) which shows western intensification due to the curvature of the rotating Earth. τ also causes Ekman upwelling in the northerly and Ekman downwelling in the southerly upper layer giving a near-surface isopycnal surface σe its typical shape: the isopycnal is shallow below the subpolar gyre and deep below the subtropical gyre. A source of newly formed deep water, D, feeds the deep ocean in which a deep western boundary current (DWBC) develops from which the deep geostrophic flow (DGF) of the interior is derived. DGF flows northward to conserve potential vorticity while slowly upwelling. This results in a vertical mass flux Q that closes the flow. In reality, $Q < D$ in this sector and the DWBC is crossing the equator setting up a global circulation [from Stocker 1999].

In the present ocean, the northern source is strong enough so that the current crosses the equator and penetrates eventually into the southern ocean. There, it mixes with the deep waters from the Weddell Sea and flows into the Indian and Pacific oceans where broad upwelling occurs. The global structure of the deep water paths was already suggested by Henry Stommel (1920-1992) in a pioneering paper (Stommel 1958); the return flow in the thermocline, preferentially via the 'warm water route' around Africa, was first described by Gordon (1986). This global flow subsequently became known as the 'conveyor belt' (Broecker 1987; Broecker 1991), but the structure is far more complicated than a simple ribbon spanning the globe (Schmitz 1995).

Evaluations of the radiation balance at the top of the atmosphere show that the oceanatmosphere system must transport heat towards the high latitudes where there is a net loss of energy over one year (Trenberth & Solomon 1994). About half of that heat is carried by ocean currents (Macdonald & Wunsch 1996). In contrast to the other ocean basins, the meridional heat transport in the Atlantic Ocean is northward at all latitudes. Evaluation of oceanographic observations (Hall & Bryden 1982) as well as model simulations (Böning et al. 1996) indicate that the meridional heat transport in the Atlantic is primarily due to the meridional overturning circulation which carries warm near-surface waters northward and cold deep water southward. This is the deep circulation of the ocean that is driven by surface buoyancy fluxes and is referred to as the "thermohaline circulation", short THC (Warren 1981). The wind-driven, near-surface circulations in the Atlantic do not transport significant amounts of heat polewards. The thermohaline circulation in the Atlantic is also often referred to as the "nordic heat pump".

The idea that the ocean has more than just a regulating or damping effect on climate changes goes back to Chamberlin (1906) who hypothesised that a reversal of the abyssal circulation could explain "some of the strange climatic phenomena of the past". Much later Stommel (1961) found two stable equilibrium states in a simple box model of the THC, and Bryan (1986) showed multiple equilibria of the THC in a 3-dimensional ocean model. This finally convinced researchers that the ocean takes a central and active role in the climate system.

F. Bryan also identified a positive feedback mechanism that maintains deep water formation at high latitudes. Surface waters that are transported by the WGC towards the sinking regions (Fig. 8.3) transport salt so that, when cooled down by atmosphere-ocean heat exchange, they loose sufficient buoyancy and sink. This mechanism operates in the northern North Atlantic but not in the North Pacific, where the water does not contain enough salt to become sufficiently dense. If now, for some reason, the flow of the dense water is diverted, cut off or even sufficiently diluted, the deep water formation can be reduced or even stopped. The THC then settles into a new state in which the meridional transport of heat is significantly reduced. Lower sea surface temperatures and cooling of the overlying air follows. Such different states could be realized in 2-dimensional thermohaline models (Marotzke et al. 1988; Wright & Stocker 1991; Stocker & Wright 1991), multi-basin 3-dimensional ocean models (Bryan 1986; Marotzke & Willebrand 1991; Mikolajewicz & Maier-Reimer 1994; Hughes & Weaver 1994; Rahmstorf

& Willebrand 1995) and coupled atmosphere-ocean models (Manabe & Stouffer 1988; Schiller et al. 1997; Fanning & Weaver 1997).

All these models essentially exhibit a universal hysteresis behaviour (Fig. 8.4). This is due to the non-linear processes that govern the atmosphere-ocean fluxes of buoyancy which are the drivers for the THC (Stocker & Wright 1991). Anomalies of sea surface temperature are removed rather efficiently by anomalies in the atmosphere-ocean heat flux because of the strong correlation between these two quantities. On the other hand, sea surface salinity anomalies are not correlated with anomalies in the net surface freshwater fluxes (evaporation-precipitation-runoff). Therefore, there is a large difference between the response time of these different anomalies (Rooth 1982). The important result of hysteretic behaviour is that certain perturbations in the climate system can induce abrupt, non-linear and sometimes irreversible reorganizations in the atmosphere-ocean system (Stocker 2000).

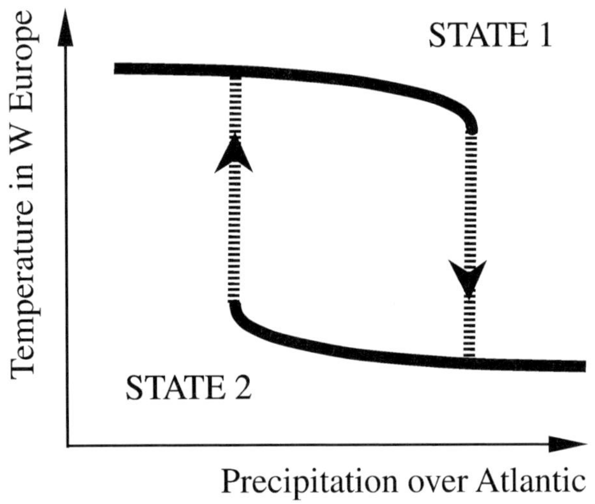

Fig. 8.4: The ocean-atmosphere system is a non-linear physical system that can exhibit hysteresis behaviour of the deep circulation in the ocean (Stocker and Wright 1991). Depending on the surface freshwater balance of the Atlantic ocean, the meridional heat flux in the Atlantic is not unique and multiple equilibria exist. Changes are linear as long as they remain on the same branch of the hysteresis loop. If certain threshold values in the atmosphere-ocean system are passed, the climate state can change abruptly by switching from one branch to the other. This is a robust feature of the climate system as demonstrated by the entire hierarchy of climate models [from Stocker & Wright 1991].

8.3.2
The Paleoclimatic Record

Abrupt warmings and gradual coolings with amplitudes of up to 10-15°C in Central Greenland (Severinghaus et al. 1998; Lang et al. 1999) characterized the glacial climate: about 24 of these, now referred to as Dansgaard/Oeschger (D/O) events, were found in the Greenland ice cores (Dansgaard et al. 1993), and in marine sediments (Bond et al. 1992; Bond & Lotti 1995; Hughen et al. 1996). Terrestrial records (Grimm et al. 1993) from the northern hemisphere show similar changes. While such climate signals appear wide-spread at least in the northern hemisphere, they are muted or absent in most records of the southern hemisphere. All D/O events exhibit a striking similarity in their temporal evolution: cooling extends generally over many centuries to about 3 kyr, while warming is abrupt and occurs within years or decades. This suggests that one common mechanism may be responsible for these climate swings. The recurrence time for the shorter D/O events is of the order of 1000 years.

Layers with abundant coarse-grain lithic fragments are found in marine sediments (Heinrich 1988), later termed "Heinrich Events", or short H-event (Broecker et al. 1992). They are thought to be detrital material transported from the bedrock underlying the great ice sheets to the North Atlantic by calving icebergs. This occurred through surging ice streams which then melted and provided an additional sediment layer. Mineralogical analysis of these layers reveal that their origin is the Canadian shield (Bond et al. 1992), but discharge can also be traced to the Fennoscandian (Fronval et al. 1995) and British ice sheets (McCabe & Clark 1998). Most of the debris material, however, originates from north of Hudson Bay (Gwiazda et al. 1996). The typical recurrence time is estimated at about 7 kyr, thus significantly longer than that for the D/O events found in the ice cores.

At first glance, D/O and H-events seem unrelated. However, Bond & Lotti (1995) shows in a high-resolution marine sediment core from the North Atlantic that smaller layers of ice rafted debris were buried between the prominent Heinrich layers, i.e. such layers also occur during or before D/O events. This is strong indication of a common cause or trigger in the climate system responsible for these changes. The different thickness of the debris layers for D/O and H events suggests that the response of the atmosphere-ocean system depends on the magnitude of the iceberg discharge and possibly involves threshold effects.

The two "fat" D/O events (number 8 and 12, 36 kyr and 45 kyr BP, respectively) are not only distinct by their duration, but they both occur after a H-event; other, shorter D/O events do not have a preceeding H event. They are clearly associated with climate change derived from the isotopic changes in the Antarctic ice cores (Blunier et al. 1998). The warming in the Antarctic record appears steady over about 2 kyr and leads the abrupt warming recorded in the Greenland ice core. At the time of the northern warming, the Antarctic signal indicates a gradual return to colder temperatures over the next 2 kyr. Interestingly, these are also the two times when a clear increase in CO_2 of about 20 ppmv is detected (see Sect. 8.4.1). Recent measurements confirm this finding and suggest two further events of Antarctic warming around 50 and 58 kyr BP which may be connected to rapid change in the north (Indermühle et al. 2000).

The above paleoclimatic evidence is consistent with and supports the hypothesis that the ocean plays the major role in causing the abrupt changes and transmitting the related signals. In particular, a collapse of the THC would stop the nordic heat pump leading to a severe cooling in the northern North Atlantic region. As the once active pump has been drawing heat from the southern ocean (Crowley 1992), this heat is now slowly accumulating in the southern ocean leading to warming. The climate system thus appears to operate as a bipolar seesaw during abrupt climate change (Broecker 1998; Stocker 1998).

8.4 Reconstructed Changes in Atmospheric CO_2

8.4.1 Evidence from Polar Ice Cores

Polar ice cores permit the reconstruction of past CO_2 concentrations over at least 4 glacial cycles (Petit et al. 1999). Prior to the formation of ice, accumulated snow is sintering into firn, a porous material which is in contact with atmospheric air (Schwander et al. 1997). Upon closure of the air-filled spaces in the firn, the composition of the air remains trapped in the bubbles within the ice matrix. Provided no further physical fractionation processes or chemical reactions occur, the air remains conserved in the bubbles. It depends on the location and the content of trace materials (dust, organic matter) whether these conditions are sufficiently satisfied. Once the ice core is retrieved from the ice sheets, the air is analyzed with various methods to provide information on past concentration of trace gases in ancient air (Raynaud et al. 1993). Carbon dioxide concentration is measured using a laser absorption technique and recent applications are described by Stauffer et al. (1998) and Indermühle et al. (1999).

For the other greenhouse gases, different analysis methods are applied. Methane is measured using gas chromatography (Blunier et al. 1998), the same technique also allows the determination of past concentrations of N_2O (Flückiger et al. 1999). In contrast to the vast majority of paleoclimatic data, CO_2, CH_4 and N_2O are among the very few variables that can be determined directly. Because they are not proxy data, one does not have to determine transfer functions or use modern analogue assumptions which may not be valid (Johnsen et al. 1995). Further direct data are isotopic and elemental concentrations of the gases trapped in the bubbles (e.g. O_2/N_2, ffi^{15}N).

Fig. 8.5 shows the most recent and complete reconstruction of past CO_2 changes. Atmospheric CO_2 concentrations vary in concert with the cycles of glaciation being around 270 to 290 ppmv during warm phases (interglacials) and around 190 ppmv during full glacial times. The continuous rise over the last 250 years is clearly unprecedented. Never in the last 420 000 years was the atmospheric CO_2 concentration as high as it is now (Petit et al. 1999). Calculations and various lines of independent evidence confirm that the recent rise, which by now is as large as the natural glacial-interglacial amplitude of CO_2, is due to the emission of CO_2 from fossil fuel burning and land use changes. Calculations which also take into account the changes of the isotopes of CO_2 (^{13}C and ^{14}C) permit the distinction of changes in the carbon inventories of the atmosphere, ocean and the terrestrial biosphere (Joos et al. 1999).

High-resolution CO_2 measurements by Stauffer et al. (1998), Indermühle et al. (1999), and Fischer et al. (1999) on Antarctic ice cores reveal a clearer picture of the dynamics of the global carbon cycle on time scales of less than a few 1000 years (Fig. 8.6). The variability of atmospheric CO_2 during the glacial does not exceed 20 ppmv, although a series of abrupt climate changes is evident in many paleoclimatic records (Broecker 1997). Some of these (the major Heinrich events and, associated with them, the longest of the Dansgaard/Oeschger events) are thought to be due to large atmosphere-ocean reorganizations in the form of complete collapses of the Atlantic THC (Stocker 2000). In order to test such hypotheses with physical-biogeochemical climate models, the measured amplitude and timing of the CO_2 changes serve as crucial constraints.

Another unexpected secular change in atmospheric CO_2 was the increase during the Holocene (the last 10 kyr) which is a relatively stable climate period (Indermühle et al. 1999). Major atmosphere-ocean reorganizations are absent with the possible exception of the brief cooling at 8200 kyr BP (Alley et al. 1997). Changes of the order of 20 ppmv can thus also occur naturally without large changes in ocean circulation.

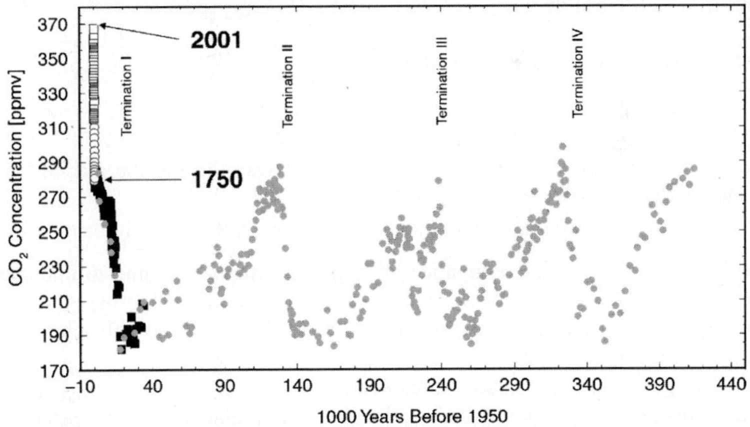

Fig. 8.5: Complete composite CO_2 record over the last 420 000 years including 4 glacial cycles from direct measurements (Keeling & Whorf 1994, white squares), and ice core measurements from Siple Station (Neftel et al. 1985, white dots), Taylor Dome (Indermühle et al. 1999, Fischer et al. 1999, black squares), and Vostok Station (Petit et al. 1999, gray dots).

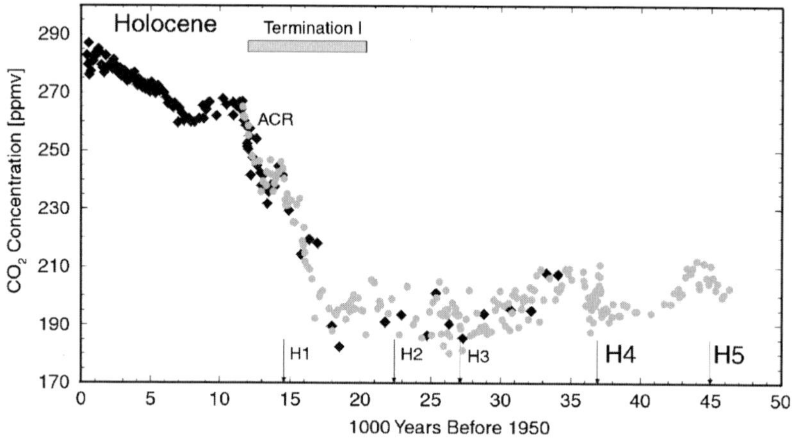

Fig. 8.6: High-resolution CO_2 record from Byrd (Stauffer et al. 1998, gray dots) and Taylor Dome (Indermühle et al. 1999, Fischer et al. 1999, black diamonds) ice cores. The timing of Heinrich events (Broecker et al. 1992) is indicated by arrows. Those events in larger letters are associated with long D/O events found in Greenland ice cores (Dansgaard et al. 1993) and corresponding smaller climate changes in Antarctica (Blunier et al. 1998).

8.4.2 Possible Mechanisms for Atmospheric CO_2 Changes

Three types of CO_2 changes can be distinguished in Fig. 8.5 and Fig. 8.6:

- glacial-interglacial changes of about 80 ppmv on time scales of about 10 000 years;
- changes of about 20 ppmv on millennial time scale during the glacial, correlated with climatic changes;
- changes of about 20 ppmv on millennial time scale during the Holocene.

It is a challenge of paleoclimatic and biogeochemical modeling to understand and quantitatively simulate these changes. There has been much progress over the last few years but some important aspects that are seen in the paleoclimatic records are not yet understood.

The longest-standing and still unresolved problem concerns the largest changes in the atmospheric CO_2 concentration, the glacial-interglacial changes. There have been a number of hypotheses explaining both the timing and the amplitude of the 80 ppmv changes but none of them is consistent with all available paleoclimatic evidence. A recent overview is given by Broecker & Henderson (1998, their Tab. 1). None of the physical mechanisms such as changes in deep water circulation, cooler sea surface temperatures, or longer residence time of the waters in high northern latitudes can explain individually the amplitude of the CO_2 drop during the glacial (Siegenthaler & Wenk 1984). Furthermore, all scenarios using sea level rise as a driver (shelf inundation and shallow-water carbonate, coral-reef hypotheses) are not consistent because sea level rise occurs after the CO_2-increase. Finally, there is no clear paleoceanographic evidence for a significant increase in marine productivity drawing down atmospheric CO_2; some of the stable isotope data, $\delta^{13}C$ in marine sediments, actually point to the contrary. Iron fertilization is another hypothesis. In some areas of the ocean, iron is a limiting nutrient and stronger dust input into the ocean by winds in a dustier atmosphere may enhance marine productivity resulting in lower atmospheric CO_2 (Martin 1990; Behrenfeld et al. 1996). As ice cores from Greenland and Antarctica indicate, the dust deposition drops significantly during the end of the glacial. However, the time scales of the changes in the dust supply are on the order of decades whereas those of CO_2 changes are millennia. At present, the most promising mechanisms appear to be linked to nutrient (through iron fertilization) and temperature changes in the Southern Ocean, with the nitrogen cycle playing an important role in explaining the relatively long time scales of CO_2 change (Broecker & Henderson 1998).

High-resolution measurements in ice cores indicate that millennial changes of CO_2 during the glacial are less than about 20 ppmv (Stauffer et al. 1998; Indermühle et al. 2000). While earlier suggestions that CO_2, much like CH_4, would change during each D/O event could not be confirmed, the CO_2 changes appear correlated to (at least some) Heinrich events and/or the longest and most prominent of the D/O events (Fig. 8.6). About 2 kyr before the rapid warming seen in Greenland ice cores, CO_2 appears to rise. This rise coincides with a rise in temperature as indicated by the stable isotopes measured on Antarctic ice cores (Blunier et al. 1998; Indermühle et al. 2000), although there is still some uncertainty in the ice-age/gas-age difference. CO_2 therefore seems to be more closely linked with the climate changes in the south, than those in the north. How then, could one explain the warming in the south? Marine sediments contain layers of ice rafted debris at or before each D/O event (Bond & Lotti 1995). This suggests the presence of pools of freshwater which could have acted as triggers of changes in the THC of the Atlantic. Indeed, meltwater discharge to the North Atlantic would be the mechanism to explain the southern warming, because a cooling in the north, caused by the disruption of the nordic heat pump, would lead to a warming in the south. This strong north-south coupling appears to be present during the few H-events and subsequent "fat" D/O events but not during the shorter D/O events (Stocker 2000).

In order to test the hypothesis of strong north-south coupling during a complete collapse of the Atlantic THC, a simplified physical-biogeochemical climate model was used (Marchal et al. 1998). The discharge of a defined amount of freshwater into the North Atlantic disrupts the THC and lead to a collapse of the circulation. The strong cooling induced in the North Atlantic is compensated by a warming in the southern ocean through the effect of the "bipolar seesaw" (Broecker 1998; Stocker 1998). The cooling in the north would lead to an increased uptake of CO_2 through a stronger solubility pump, while the opposite is true for the southern ocean. While the global effect of changes in sea surface temperature remains less than about 5 ppmv with the warming in the south dominating the cooling in the North Atlantic, the combined effect of changes in DIC and alkalinity due to the discharge of the freshwater is an increase of atmospheric CO_2 a few 100 years after the full collapse of the THC in the North Atlantic. The net effect is thus an increase in atmospheric CO_2 between 7 and 30 ppmv on a timescale of 100 to 2000 years depending on the intensity of the THC change. At the time of abrupt warming in the north (resumption of the circulation), CO_2 is decreasing again. This is in qualitative agreement with the information from the ice cores. A recent study of the Younger Dryas confirms these findings (Marchal et al. 1999). Hence, the model suggests that the millennial CO_2 changes are linked with ocean-atmosphere reorganizations triggered by freshwater pulses. It is important to note that the "chicken-and-egg" problem is not yet solved: did changes in the south trigger collapses of the Atlantic THC, or did ice sheet disintegration in the north trigger changes in ocean circulation? It is very likely that the sequence of H and D/O events is a truly coupled ice-ocean-atmosphere phenomenon.

Changes of atmospheric CO_2 during the Holocene came as a surprise (Indermühle et al. 1999). It was long thought that the relatively stable climate period of the Holocene would also be reflected in the concentration of greenhouse gases. However, CH_4 showed a pronounced minimum at around 5 kyr BP indicating a weaker hydrological cycle in the low latitudes during that time (Blunier et al. 1995). Also CO_2 is not constant the Holocene but rises slowly by about 20 ppmv from about 8 kyr BP to the preindustrial level of 280 ppmv (Fig. 8.6). Here, ocean-atmosphere reorganizations cannot be invoked because rapid coolings and warmings, unlike during the glacial, have not occurred in this time period. Although sparse, data from the stable carbon isotope (^{13}C) of atmospheric CO_2 and model calculations indicate that most of the atmospheric carbon inventory increase comes from the terrestrial biosphere; a minor part would be due to a slow increase in global mean sea surface temperature during several thousand years. This highlights the importance of the terrestrial biosphere for millennial time scale changes of atmospheric CO_2. In consequence, more data of $^{13}CO_2$ are needed from polar ice cores.

8.5 Future Changes and Feedback Processes

It has been shown that the surface freshwater balance exerts a strong control on the THC in the North Atlantic (Manabe & Stouffer 1988; Stocker & Wright 1991; Mikolajewicz & Maier-Reimer 1994) and that multiple equilibria can result (Fig. 8.4). Warmer air temperatures, such as projected for the next century due to a continuing emission of greenhouse gases, are likely to enhance the hydrological cycle. This link between temperature and hydrological cycle is exhibited in the methane changes during each D/O event: the warming is associated with a 50% increase in the CH_4 concentration (Chappellaz et al. 1993). Warming would tend to strengthen evaporation at low latitudes, increase precipitation at higher latitudes and thus accelerate the hydrological cycle. As ocean models indicate, this would lead to a reduction of deep water formation in the northern North Atlantic due to a freshening of the surface waters. A reduction of sea surface density is also caused by the increased surface air temperatures further reducing the thermohaline circulation. While the relative strength of these two mechanisms is under debate (Dixon et al. 1999), a general reduction of the Atlantic THC in response to global warming appears to be a robust result found by the entire hierarchy of climate models (Manabe & Stouffer 1993; Manabe & Stouffer 1994; Stocker & Schmittner 1997; Wood et al. 1999).

As already Fig. 8.4 suggests, threshold values of key climate variables exist beyond which the THC can no longer be sustained. Among these are the level of greenhouse gases in the atmosphere (Manabe & Stouffer 1993) as well as the rate of increase of greenhouse gas concentration (Stocker & Schmittner 1997). The few model simulations show, that the critical level is somewhere between double and fourfold preindustrial CO_2 concentration, but this depends critically on various model parameterizations (Schmittner & Stocker 1999; Knutti et al. 2000). The threshold lies lower if the CO_2 increase is faster. If the threshold is crossed, a complete collapse of the Atlantic THC ensues which results in regional cooling and water mass reorganization very similar to the paleoclimatic experiments with the same model.

Model simulations using 3-dimensional ocean general circulation models with prescribed boundary conditions predicted a minor (Maier-Reimer et al. 1996) or a rather strong (Sarmiento & Le Quéré 1996) feedback between the circulation changes and the uptake of anthropogenic CO_2 under global warming scenarios. However, the complete interplay of the relevant climate system components was only taken into account in the recent study by Joos et al. (1999). A series of CO_2 stabilization profiles were prescribed for the next 500 years along with a specific climate sensitivity, i.e. the global mean temperature increase due to a doubling of CO_2: typically 1.5-4.5°C (Fig. 8.7a). As expected this leads to a reduction of the Atlantic THC. Again, a threshold value is between 750 and 1000 ppmv for a complete cessation of the THC (Fig. 8.7b).

With this model different experiments with the ocean carbon pumps operating or suppressed can be performed. Such experiments are essential for a better understanding of the various processes influencing ocean uptake of CO_2. A maximum uptake of 5.5 Gt yr^{-1} is simulated if there is no change in ocean circulation nor sea surface temperature. The full simulation including all feedbacks (sea surface temperature, circulation and biota) shows a long-term reduction of almost 50% in the uptake flux provided the Atlantic THC collapses (Fig. 8.8a). The solubility effect is important in the first 100 years (curve C) but later, the circulation effect takes over (curve D). If the circulation does not break down, circulation and biota feedback compensate each other, and the solubility effect remains the only significant feedback effect (Fig. 8.8b). The reduction of strength of the ocean as a major carbon sink appears a robust result, but the model also shows that dramatic feedback effects (such as a runaway greenhouse effect) are very unlikely. The maximum increase of CO_2 in the case of a collapsed Atlantic THC is estimated at about 20%. This result is entirely consistent with the evidence from the paleoclimatic records:

major atmosphere-ocean reorganizations such as during H or D/O-events appeared to have a relatively small influence on atmospheric CO_2.

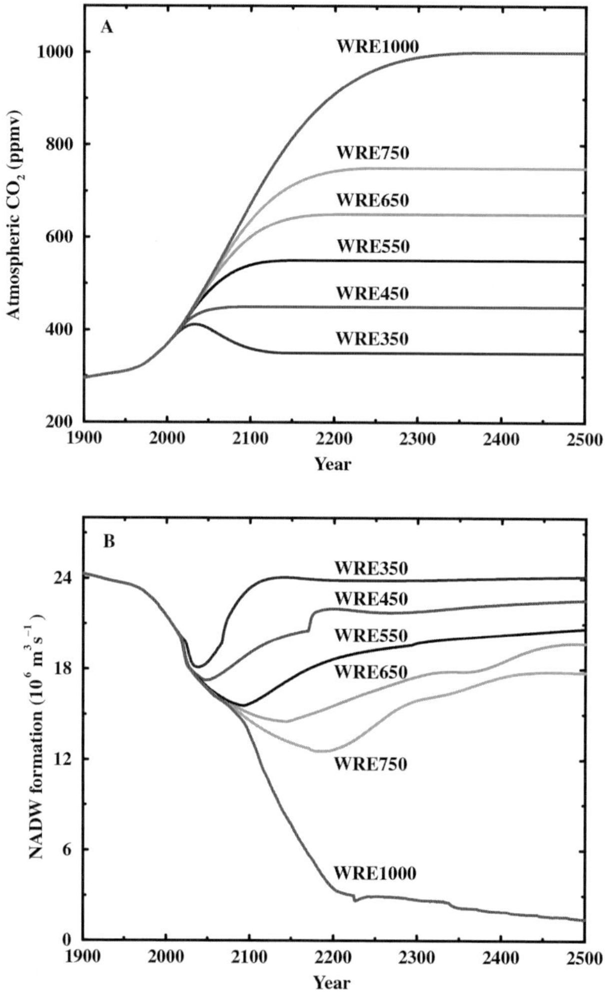

Fig. 8.7.: **a**) Scenarios for the stabilization of atmospheric CO_2 concentration over the next 500 years; **b**) evolution of the overturning circulation in the Atlantic in response to global warming caused by increasing concentration of CO_2. The climate sensitivity of the current experiments is 3.7°C for a doubling of CO_2 [From Joos et al. 1999].

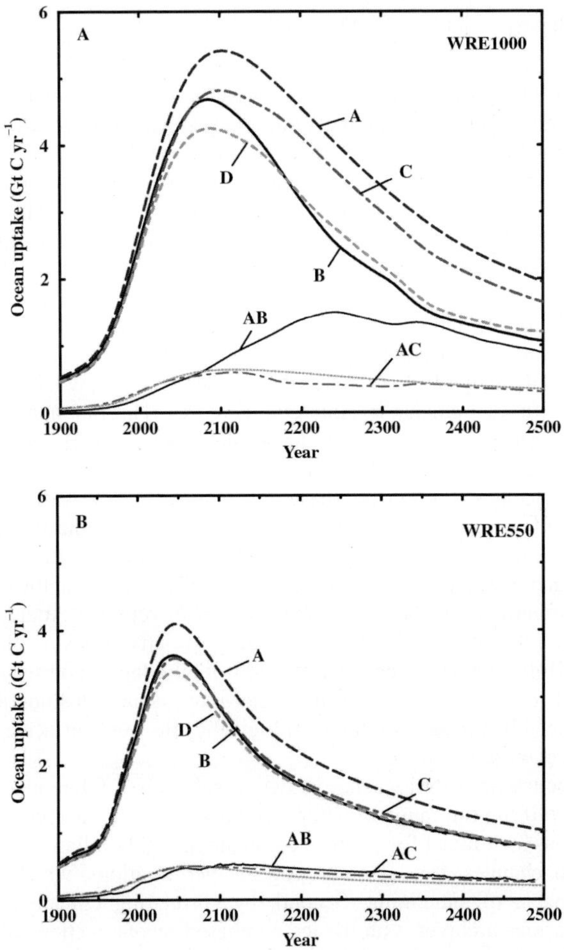

Fig. 8.8.: Evolution of the oceanic uptake of carbon dioxide for the different stabilization scenarios of Fig. 8.7. In WRE1000 (**a**), the Atlantic thermohaline circulation collapses completely by year 2500, in WRE500 (**b**) it remains close to the initial strength. The different simulations are labelled A-D. In A, all feedbacks are neglected (constant ocean), B is the full simulation (temperature, circulation, biota changes), C is only sea surface temperature changes (circulation held constant), D (temperature and circulation changes). Further explanations on regarding the different scenarios are in the text [From Joos et al. 1999].

8.6 Conclusions

The physical-biogeochemical state of the ocean is strongly influencing the atmospheric concentration of CO_2, the most important greenhouse gas after water vapor. The paleoclimatic record exhibits three different types of CO_2-changes, two of which are strongly associated with ocean circulation changes. By far the largest changes are the glacial/interglacial cycles of about 80 ppmv which still defy a complete, quantitative explanation. While the ocean plays a significant role through variations in the carbon pump strengths, most probably only a combination of different effects can explain the reconstructed changes. These include changes in the sea surface temperature, marine productivity, nitrogen fixation and in the interaction with carbonate sediments.

Two types of smaller CO_2-changes of the order of 20 ppmv have emerged from highresolution measurements in Antarctic ice cores. The first appear to be correlated to the largest climate changes during the glacial, the H-events or the long D/O events of the Greenland ice cores. These climate changes are distinguished by the fact that changes in opposite phase are recorded in the Antarctic cores. This suggests an interhemispheric seesaw which is in operation during abrupt climate change. Simulations with coupled physical-biogeochemical climate models lend support to such a scenario. The driving force is afforded by meltwater discharges from the northern hemisphere ice sheets which disrupt the Atlantic THC. CO_2 changes of similar magnitude also occur during relatively stable climate periods such as the Holocene. Here, changes are most likely due to a terrestrial biosphere which is still responding to the recent glacial-interglacial transition and continues to be influenced by changes in the hydrological cycle, land surface conditions (retreating ice cover) and climate.

The paleoclimatic record teaches us two lessons. First, future changes can only be understood if the full range of natural climate variability is reconstructed. Such reconstructions are crucial for a sensible climate model building and development. It is the past changes, free of anthropogenic perturbations, which these models must be capable of simulating. Second, by correlating climate changes documented in various archives with the ice corebased reconstructions of atmospheric CO_2, the link between the global carbon cycle and the atmosphere-ocean-biosphere system can be quantified. Model simulations suggest, that there is a potential in large non-linear changes in the physical climate system. These are shown to have an influence on the carbon cycle, on the uptake capacity of the world ocean and ultimately on the atmospheric concentration of CO_2.

Acknowledgments

I thank the Science Museum of Barcelona for organizing a stimulating conference. Comments by A. Indermühle, O. Marchal and G.-K. Plattner are appreciated. Financial support of the Swiss National Science Foundation and the Swiss Priority Program of Environment is acknowledged.

References

Alley, R. B., P. A. Mayewski, T. Sowers, M. Stuiver, K. C. Taylor, & P. U. Clark (1997). Holocene climatic instability: A prominent, widespread event 8200 yr ago. Geology 25, 483-486.

Behrenfeld, M. J., A. J. Bale, Z. S. Kolber, J. Aiken, & P. G. Falkowski (1996). Confirmation of iron limitation of phytoplankton photosynthesis in the equatorial Pacific Ocean. Nature 383, 508-511.

Berger, A., M.-F. Loutre, & C. Tricot (1993). Insolation and Earth's orbital periods. J. Geophys. Res. 98, 10341-10362.

Blunier, T., J. Chappellaz, J. Schwander, A. Dällenbach, B. Stauffer, T. F. Stocker, D. Raynaud, J. Jouzel, H. B. Clausen, C. U. Hammer, & S. J. Johnsen (1998). Asynchrony of Antarctic and Greenland climate change during the last glacial period. Nature 394, 739-743.

Blunier, T., J. Chappellaz, J. Schwander, B. Stauffer, & D. Raynaud (1995). Variations in the atmospheric methane concentration during the Holocene. Nature 374, 46-49.

Bond, G., H. Heinrich, W. Broecker, L. Labeyrie, J. McManus, J. Andrews, S. Huon, R. Jantschik, S. Clasen, C. Simet, K. Tedesco, M. Klas, G. Bonani, & S. Ivy (1992). Evidence for massive discharges of icebergs into the North Atlantic ocean during the last glacial period. Nature 360, 245-249.

Bond, G. C. & R. Lotti (1995). Iceberg discharges into the North Atlantic on millennial time scales during the last glaciation. Science 267, 1005-1010.

Böning, C., F. O. Bryan, W. R. Holland, & R. Döscher (1996). Deep-water formation and meridional overturning in a high-resolution model of the North Atlantic. J. Phys. Oceanogr. 26, 1142-1164.

Broecker, W. S. (1987). The biggest chill. Natural Hist. 96, 74-82.

Broecker, W. S. (1991). The great ocean conveyor. Oceanography 4, 79-89.

Broecker, W. S. (1997). Thermohaline circulation, the Achilles heel of our climate system: will manmade CO_2 upset the current balance? Science 278, 1582-1588.

Broecker, W. S. (1998). Paleocean circulation during the last deglaciation: a bipolar seesaw? Paleoceanogr. 13, 119-121.

Broecker, W. S., G. Bond, M. Klas, E. Clark, & J. McManus (1992). Origin of the northern Atlantic's Heinrich events. Clim. Dyn. 6, 265-273.

Broecker, W. S. & G. M. Henderson (1998). The sequence of events surrounding Termination II and their implications for the cause of glacial-interglacial CO_2 changes. Paleoceanogr. 13, 352-364.

Bryan, F. (1986). High-latitude salinity effects and interhemispheric thermohaline circulations. Nature 323, 301-304.

Chamberlin, T. C. (1906). On a possible reversal of deep-sea circulation and its influence on geologic climates. J. Geology 14, 363-373.

Chappellaz, J., T. Blunier, D. Raynaud, J. M. Barnola, J. Schwander, & B. Stauffer (1993). Synchronous changes in atmospheric CH_4 and Greenland climate between 40 and 8 kyr BP. Nature 366, 443-445.

Crowley, T. J. (1992). North Atlantic deep water cools the southern hemisphere. Paleoceanogr. 7, 489-497.

Crowley, T. J. & S. K. Baum (1997). Effect of vegetation on an ice-age climate model simulation. J. Geophys. Res. 102, 16463-16480.

Crowley, T. J. & G. R. North (1991). Paleoclimatology. Number 18 in Oxford Monographs on Geology and Geophysics. Oxford University Press. 339 pp.

Dansgaard, W., S. J. Johnsen, H. B. Clausen, D. Dahl-Jensen, N. S. Gundestrup, C. U. Hammer, C. S. Hvidberg, J. P. Steffensen, A. E. Sveinbjornsdottir, J. Jouzel, & G. Bond (1993). Evidence for general instability of past climate from a 250-kyr ice-core record. Nature 364, 218-220.

Dixon, K. W., T. L. Delworth, M. J. Spelman, & R. J. Stouffer (1999). The influence of transient surface fluxes on North Atlantic overturning in a coupled GCM climate change experiment. Geophys. Res. Let. 26, 2749-2752.

Fanning, A. F. & A. J. Weaver (1997). Temporal-geographical meltwater influences on the North Atlantic conveyor: implications for the Younger Dryas. Paleoceanogr. 12, 307-320.

Fischer, H., M. Wahlen, J. Smith, D. Mastroianni, & B. Deck (1999). Ice core records of atmospheric CO_2 around the last three glacial terminations. Science 283, 1712-1714.

Flückiger, J., A. Dällenbach, T. Blunier, B. Stauffer, T. F. Stocker, D. Raynaud, & J.-M. Barnola (1999). Variations in atmospheric N_2O concentration during abrupt climatic changes. Science 285, 227-230.

Fronval, T., E. Jansen, J. Bloemendal, & S. Johnsen (1995). Oceanic evidence for coherent fluctuations in Fennoscandian and Laurentide ice sheets on millennium timescales. Nature 374, 443-446.

Gordon, A. L. (1986). Interocean exchange of thermocline water. J. Geophys. Res. 91, 5037-5046. Grimm, E. C., G. L. Jacobson Jr., W. A. Watts, B. C. S. Hanson, & K. A. Maasch (1993). A 50,000-year record of climate oscillations from Florida and its temporal correlation with the Heinrich events. Science 261, 198-200.

Gwiazda, R. H., S. R. Hemming, & W. S. Broecker (1996). Provenance of icebergs during Heinrich event 3 and the contrast to their sources during other Heinrich episodes. Paleoceanogr. 11, 371-378.

Hall, M. M. & H. L. Bryden (1982). Direct estimates and mechanisms of ocean heat transport. Deep Sea Res. 29, 339-359.

Heinrich, H. (1988). Origin and consequences of cyclic ice rafting in the Northeast Atlantic Ocean during the past 130,000 years. Quat. Res. 29, 142-152.

Hughen, K. A., J. T. Overpeck, L. C. Peterson, & S. Trumbore (1996). Rapid climate changes in the tropical Atlantic region during the last deglaciation. Nature 380, 51-54.

Hughes, T. M. C. & A. J. Weaver (1994). Multiple equilibria of an asymmetric two-basin ocean model. J. Phys. Oceanogr. 24, 619-637.

Indermühle, A., E. Monnin, B. Stauffer, T. F. Stocker, & M. Wahlen (2000). Atmospheric CO_2 concentration from 60 to 20 kyr BP from the Taylor Dome ice core, Antarctica. Geophys. Res. Let. 27, 735-738.

Indermühle, A., T. F. Stocker, F. Joos, H. Fischer, H. J. Smith, M. Wahlen, B. Deck, M. D., J. Tschumi, T. Blunier, R. Meyer, & B. Stauffer (1999). Holocene carbon-cycle dynamics based on CO_2 trapped in ice at Taylor Dome, Antarctica. Nature 398, 121-126.

Johnsen, S. J., D. Dahl-Jensen, W. Dansgaard, & N. Gundestrup (1995). Greenland palaeotemperatures derived from GRIP bore hole temperature and ice core isotope profiles. Tellus 47B, 624-629.

Joos, F., R. Meyer, M. Bruno, & M. Leuenberger (1999). The variability in the carbon sinks as reconstructed for the last 1000 years. Geophys. Res. Let. 26, 1437-1440.

Joos, F., G.-K. Plattner, T. F. Stocker, O. Marchal, & A. Schmittner (1999). Global warming and marine carbon cycle feedbacks on future atmospheric CO_2. Science 284, 464-467.

Joos, F. & J. L. Sarmiento (1995). Der atmosphärische CO_2-Anstieg. Physikal. Blätter 51, 405-411.

Keeling, C. D. & T. P. Whorf (1994). Atmospheric CO_2 records from sites in the SIO network. In T. Boden, D. Kaiser, R. Sepanski, & F. Stoss (Eds.), Trends '93: A Compendium of Data on Global Change, pp. 16-26. Carbon Dioxide Information Analysis Center.

Killworth, P. D. (1983). Deep convection in the world ocean. Rev. Geophys. Space Phys. 21, 1-26.

Knutti, R., T. F. Stocker, & D. G. Wright (2000). The effects of sub-grid-scale parameterizations in a zonally averaged ocean model. J. Phys. Oceanogr. 30, 2738-2752.

Lang, C., M. Leuenberger, J. Schwander, & S. Johnsen (1999). 16°C rapid temperature variation in Central Greenland 70,000 years ago. Science 286, 934-937.
Macdonald, A. M. & C. Wunsch (1996). An estimate of global ocean circulation and heat fluxes. Nature 382, 436-439.
Maier-Reimer, E., U. Mikolajewicz, & A. Winguth (1996). Future ocean uptake of CO_2: interaction between ocean circulation and biology. Clim. Dyn. 12, 711-721.
Manabe, S. & R. J. Stouffer (1988). Two stable equilibria of a coupled ocean-atmosphere model. J. Clim. 1, 841-866.
Manabe, S. & R. J. Stouffer (1993). Century-scale effects of increased atmospheric CO_2 on the ocean-atmosphere system. Nature 364, 215-218.
Manabe, S. & R. J. Stouffer (1994). Multiple-century response of a coupled ocean-atmosphere model to an increase of atmospheric carbon dioxide. J. Clim. 7, 5-23.
Marchal, O., T. F. Stocker, & F. Joos (1998). Impact of oceanic reorganizations on the ocean carbon cycle and atmospheric carbon dioxide content. Paleoceanogr. 13, 225-244.
Marchal, O., T. F. Stocker, F. Joos, A. Indermühle, T. Blunier, & J. Tschumi (1999). Modelling the concentration of atmospheric CO_2 during the Younger Dryas climate event. Clim. Dyn. 15, 341-354.
Marotzke, J., P. Welander, & J. Willebrand (1988). Instability and multiple equilibria in a meridional plane model of the thermohaline circulation. Tellus 40A, 162-172.
Marotzke, J. & J. Willebrand (1991). Multiple equilibria of the global thermohaline circulation. J. Phys. Oceanogr. 21, 1372-1385.
Marshall, J. & F. Schott (1999). Open-ocean convection: observations, theory and models. Rev. Geophys. 37, 1-64.
Martin, J. H. (1990). Glacial-interglacial CO_2 change: the iron hypothesis. Paleoceanogr. 5, 1-13.
McCabe, A. M. & P. U. Clark (1998). Ice-sheet variability around the North Atlantic Ocean during the last deglaciation. Nature 392, 373-377.
Mikolajewicz, U. & E. Maier-Reimer (1994). Mixed boundary conditions in ocean general circulation models and their influence on the stability of the model's conveyor belt. J. Geophys. Res. 99, 22633-22644.
Neftel, A., E. Moor, H. Oeschger, & B. Stauffer (1985). Evidence from polar ice cores for the increase in atmospheric CO_2 in the past two centuries. Nature 315, 45-47.
Neftel, A., H. Oeschger, T. Staffelbach, & B. Stauffer (1988). CO_2 record in the Byrd ice core 50,000-5,000 years BP. Nature 331, 609-611.
Pedlosky, J. (1996). Ocean Circulation Theory. Springer. 453 pp.
Peixoto, J. P. & A. H. Oort (1992). Physics of Climate. American Institute of Physics. 500 pp.
Petit, J. R., J. Jouzel, D. Raynaud, N. I. Barkov, J.-M. Barnola, I. Basile, M. Bender, J. Chappellaz, M. Davis, G. Delaygue, M. Delmotte, V. M. Kotlyakov, M. Legrand, V. Y. Lipenkov, C. Lorius, L. P'epin, C. Ritz, E. Saltzman, & M. Stievenard (1999). Climate and atmospheric history of the past 420,000 years from the Vostok ice core, Antarctica. Nature 399, 429-436.
Rahmstorf, S. & J. Willebrand (1995). The role of temperature feedback in stabilizing the thermohaline circulation. J. Phys. Oceanogr. 25, 787-805.
Raynaud, D., J. Jouzel, J. M. Barnola, J. Chappellaz, R. J. Delmas, & C. Lorius (1993). The ice record of greenhouse gases. Science 259, 926-934.
Rooth, C. (1982). Hydrology and ocean circulation. Prog. Oceanogr. 11, 131-149.
Sarmiento, J. L. & C. Le Quéré (1996). Oceanic carbon dioxide in a model of century-scale global warming. Science 274, 1346-1350.
Schiller, A., U. Mikolajewicz, & R. Voss (1997). The stability of the North Atlantic thermohaline circulation in a coupled ocean-atmosphere general circulation model. Clim. Dyn. 13, 325-347.
Schmittner, A. & T. F. Stocker (1999). The stability of the thermohaline circulation in global warming experiments. J. Clim. 12, 1117-1133.
Schmitz, W. J. (1995). On the interbasin-scale thermohaline circulation. Rev. Geophys. 33, 151-173.

Schwander, J., T. Sowers, J.-M. Barnola, T. Blunier, A. Fuchs, & B. Malaizé (1997). Age scale of the air in the Summit ice: Implication for glacial-interglacial temperature change. J. Geophys. Res. 102, 19483-19493.

Severinghaus, J. P., T. Sowers, E. J. Brook, R. B. Alley, & M. L. Bender (1998). Timing of abrupt climate change at the end of the Younger Dryas interval from thermally fractionated gases in polar ice. Nature 391, 141-146.

Siegenthaler, U. & J. L. Sarmiento (1993). Atmospheric carbon dioxide and the ocean. Nature 365, 119-125.

Siegenthaler, U. & T. Wenk (1984). Rapid atmospheric CO_2 variations and ocean circulation. Nature 308, 624-626.

Stauffer, B., T. Blunier, A. Dällenbach, A. Indermühle, J. Schwander, T. F. Stocker, J. Tschumi, J. Chappellaz, D. Raynaud, C. U. Hammer, & H. B. Clausen (1998). Atmospheric CO_2 and millennial-scale climate change during the last glacial period. Nature 392, 59-62.

Stocker, T. F. (1998). The seesaw effect. Science 282, 61-62.

Stocker, T. F. (1999). Climate changes: from the past to the future - a review. Int. J. Earth Sci. 88, 365-374.

Stocker, T. F. (2000). Past and future reorganisations in the climate system. Quat. Sci. Rev. 19, 301-319.

Stocker, T. F. & A. Schmittner (1997). Influence of CO_2 emission rates on the stability of the thermohaline circulation. Nature 388, 862-865.

Stocker, T. F. & D. G. Wright (1991). Rapid transitions of the ocean's deep circulation induced by changes in surface water fluxes. Nature 351, 729-732.

Stommel, H. (1948). The westward intensification of wind-driven ocean currents. Trans. Am. Geophys. Union 29, 202-206.

Stommel, H. (1958). The abyssal circulation. Deep Sea Res. 5, 80-82.

Stommel, H. (1961). Thermohaline convection with two stable regimes of flow. Tellus 13, 224-241.

Stommel, H. & A. B. Arons (1960). On the abyssal circulation of the world ocean - I. Stationary planetary flow patterns on a sphere. Deep Sea Res. 6, 140-154.

Takahashi, T., J. Olafsson, J. G. Goddard, D. W. Chipman, & S. C. Sutherland (1993). Seasonal variation of CO_2 and nutrients in the high-latitude surface oceans: a comparative study. Global Biogeochem. Cyc. 7, 843-878.

Trenberth, K. E. & A. Solomon (1994). The global heat balance: heat transports in the atmosphere and ocean. Clim. Dyn. 10, 107-134.

Volk, T. & M. I. Hoffert (1985). Ocean carbon pumps: analysis of relative strengths and efficiencies in ocean-driven atmospheric CO_2 changes. In E. T. Sundquist & W. S. Broecker (Eds.), The Carbon Cycle and Atmospheric CO_2: Natural Variations Archean to Present, Volume 32 of Geophysical Monograph, pp. 99-110. Am. Geophys. Union.

Warren, B. A. (1981). Deep circulation of the world ocean. In B. A. Warren & C. Wunsch (Eds.), Evolution of Physical Oceanography - Scientific Surveys in Honor of Henry Stommel, pp. 6-41. MIT Press.

Wood, R. A., A. B. Keen, J. F. B. Mitchell, & J. M. Gregory (1999). Changing spatial structure of the thermohaline circulation in response to atmospheric CO_2 forcing in a climate model. Nature 399, 572-575.

Wright, D. G. & T. F. Stocker (1991). A zonally averaged ocean model for the thermohaline circulation, Part I: Model development and flow dynamics. J. Phys. Oceanogr. 21, 1713-1724.

9 Understanding Future Climate Change Using Paleorecords

Keith Alverson[1], Christoph Kull[2]

PAGES International Project Office. Bärenplatz 2, Bern, CH-3011, Switzerland
[1]alverson@pages.unibe.ch
[2]kull@pages.unibe.ch

9.1 Introduction: Why Study the Past?

Warming has been measured over most parts of the globe during the late 20th century. The instrumental data that record this warming provide only a limited perspective on either its nature or its cause. Because instrumental measurements cover mainly the period of industrialization, for example, they offer little information with which to distinguish between natural and anthropogenic effects. Furthermore, the instrumental record does not capture the spatial and temporal range of decadal to centennial scale variability which models and paleodata suggest is inherent to the climate system. An understanding of these decadal to centennial scale modes of variability must be an inherent component of any attempt at climate prediction. Furthermore, the paleorecord provides numerous examples of abrupt shifts in climate, and the ecosystem responses to these. Because such changes are absent in the instrumental period, societal infrastructure has largely been built without consideration of such possibilities, leaving many societies highly vulnerable to the types of climate changes which we know have occurred in the past.

This chapter seeks to highlight some of the ways in which paleoenvironmental data and models can contribute to three areas of global climate research: (1) detecting and attributing global anthropogenic change, (2) understanding processes at work in the Earth System, and (3) assessing human and ecosystem vulnerability to potential future changes. The remainder of this first section (Sect. 9.1) provides a short overview of the paleoenvironmental archives and models which serve as the primary tools used to address these issues, including a call to include these paleoarchives in international global climate observing systems. Sect. 9.2 addresses the question of detection and attribution of anthropogenic global climate change, with particular emphasis on high mountains. Sect. 9.3 provides a few examples of climate processes - the past operation of which provide information of relevance to future concerns. Sect. 9.4 tackles the question of human vulnerability to future change, from the perspective of the paleorecord. None of these sections contains a comprehensive list of the wealth of paleoenvironmental material relevant to these topics. Rather, they provide isolated glimpses which can hopefully serve as an entrée to the much wider range of paleoenvironmental science with great relevance to future concerns. In Sect. 9.5, we conclude with a synopsis of why and how the past is significant for the future.

9.1.1 The Need for a Global Paleoclimate Observation System

Paleoarchives extend the quantitative climate record to millennial time scales. In addition, paleodata inform us about the nature of ecosystem responses to climate change. Climate-related variables such as air and water temperature, salinity, pH, oxygen concentrations, CO_2 concentration, moisture balance and circulation strength can be quantitatively reconstructed using various biological indicators, isotopes, and other proxy measurements in lake and ocean sediments, corals, speleothems, ice cores from polar regions and high altitudes, tree rings, documentary records and other paleoarchives. Tapping the information contained in these paleoarchives has an enormous potential both to reconstruct past climate variability and contribute to our understanding of linkages between ecosystems and climate (Alverson et al. 2000; Alverson et al. 2001).

All natural archives contain many lines of evidence which serve as proxies for climatic and environmental change. For example, a single ice core contains indications of air temperature, atmospheric gas composition, volcanic activity and dust deposition rates. Similarly, within a single lake, sediments can record temperature, erosion rates and the makeup of terrestrial and aquatic ecosystems. Paleoenvironmental reconstruction requires that the properties measured in natural archives (proxies) be quantitatively translated into environmental parameters. For this approach to work the proxies must be rigorously calibrated against relevant direct observations, for example, July air temperature, sea-surface salinity or the makeup of vegetation cover. Thus, a period of overlap between the proxy record and contemporary data is vitally important for deriving well calibrated, quantitative paleo-reconstructions.

Recently initiated international global climate observation programs will need to be continuously operated for at least 50 years before they begin to provide information that is relevant to producing reliable predictions of climate change and its ecological impacts. Natural archives of past climate variability, on the other hand, can provide relevant information now. Unfortunately, recent anthropogenic factors are leading to the destruction or alteration of many of these paleoarchives, lending urgency to the task of retrieving the records they contain. An internationally coordinated effort to expedite large-scale observational and experimental campaigns to investigate the processes recorded in these endangered natural archives is the only way to rescue these endangered natural archives of past environmental variability (Alverson et al. 2001; Alverson and Eakin 2001).

A prime example of destruction of paleoarchives is the ongoing rapid retreat of alpine glaciers in both tropical and temperate latitudes. Ice cores from such glaciers have been used to reconstruct temperature, precipitation and atmospheric dust levels (Thompson 1998). Furthermore, they provide annual to millennial scale records of climate dynamics, including changes in the strength of the Asian monsoon (Thompson et al. 2000) and the strength and frequency of ENSO (Henderson et al. 1999; Moore et al. 2001). Over longer time scales, these ice cores contain archives of decadal to millennial scale climatic and environmental variability and provide unique insight into regional and global scale events including the so-called "Little Ice Age" and rapid climatic transitions such as that which occurred at the end of the Younger Dryas cold phase at the end of the last glacial period.

As shown in Fig. 9.1, the total area of the summit glacier on Kilimanjaro has decreased by 82% between 1912 and 2000. If this trend continues, there will soon be no paleoclimatic information stored in ice at high altitude, low-latitude sites. The only information from the Kilimanjaro ice will be what is left of the cores extracted in the year 2000 by Lonnie Thompson and his group and now stored in freezers at Ohio State University. Given the critical importance of the low latitude regions as drivers of climate on Earth, this loss will irreplaceably hobble our ability to use the past to predict the future

The situation on Kilimanjaro is not an isolated one. Tropical warming is causing the rapid retreat and, in some cases, the disappearance, of ice caps and glaciers at high elevations in the tropics and subtropics around the world. The retreat of the Quelccaya ice cap in Peru, for example, is also well documented. On Quelccaya ice cap in Peru, for example, the seasonally resolved paleoclimatic record, formerly (1983) preserved as $\delta^{18}O$ variations, is no longer being retained within the currently accumulating snow (Thompson 2000). The percolation of meltwater throughout the accumulating snowpack is homogenizing the stratigraphic record of $\delta^{18}O$ (Thompson 2001). These ice masses record thousands of years of local climate history, and are at the same time particularly sensitive to recent changes in ambient temperatures since they exist very close to the melting point. This makes them one of several high altitude long term climate recorders which serve as potential harbingers of recent anthropogenic climate change (Sect. 9.2.2)

Of course, mountain glaciers have always waxed and waned. Indeed there are many cases, for example in the Swiss Alps, where the dramatic modern day retreat has exposed dated wood only a few thousand years old, indicating that not very long ago, these glaciers were still more restricted than today (Hormes et al. 2001). However, at the time these European glaciers were in strong retreat, those in South America were not. The novel element today is that mountain glaciers everywhere in the world, with the minor exception of a few in Scandinavia where temperature is not a dominant control on glacial mass balance, are rapidly retreating (Thompson 2000). An extensive overview of the status of glaciers around the world, is available from the world glacier monitoring service web site (http://www.geo.unizh.ch/wgms/).

Glaciers are not the only paleoarchives that are disappearing, a second example is the ongoing widespread bleaching of corals. Measurements in corals have been used successfully to reconstruct sea surface temperature and salinity and even the surface circulation of the tropical oceans for the past several hundred years, and for isolated windows during the more distant past, often with a temporal resolution of only a few weeks (Gagan et al. 2000; Tudhope et al. 2001). Furthermore, corals have recorded past changes in the frequency and magnitude of ENSO events with direct consequences for predictability. Coral based reconstructions, for example, indicate a shift from primarily decadal tropical pacific sea surface temperature variability in the 1800's to the interannual variability of the last century (Urban et al. 2000). Such past changes in the dominant frequency of tropical pacific variability are not simply an artifact of coral records. They are corroborated by independent data from tree rings (Villaba et al. 2001) and ice cores (Henderson et al. 1999; Moore et al. 2001). Long coral based records also quantify the statistical significance of the 1976 ENSO regime shift, which has been questioned in the framework of the instrumental record alone (Wunsch 1999), by providing the required longer time series.

Large living corals more than 100 years old are relatively rare in most reef areas of the world. Unfortunately, a significant number of these corals have been killed in recent years, and the outlook for many others is bleak. Corals are under threat from localized stresses related to coastal development and population pressure. Dynamite and cyanide fishing, dredging for engineering works, nutrient overloading, the disruption of grazing fish populations, coastal pollution, and unregulated development all contribute to coral mortality, particularly in populated regions of the tropics. In addition to these local processes, there is a widespread mortality of corals due to coral bleaching associated with exceptionally warm ocean temperatures . Such extremes have been observed in all the world's tropical oceans. Once corals die, the potential for climate reconstruction from the skeletons is severely reduced. The dead coral skeletons are prone to rapid physical and biological erosion, and the absolute chronology, a key factor for any paleoclimate record, is lost. Some scientists predict a global demise of coral reefs within the next few decades due to global warming (Hoegh-Guldberg 1999).

Until the last few years, the use of tree rings as records of past environmental variability has been limited, almost entirely, to the middle and high latitudes. Real progress is being made with developing methods for decoding the climatic signals in tropical trees. At the same time, many old growth tropical trees are under severe threat. For example, teak trees in South Asia have annual rings that can be used as records of climate. At the same time as this progress has been made, massive felling of old-growth teak continues, raising the possibility that many of the oldest trees, those containing the most valuable information, will have been fed to the sawmills long before the climatic information they contain can be extracted (D'Arrigo 1998).

Though their existence is not under threat, the interpretation of other palaeodata is increasingly complicated by human activities. The relationships between tree-ring properties and regional climate parameters, for example, are widely used for reconstructing past climate. At high northern latitudes, tree-ring densities show a strong correlation with summer temperature. Transfer function based estimates of temperature from trees in this region are accurate recorders of large-scale temperature on short (interannual and decadal) and long (multidecadal) timescales: demonstrably during the early part of the twentieth century. During the second half of the twentieth century, tree density averaged around the Northern Hemisphere, still mirrors the year-to-year change in hemispheric temperatures accurately, but the density and temperature trends have increasingly diverged (Briffa et al. 1998). When human influences become so strong that the trees are no longer reacting directly to climate but responding instead to other anthropogenic factors, transfer functions based on comparison with instrumental data during the most recent period, can no longer be applied. The evidence on more local geographic scales is varied but, at least in subarctic Eurasia, these changes might be linked to an increase in winter precipitation that delays the onset of tree growth (Vaganov et al. 1999). Other speculative possibilities include some link with a general fertilization effect seen in other tree-growth parameters, possibly due to high atmospheric CO_2 concentrations or nitrogen-bearing precipitation, or the effects of acid rain or enhanced ultraviolet radiation (Briffa 2000).

Whatever the cause, this phenomenon points toward the need for more integrated process studies that involve the use of multiple archives, observations, models and experiments capable of providing an understanding of the time-dependent interrelationships between different biological and physical processes.

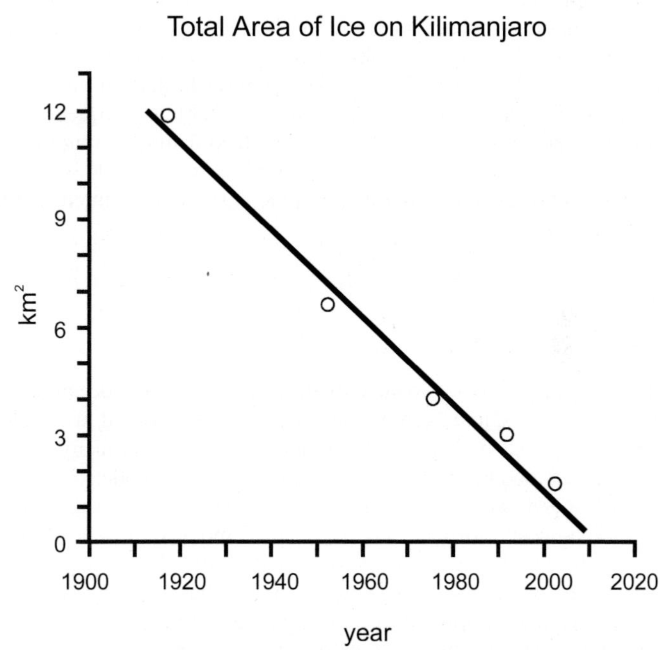

Fig. 9.1. Total area cover of the ice cap on Kilimanjaro from 1912 to present. Should the measured rate of retreat continue unchanged, the ice cap will have vanished by around 2015. Data: Lonnie Thompson (Alverson et al. 2001).

9.1.2
Chronology

Understanding climate history requires a clear chronology of the past. Records must be dated so that the timing of events, rates of change and relationships among different records can be established. Only in this way is it possible to build a coherent picture of the true temporal and spatial evolution of earth system processes. Chronologies are developed using a wide variety of methods. Several of these, radiocarbon dating for example, depend on radioactive decay. Others rely on counting annual layers, whether of snow accumulation, tree rings or seasonally deposited sediments. Models of physical processes such as ice accumulation and flow can also be used. Markers of known age such as volcanic ash layers are especially useful for inter-site comparisons. These approaches all require a great deal of research into the environmental processes that make them useful as chronometers. The best chronologies are often established using several independent methods.

Paleoresearchers have often engaged in highly elaborate tuning of their chronologies. This has usually been done for reasons of expedience, insufficient financing, or simply lack of datable materials within a given archive. None of these reasons can continue to justify the practice. Tuned chronologies have lead to a plethora of paleoenvironmental records which appear to support clear climatic correlations, but which in fact often confuse attempts to discern the temporal and spatial patterns of past climatic change. An unavoidable consequence of the advances that the paleoclimatic community has made in demonstrating the relevance of their research to modern and future concerns, is that records without precise and independent chronological control are of increasingly little interest at the forefront of the field.

9.1.3
Paleoclimate Modeling

Modeling is an integral part of paleoclimate research. Winds, ocean currents, the movement of ice sheets, and even biological processes involved in the global carbon cycle are governed by basic physical and chemical laws, many of which are well enough understood to serve as the basis for numerical models.

Paleoclimatologists employ a range from simple analytical equations to coupled atmosphere-ocean-biosphere GCMs. Simple models allow a wide range of hypotheses to be tested. Coupled GCMs, on the other hand, require enormous amounts of computer time, severely limiting the number and length of simulations that can be carried out. Simplified dynamical models of various levels of complexity fill in the range between these two extremes in the Earth model hierarchy.

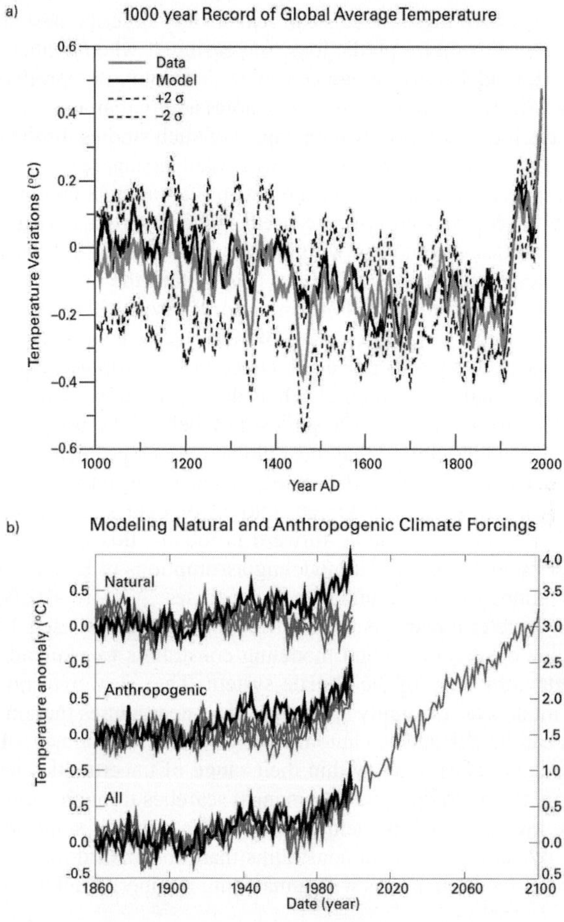

Fig. 9.2. a) A comparison of a decadally-smoothed Northern Hemisphere mean annual temperature record covering the past millennium (Mann et al, 1999) and an energy balance model calculation employing reconstructed natural solar and volcanic forcings as well as recent greenhouse gas level increases (Crowley, 2000). The dashed lines indicate the 95% confidence level for the temperature reconstruction. b) Annual-mean global near surface (1.5 m) temperature anomalies (relative to 1881-1920) for a series of model experiments using a coupled ocean-atmosphere GCM. The black line shows instrumental records over the same period. The combination of natural (top) and anthropogenic (middle) forcings employed by this model reproduce observed decadal to century scale climate variability in the instrumental record. The bottom graph shows the results when both natural and anthropogenic effects are included as well as a future prediction based on greenhouse gas increases from the IPCC scenario B2 with no changes from 1999 solar and volcanic forcings.

Since this same models are used to assess how climate will evolve in the future, the extent to which these models can realistically describe past changes is of immediate relevance to future predictions. Increasingly, when applied to the relatively recent past, models have shown considerable fidelity in reproducing past records, but only when both natural (solar, volcanic) and human induced (CO_2) forcings are included. Fig. 9.2 shows two examples of such studies. In the upper panel an EMIC calculation (Crowley 2000) is evaluated against proxy reconstructed Northern Hemisphere temperatures for the past millennium (Mann et al. 1999). In the lower panel, a coupled atmosphere-ocean general circulation model experiment (Stott et al. 2000) is compared with instrumental data for the past century.

A relatively new modeling technique, at least within the paleoclimate community, is the use of inverse models (LeGrand and Wunsch 1995; LeGrand and Alverson in press). Inverse procedures seek solutions that are consistent, within prescribed uncertainties, with both paleodata constraints and model equations. Such models allow traditional assumptions, such as the precise prescription of uncertain model parameters, to be relaxed. The philosophy behind the use of inverse techniques comes from a recognition that most global climate problems, especially paleo ones, are fundamentally underdetermined. That is, in the state of the climate system must be inferred from a limited amount of data and imperfect models. Usually, in order to obtain tractable forward problems, this underdeterminacy is removed, either by making implicit modeling assumptions or by incorporating exact parameterizations of poorly understood processes. A more satisfying way to handle such underdeterminacy is to use inverse methods (Wunsch 1996). These methods combine observational and modeling constraints to estimate, to the best of our knowledge, the state of the climate system. They deal with imperfect data and imperfect models by explicitly specifying the uncertainties therein. Moreover, these methods can be designed to automatically search for optimal solutions consistent with available constraints within their range of uncertainties instead of requiring modelers to employ painstaking manual searches through parameter space. In many cases, inverse methods yield a much wider range of solutions consistent with available dynamical and data constraints than do forward methods based on the resolution of a set of equations with a matching number of unknowns.

9.2 Detecting Anthropogenic Change.

9.2.1 Greenhouse Gasses and Global Warming

Aspects of ongoing and predicted future environmental change are, in many cases, outside of the envelope of natural variability. Some of the most compelling evidence supporting this statement comes when paleorecords are set alongside a range of predictions. In this section we present two examples, one showing the increase in greenhouse gas concentrations over the past half million years and the other Northern Hemisphere temperatures over the past millennium, both alongside a range of 'best guess' estimates for the year 2100.

Fig. 9.3 shows the past record of two important greenhouse gasses, carbon dioxide and methane, as recorded in air bubbles from the Vostok ice core. This record spans nearly half a million years. Over this time these greenhouse gas levels correlate very highly with inferred temperature reconstructions (not shown). The controls on this long term variability in atmospheric greenhouse gas concentrations, and their relationship to global climate, are not fully understood. Nonetheless, it is clear that modern levels of these gasses, as indicated by the asterisks, are already well outside of the long term historical range. The range of concentrations available from a series of 'best guess' future scenarios are indicated by the vertical bars. Ongoing anthropogenic greenhouse gas forcings clearly dwarf those associated with glacial cycles, which we know to have fundamentally altered the global climate and environment.

Fig. 9.4 shows an estimate of Northern Hemisphere average temperature change over the past millennium (Mann et al, 1999) set against the range of estimated temperature changes extrapolated to the year 2100 (Houghton et al. 2001). The past decade was the warmest of the last millennium, and the range of future estimates lies well beyond the envelope of natural variability over the past millennium. The environmental, societal, and economic consequences of regional climatic change associated with this rapid global warming will be large.

9.2.2 High Altitude Harbingers

Mountain archives, including glaciers, high altitude lakes, and trees near the limits of their habitable range provide ideal archives to study past climatic changes. They are also sensitive harbingers of recent global change. Because these archives are often relatively remote and insulated from direct human influences, they can serve as early indicators of global climate change.

Mountainous topography affects climate by modifying regional and global atmospheric circulation. As a consequence, myriad local climates occur at different altitudes and on lee or windward sides of mountain chains, providing a basis for a variety of ecosystems. These ecosystems are concentrated in small areas which are strongly linked by their geographic proximity and thus interact. Changes in one part of a mountain ecosystem thus influencing other ecosystems providing one reason for the great sensitivity of mountain regions to global change. For example, a relatively minor temperature rise near the freezing line may lead to pronounced snow melt, slope instability, enhanced river runoff, changes in vegetation patterns and biodiversity, thereby influencing both local ecosystems and highland–lowland interactions on a larger scale.

The vertical structure of mountains provides a unique basis to study impacts of climate change on the different local ecosystems. In the uppermost regions, ice cores provide long term records of past conditions. At lower altitudes similar information is stored in lake sediments and peat bogs. Trees growing at the upper limits of their range are strongly influenced by climatic variability and provide high resolution information about growing conditions. In all altitudinal belts the geomorphology of the landscape provides information on past climate conditions. The locations of moraines, rock glaciers, slope sediments, fluvial deposits and soils are strongly related to the history of climate conditions.

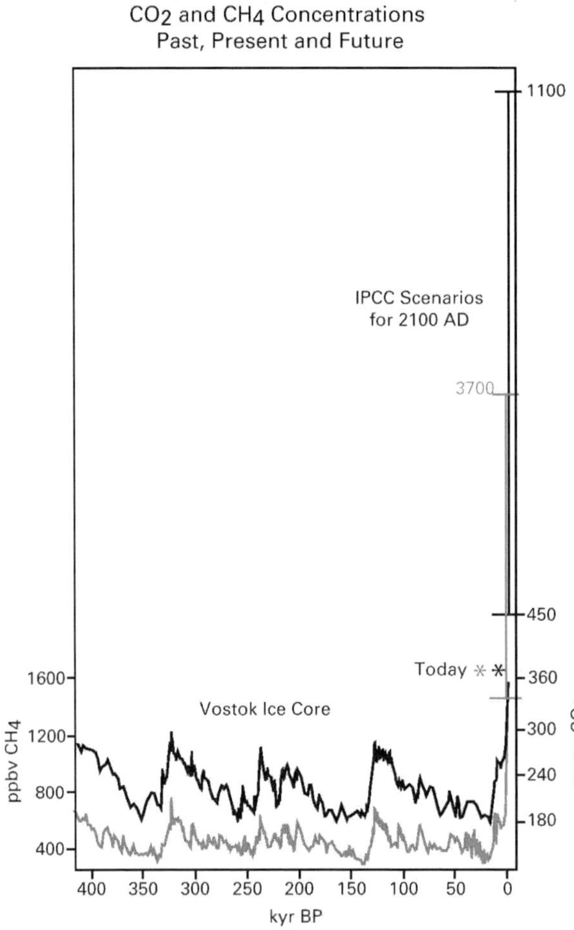

Fig. 9.3. Greenhouse trace gas (CO_2 and CH_4) changes over the last four glacial periods as recorded in the Vostok ice core. Present day values and the range of estimates for the year 2100 from (Houghton et al. 2001) are also indicated.

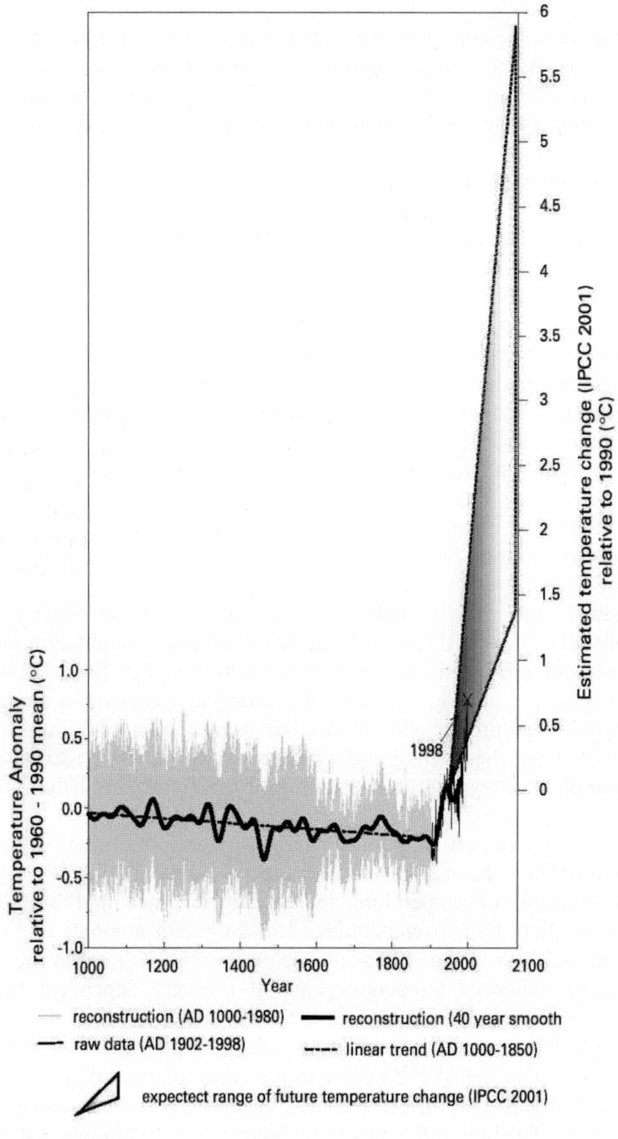

Fig. 9.4. A multiproxy reconstruction of mean annual Northern Hemisphere temperature extrapolated to the range of estimates for the year 2100 from (Houghton et al. 2001). A statistical confidence interval for the reconstruction is also shown.

Glaciers

In glaciated areas, ice cores provide a unique archive of past climate conditions at high elevations. Accumulated snow and ice provide information about past precipitation. The chemical and isotopic composition of the ice provide information about past atmospheric conditions including local and remote temperature as well as mechanisms of moisture transport to the site. Numerous high altitude ice cores have been drilled worldwide in the Americas (Andes, Rockies), Asia (Himalayas), Africa (Kilimanjaro) and Europe (Alps).Taken together, these provide an integral picture of high altitude climate variability during the Holocene and late Pleistocene.

Climatic interpretation of ice cores, as with any archive, relies first and foremost on an accurate chronology. Visual identification of the annual accumulation and ablation layers is limited, due to compacting, to the uppermost portion of the ice. Independent counting can be done with annual oscillations of $\delta^{18}O$ or other chemical species. Time markers such as datable organic material, chemical signatures from known volcanic eruptions and atmospheric nuclear weapon tests serve as independent dating checks and help demonstrate that a given ice core does not contain time hiatuses due to ice motion or negative mass balance. Mass loss by sublimation plays an important factor in calculating net accumulation especially in arid regions where up to 50% of the accumulation may be lost again by sublimation.

Sublimation rates vary through time. Although this is not directly obvious in the ice core it certainly influences the accuracy of any accumulation or precipitation reconstruction. Evidence of the presence or lack of significant sublimation effects can often be found by analyzing the chemical composition along the core. Reconstructed paleo-precipitation as derived from accumulation rate (the sum of precipitation, avalanche feeding and wind deposition) therefore strongly depends on a precise chronology and understanding of the net accumulation and ablation history.

Reconstructions of past temperature variability are often based on the isotopic composition of the accumulated water in the ice core. ^{16}O and ^{18}O isotopes fractionate as a function of temperature, and therefore latitude, and rainfall intensity. During evaporation, higher temperatures lead to higher amounts of ^{18}O in water vapor. However there is also a dependence on the $^{16}O/^{18}O$ ratio in the source water. The most important temperature/rainfall intensity dependent fractionating process occurs as water precipitates in the atmosphere. In initial precipitation the heavy isotope ^{18}O is enriched leaving the remaining atmospheric water vapor depleted in ^{18}O. The recorded $\delta^{18}O$ value in the accumulation of an ice core is thus influenced by at least two different remote variables: moisture source and atmospheric transport. The link to the prevailing temperature conditions at the site of the core is therefore difficult to construct. Such reconstructions often seem most plausible on centennial and longer timescales. Many ice cores from both polar regions and mountain glaciers indicate a similar shift in the $\delta^{18}O$ at the relatively abrupt end of the last glacial period, lending some credibility to the interpretation of these values as temperature proxies. But the values may also be altered by changed moisture supply and transport processes occurring contemporaneously with the warming.

Ice core chemical composition contains further information. In dry environments many chemical species may be dry deposited with aeolian particles (dry deposition).Relative aridity is indicated by dust enrichment in ice core segments, for example for most continental areas during the last glacial period. However such enrichment can also be produced by massive ice losses due to sublimation which does not remove any dust or irreversibly deposited chemical species. The effect of sublimation is more related to local climatic conditions such as cloudiness, radiation, water vapor deficit and wind than to continental scale aridity (Kull and Grosjean 2000; Ginot et al. in press).

Climate reconstruction based on ice cores can be made more precise by measuring and modeling mass balance at the drilling site and by comparing the results with independent data from other local archives (Kull et al. in press). A quantitative, global picture of past changes recorded in these cores allows the ongoing rapid retreat of glaciers around the world (Sect. 9.1, Fig. 9.1) to be put in the perspective of long term records. The result of such comparisons appears to be that modern high altitude glacial retreat, globally, is unprecedented over the last 10 000 years.

Lake Sediments

At slightly lower elevations, lakes and peat bogs provide information in the accumulation of organic material in sediments. These records also require an accurate chronology. Lakes which freeze in winter or with anaerobic bottom waters are sometimes characterized by varves, a visually or otherwise identifiable marker of the annual cycle. Dating of the sediment core by varve counting and ^{14}C dating of the organic material is easier than in ice cores. Problems can arise, for example, due to fossil water supply, which reduces ^{14}C values in the lake water and plants growing in it. Bioturbation may also disturb the sedimentary stratigraphy and affect dating precision.

Sediment composition varies due to the source of the water supply, organic production and prevailing climate conditions. Fine gravel and sand input by mountain rivers (clastic material) points to enhanced river discharge and can thus be used for climatic interpretation. Fine organic and siliclastic deposition of diatomeen algae are typical in deeper open water bodies. Higher deposition rates of evaporitic minerals such as carbonate and gypsum, on the other hand, provide evidence for shallow waters and semiarid climatic conditions. The composition and distribution of the different species, as reconstructed pollen and plant macrorests in sediments provides evidence as to former environmental conditions. Transfer functions, which express modern as a function of the plant distributions in different lakes, are used to reconstruct past climate conditions in a given lake as a function of time. One problem with the interpretation of pollen composition is the fact that wind direction and strength may be responsible for the observed composition and overprint the local climate signal.

Because of their remote location, high mountain lake and bog ecosystems are less likely than those in the lowlands to have been influenced by direct human activities. As such, they are particularly useful for the detection of any indirect influences of global anthropogenic climate change.

Trees

Trees living at high altitudes often exist at the extreme, either in terms of temperature or moisture availability, of their tolerated range. Given such conditions, relatively small climatic variations thus strongly influence tree growth, and their annual growth rings are generally characterized by variable width and density. Climate chronologies from high altitude trees, often extending for several thousand years, have been constructed for many regions of the world. These records provide an invaluable, high temporal resolution, component to multiproxy climate reconstructions. They also allow ongoing changes to be better understood, and compared to a long term record. For example, with the help of past records and in situ modern ecological studies, it is clear that vertical movement of mountain treelines is not significantly influenced, as has been suggested, by CO_2 concentration-related fertilization effects (Körner 2001). Recent tree line advances, when not influenced by direct human activities, thus reflect climatic changes. For example comparison of contemporary vegetation with ~100 year old photographs indicates vertical treeline displacements of 20-80m (200-900m horizontally) along a continuous transect from the southern to polar Urals (Shiyatov et al. 2001). Analysis of rings with fluctuations from trees growing within the forest-tundra ecotone along this transect, together with sparse instrumental data, corroborate that increasingly warm temperatures favorable to tree growth are the primary cause of these major ecosystem changes (Shiyatov et al. 2001).

Geomorphology

Mountain landscapes extend through many altitudinal belts and are formed under a broad variety of local climate conditions including permafrost, frost alternating and frost free conditions. It is occasionally possible to provide quantitative constraints, within the broader framework of a multiproxy climatic reconstruction, from geomorphological evidence. Moraine exposure age dating by cosmogenic radionuclides, for example, can supply dates of maximum glacial advance. Modeling allows reconstruction of past climate conditions compatible with known glacial advances (Kull and Grosjean 2000). Often, sediments in proglacial lakes and peat bogs behind the terminal moraine allow a minimum age for the onset of glacial retreat to be calculated. Relict permafrost features in currently permafrost free regions can provide quantitative estimates for maximum temperature decrease during glacial times (Porter et al. 2001; Vandenberghe and Lowe 2001). In valleys, river and slope activity is recorded in fans, which often provide a detailed history of accumulation and erosion periods. Climatic interpretation is made difficult due to difficulty constructing high resolution chronologies and because these features are influenced by many local parameters, including the magnitude and frequency of precipitation events, vegetation coverage, geology and relief (material availability) and local runoff conditions. Thus, such features generally provide qualitative information about past climatic and environmental changes.

High Altitude Temperature Change

Average global surface temperature has increased since the late 19th century by approximately 0.6°C with a 95% confidence interval between 0.4 and 0.8°C (Houghton et al. 2001). Most of this warming occurred during two periods, between 1910 and 1945, and since 1975. Whereas tropical temperature trends are more uniform than extratropical trends before 1975, the most recent decades are marked by globally synchronous warming. This warming is unique within the last 1000 years, as shown by multiproxy reconstructions (eg. Fig. 9.4). The increase in temperature during recent decades is marked by a faster rise in daily minimum than daily maximum temperature, and thus a decrease in the diurnal temperature range.

At high elevations, there is an apparent discrepancy in temperature trends as observed near the land surface and in free air at equivalent heights in the mid troposphere. Temperatures near the land surface appear to have risen faster than in the free atmosphere as measured by either satellites or radiosondes and recorded in reanalysis datasets (Houghton et al. 2001). These discrepancies are not yet fully understood, but changing high altitude land surface characteristics play a likely role. In agreement with the near land surface instrumental records, temperature proxies such as glaciers, vegetation and permafrost all indicate a strong warming at high elevation sites. Several studies in different mountain regions show a greater warming at high elevation sites than at low elevations (e.g. Diaz and Graham, 1996; Beniston et al., 1997) Because mountain climate is more variable than in other regions, for example over the ocean, the stronger recent warming signal does not necessarily provide a stronger signal to noise ratio. Thus, a relatively small climatic shift in a naturally less variable region might be statistically more significant than a large change in a highly variable mountain region (Fyfe and Flato 1999). Despite these qualifications, the strong warming that has been seen in mountain regions around the world in recent decades appears to be a detectable signal of global climate warming, probably enhanced locally relative to the free troposphere due to local land and snow cover change or other as yet unknown processes.

Changes in Mountain Ecosystems

Climate change has wide implications for ecosystems. In mountain regions, many different ecosystems are situated in close proximity, existing in different altitudinal belts or exposures. This provides unique areas for studying the impacts of climate change and ecosystem behavior and interaction.

Vegetation in alpine environments is often located at the climatological limit of its range and thus strongly influenced by climatological factors. Even changes in non-growing season climate, for example the duration of the winter snow cover may have massive influences on mountain ecology. Different species respond or adapt to climatic changes as individuals, not as coherent communities or ecosystems (Colinvaux et al. 2000). Therefore, it is expected, and indeed observed, that changes in flora and fauna will respond differentially during periods of climatic change. Some species are replaced by others, some disappear - especially due to warming conditions in the highest environments, when there is no possibility of upward movement to cooler climes. Climate change thus clearly influences biodiversity. Regionally amplified climate variability, the ability to support isolated pools of cold tolerant species seeking refuge from the relatively unusual warmth of the Holocene, and the sometimes reduced direct human influences, are some of the reasons why the regions of greatest biodiversity around the globe tend to be in mountainous areas (Fig. 9.5).

Both past climatic changes and ecosystem responses to these are recorded in the paleorecord. Understanding such interactions in the past provides one way to predict likely future changes. The Cape Floral Kingdom in the highlands of southwestern South Africa, for example, covers only 90000 square kilometers, but hosts approximately 9000 species, roughly 2/3 of which are endemic to the region (Alverson et al. 2001). Understanding the basis for the persistence of such unique biodiversity rich mountain areas in the past is part of the key to ensuring their future survival.

One key question is how ecosystems can respond to rapid climatic change. The rate and magnitude of projected climate change will require many species to shift their ranges at rates at least 10 times faster than has been observed during the Holocene. During the last period of similar rapid climatic change, at the end of the last glacial period, some ecosystems, such as in the Swiss Alps, appear to have responded over a few decades at most, whereas in others, such as northern Sweden, establishment of forest ecosystems lagged behind temperature changes by several thousand years, depending on altitude (Amman and Oldfield 2000). In the modern day, rapid climatic change of a similar magnitude is superimposed on large scale systemic changes due to human activities which have lead to fragmentation of landscapes globally. These changes will exacerbate climatic disruption of ecosystems and complicate attempts to understand and predict ecosystem functioning. Ecological instability and the occurrence of disturbances such as wildfires, will probably increase.

Changes in the Cryosphere

Climate change has direct and wide impact on the cryosphere. The cryosphere is marked by a strong inherent nonlinearity associated with the freezing point of water. This threshold value is at the root of many interesting feedbacks in regional and global climate dynamics. Except in extremely cold regions where temperatures are always well below freezing, rising temperatures will tend to reduce the amount and duration of snow cover. This has been observed in the mid- to late 1980s over wide parts of the northern hemisphere (Houghton et al. 2001).

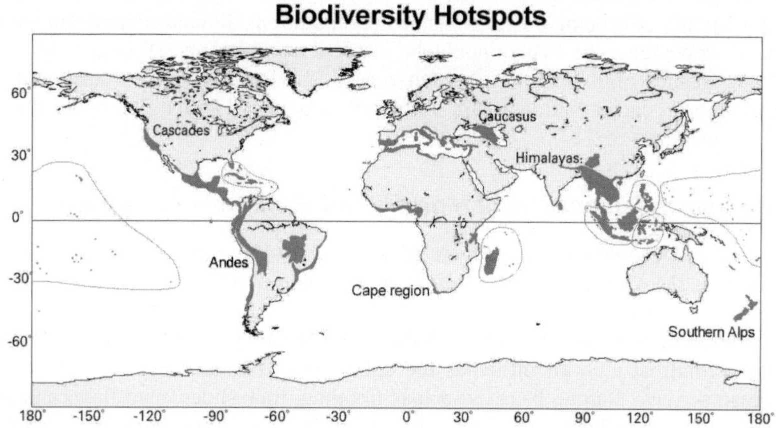

Fig. 9.5. A map of global biodiversity hotspot map with mountain regions indicated. Many high biodiversity regions are in mountainous zones.

Evidence of recent changes in mountain cryosphere systems are provided by measured warming of the permafrost (e.g. Wang et al., 2000). Permafrost is influenced not only by ambient temperature but also by radiation and snow coverage, but it is certainly a direct indicator of climate change in alpine environments. Warming of permafrost systems can have wide consequences for geomorphologic activity and the stability of high mountain environments. Direct effects include endangering intensely-used structures, such as cablecars, that are built on permafrost, as well as influence on human activity and ecosystems in downslope areas due to increasing instability of mountain slopes.

As described in Sect. 9.1.1 (Fig. 9.1), another sensitive indicator of climate change is mountain glaciers around the world. Historical data (for example old maps and paintings), ice cores, dateable moraines, and recently exposed glacier-trapped dateable wood, allow quantitative records of glacial geometry (advance or retreat phases) to be established. It is evident, that glacier retreat in the 20th century is a worldwide phenomena direct linked to anthropogenic climate change. The available data suggest that the dramatic retreat began at low and mid-latitudes around 1850 and later at high latitudes. The rise in the mean ELA of alpine glaciers and their retreat since the late 19th century is consistent with the measured temperature rise of about 0.6 to 0.8°C. (Oerlemans, 1994). The response times of mountain glaciers to climate change depends on their extension but is generally between 10 and 100 years. Therefore, globally significant high altitude warming probably started no later than the mid 19th century. Further evidence for increasing temperatures is provided by enhanced $\delta^{18}O$ values and bore hole temperatures in the upper (younger) portions of ice cores from around the world (Thompson et al., 2000). Glaciers do not only react to temperature. Precipitation, radiation and humidity are key parameters in determining mass balance and therefore advance and retreat. In very few regions, such as western Norway and New Zealand, glaciers are currently advancing due to enhanced precipitation. Climate change will lead to the disappearance of many mountain glaciers over the next few decades. The iconic glacier of Kilimanjaro, which served as Hemmingway's muse, and the eponymous Glacier National Park in the USA, may well be ice free by the middle of this century.

The Human Dimensions of High Altitude Environmental Change

According to one recent estimate 26% of the world's population lives in mountain areas (Meybeck et al. 2001). Human activity in these regions and surrounding lowlands is directly affected by climatic and ensuing environmental changes. Changes in the distribution and magnitude of extreme precipitation events, runoff and permafrost area all influence the entire range of mountain ecosystems and human activity. Natural hazards such as flooding, rock slides, mud flows and avalanches are all influenced by changing frequency of extreme events coupled with geomorphological changes such as reduced cohesion by permafrost or vegetation.

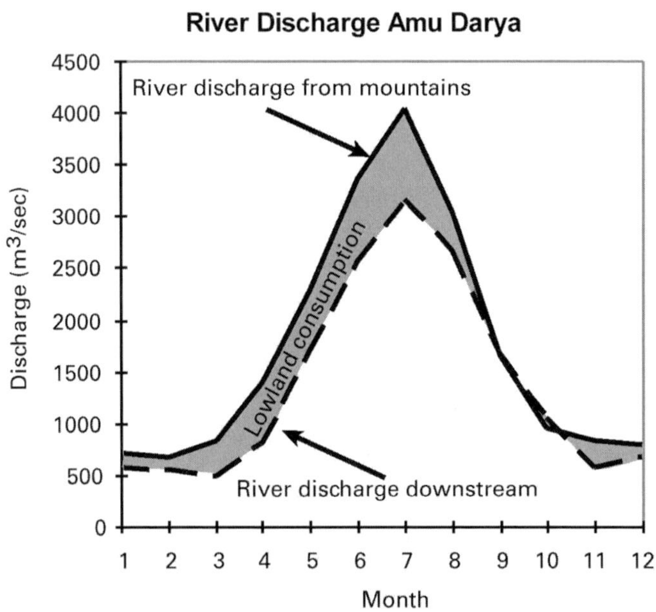

Fig. 9.6. In many parts of the arid and semiarid lowlands (e.g. Central Asia) populations are entirely dependent on the mountain waters, and are overusing local water resources. Shown are the mountain river discharge, the lowland discharge and the resulting negative water balance due to the high water consumption in the Amu Darya basin (Source: GRDC Koblenz, Germany).

Changes in the mountain hydrosphere are of major importance for lower lying regions which depend on mountain runoff for power generation, potable water and agricultural irrigation. One recent estimate suggests that 32% of the world's surface freshwater runoff is derived from mountain areas (Meybeck et al. 2001). Taking into account that much of the global total surface runoff is in low lying areas where water is not an especially critical resource, such as tropical rainforests, the percentage of surface runoff that is important for human activities derived from mountain regions must be much larger still. In many subtropical areas, water supply provided by ice and snow melt during the dry summer months is crucial for the irrigation of crops and energy production. As shown in Fig. 9.6, human populations, in large swathes of the subtropical lowlands, depend entirely on mountain waters (Liniger et al. 1998).

In addition to the agriculture and power industries, winter tourism is a global multi billion dollar industry almost entirely dependent on the timing and amount of winter snowfall. A rise in the mean snowline and a reduced snow cover period has already begun to force tourist resorts at lower elevations out of business. The development of large resorts at ever higher elevations leads to increased infrastructure expenses, and often also to dangers associated with less stable local geomorphology. Global climate change, and its expression in mountain regions, is thus of crucial concern to a large segment of the world's population as well as major sectors of the global economy.

9.3 Understanding Earth System Processes

The past record of earth system changes provides more than a simple extension of instrumental records. Although useful for helping detect anthropogenic change, as described in the previous section, paleodata also contain a wealth of information about processes. The paleorecord contains an imprint of a much more robust suite of biological, geological, physical and chemical processes than even the most powerful numerical model can possibly capture. Thus, a combination of data reconstruction, statistical techniques, and dynamical modeling set in the past, can provide information highly relevant to the potential future evolution of the system. In this section we present two examples of such process based studies. The first examines changing frequency and teleconnection patterns associated with ENSO and the second the tight correlation between atmospheric CO_2 concentration and global temperature on glacial-interglacial timescales. The full range of process studies being carried out is of course far wider than the examples presented here.

9.3.1 ENSO in the Past and Future

The El Nino Southern Oscillation (ENSO) is the interannual, large scale climatic mode that has shown the most potential for society and economically beneficial climate forecasting. However, the changing frequencies, amplitudes, and teleconnection patterns that we know to have occurred in the past need to be understood if predictions are to become reliable.

Regional climate predictions and climate reconstructions often both depend on the same underlying assumption - that climatic modes will be, or were, relatively unchanged outside the instrumental reference period. Usually, the instrumental period is too short to capture the full range of decadal scale variability for a given climate variable at a given location, and where longer proxy-based records do exist they often indicate that this assumption of statistical stationarity is a tenuous one. Occasionally, one finds an apparent change in mode in a given climatic timeseries. The degree to which such a shift can be shown to be statistically significant, and thus be argued to reflect a fundamental change in the climate system, is directly proportional to the length of the timeseries. The longer the record, the more confident one can be that the changes one sees are not simply random fluctuations.

Although the basic coupled ocean-atmosphere dynamics involved in ENSO within the equatorial Pacific are fairly well understood, our knowledge of climatically important ENSO teleconnections outside this region are, with a few exceptions, primarily based on statistical analyses. Similarly, the role of decadal scale variability in modulating ENSO is primarily based on statistical studies. To first order, it seems clear that when decadal variability is acting to cause an overall background warming in the tropical Pacific, one might expect 'enhanced' ENSO warming, and vice versa. However, an understanding of the detailed dynamical nature of the interaction between ENSO and decadal variability, remains somewhat elusive.

In Fig. 9.7 two climatic timeseries are presented. The first is the Niño 3,4 equatorial Pacific sea surface temperature (SST) index as extended using proxy data (Kaplan et al. 1998). Note that the extension of the SST index back in time is based on proxy data from remote sites which are highly correlated with SST in the modern instrumental record. The validity of the extended SST record thus rests, to some degree, on the assumption that this correlation was not different in the earlier period. The second curve is a record of precipitation derived from snow accumulation recorded in an ice core taken from Mt. Logan in the Yukon, Canada, at approximately 60.5°N (Moore et al. 2001). Both timeseries have been low pass filtered with a cut-off period of 15 years to show decadal scale variability and normalized by subtracting the mean and dividing by the standard deviation.

During the period, from 1900 to the present, the two records show a strong positive correlation (+.55), statistically significant at the 95% level. This provides statistical evidence for an ENSO teleconnection influencing precipitation at this location. As is customary, the dynamical reason for this teleconnection is not fully understood, although it is clearly related to the Pacific Decadal Oscillation (PDO), which is marked by changes in the synoptic scale flow regime associated with the strength of the Aleutian low. Analysis during the period of overlap with the NCEP/NCAR reanalysis (1948-87) support the contention that there is a coherent upper tropospheric circulation anomaly extending over much of the North Pacific Ocean and North America influencing snow accumulation at the ice core site (Moore et al. 2001; Moore et al. Submitted). During this period, heavy snow accumulation on Mt. Logan is associated with an enhancement of the Aleutian low. This pressure anomaly is associated with a southward movement of the upper level jet stream over eastern North Pacific and northward displacement along the West coast of North America, leading to enhanced advection of tropical moisture to the vicinity of Mt. Logan, and consequently higher snowfall.

Given the strong statistical correlation between these two records for the past century, and even some indication of the underlying dynamical causality during the period of overlap with the NCEP/NCAR reanalysis, one might be tempted to use this correlation either for reconstruction or prediction. Interestingly, in the period prior to 1900 the correlation is still significant (99%), but is of the opposite sign. Thus, climate predictions based on teleconnection patterns established during the short instrumental period are not straightforward.

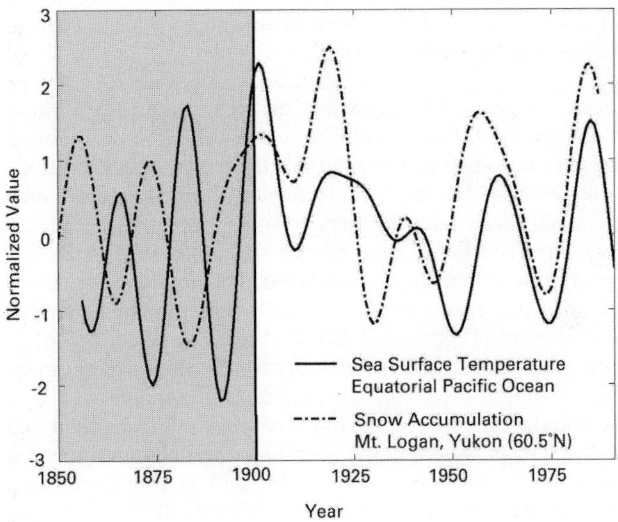

Fig. 9.7. The Niño 3,4 equatorial Pacific sea surface temperature (SST) index as extended using proxy data (Kaplan et al. 1998) and the snow accumulation record from Mt. Logan at 60.5 °N in Western Canada (Moore et al. 2001), both low pass filtered with a cutoff of 15 years. Around 1900 there was a statistically significant change in the sign of the correlation between the low frequency equatorial variability and the extra-tropical response at Mt. Logan.

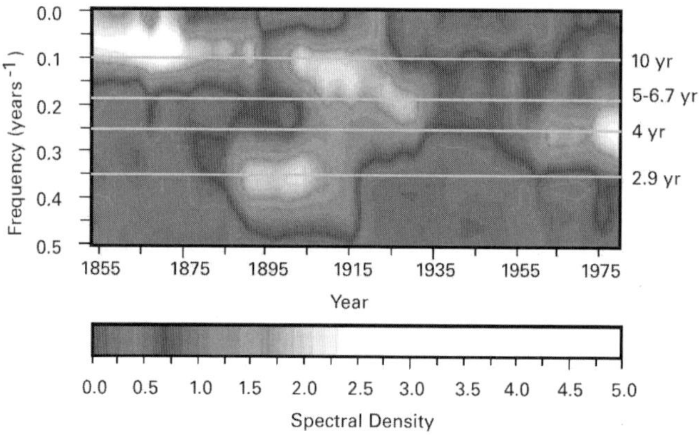

Fig. 9.8. Evolutionary spectral analysis of sea surface temperature estimates as recorded in a coral from Maiana Atoll in the equatorial Pacific Ocean (Urban et al. 2000). Colors showing the dominant frequencies with which sea surface temperatures have varied over the past 150 years. In recent years, power is concentrated in the well known 4 year ENSO band. Prior to 1900 however, there is a near absence of power in the 4 year band and a much stronger decadal signal.

It is unclear what caused the change in the sign of the ENSO-Mt. Logan teleconnection around 1900. One possibility is that the Mt. Logan core lies near a 'node' in the teleconnection pattern, and that there was simply a small shift in the location of the node. In this case, the dramatic shift in the ice core record, would not represent a large scale climatic change. Another possibility is that the ice core has indeed captured a major change in regional climate expression. Fig. 9.8 presents the opportunity to engage in speculation. This figure shows an evolutionary spectral SST analysis derived from an annually resolved coral from Maiana atoll in the equatorial Pacific (Urban et al. 2000). According to this record, 1900 was a period of time when tropical SST variability was shifting from primarily decadal power to higher frequencies. This shift appears to correlate with a shift in the phase of the correlation between the low frequency equatorial and extra-tropical response at Mt. Logan. Interestingly, tree-ring chronologies from high latitude sites in Alaska and Patagonia show a similar shift in coherence on the decadal time-scale around the same time (Villaba et al. 2001).

What can we conclude about decadal variability and ENSO prediction? From the evidence presented here, not much more than that understanding decadal variability of climate modes is difficult. Finding statistical significance in changes in interannual to decadal scale climate modes absolutely requires proxy based paleo-records that extend timeseries from the instrumental period. However, it is clear that no single proxy is sufficient. It is only with a wide geographical range of accurately dated, independent proxies including documentary evidence, tree rings, varved lake sediments, corals and ice cores that an understanding of decadal variability in climate teleconnections such as those associated with ENSO might evolve. And only then, will regional climate predictions based on these climate modes reliably transcend the assumption of statistical stationarity.

9.3.2 The Global Carbon Cycle in the Past and Future

Atmospheric CO_2 has varied in tight correlation with temperature over the past several hundred thousand years. Over the past 150 000 years this correlation is estimated to have an r^2 value of 0.89 (Cuffey and Vimeaux 2001). The variability in atmospheric CO_2 on long timescales (Fig. 9.2) and its correlation with inferred global temperature presents an intriguing set of questions about the operation of the Earth System, and in particular the global carbon cycle.

The attention of some process studies has been focused on what sets the upper and lower limiting values of order 180 and 300 ppm respectively (Falkowski et al. 2000). Terrestrial ecosystem slowdown leading to reduced terrestrial sinks has been implicated as a possible explanation for the lower bound, while the ocean is clearly implicated as the potential glacial period sink. The upper bound might simply be a response to, rather than a forcing of, global cooling associated with glacial cycles, with oceanic processes probably setting the range and frequency of variability on timescales longer than centuries.

Several oceanographic mechanisms have been proposed to explain, either individually or in concert, the decrease in atmospheric CO_2 associated with glacial states. These include: increased efficiency of the "biological pump", in which net carbon uptake, especially by the high latitude marine biosphere, is increased relative to upwelling fluxes (Sarmiento and Toggweiler 1984; Sigman and Boyle 2000), changes in the rate of ventilation of deep water in the Southern Ocean (Francois et al. 1997; Toggweiler 1999), decreased global thermohaline circulation (Sarmiento and Toggweiler 1984; Siegenthaler and Wenk 1984), enhanced air-sea exchange in the northern Atlantic (Keir 1993), decreased air-sea exchange in the Southern ocean due to enhanced sea ice cover (Elderfield and Rickaby 2000; Stephens and Keeling 2000), increased solubility of CO_2 in colder seawater (Bacastow 1996), and a mid-depth chemical divide separating watermasses with low and high CO_2 concentrations (Toggweiler 1999).

Some of these studies suggest simple, albeit highly specific, mechanisms which explain decreased CO_2 during glacial times (Toggweiler 1999; Stephens and Keeling 2000), while others claim that all simple mechanisms can be eliminated from consideration (Archer et al. 2000; Pedersen and Bertrand 2000) and that multiple or unknown processes must be at work. In contrast, (LeGrand and Alverson in press), employing an inverse solution procedure (see Sect. 9.1.3), find several simple solutions wherein a single change in the Southern Ocean relative to their modern control simulation, either reduced air-sea gas flux, reduced deep water ventilation, or enhanced biological productivity could explain all of the model and data constraints that they employ, within the uncertainties that they prescribe.

What can we conclude about the global carbon cycle? Clearly, the upper bound of around 300 ppm that has marked atmospheric CO_2 variability over at least the past 400 000 years, is not an impenetrable one. It has already been shattered by modern anthropogenic inputs of fossil carbon to the atmosphere, and is rising fast. Will this perturbation disrupt the system sufficiently to drag the Earth out of the attractor, marked by glacial oscillations, that it has been in for millions of years? Unless we understand how the system has operated in the past, it will remain impossible to move beyond simple conjecture based on correlation. Such conjecture is particularly tenuous when extrapolated to atmospheric CO_2 levels, such as those anticipated for 2100, for which there is no analogue during the period for which well dated, relatively precise paleorecords exist.

9.4
Assessing the Risks of Future Global Climate Change

Risk assessment is a rapidly developing field. Integrated assessment typically includes social, economic, institutional, and environmental dimensions forced by, or sometimes interacting with, predicted climatic drivers. These models are called 'integrated' because they draw on a wider range of fields than just physical climatology or 'participatory' because they include viewpoints from numerous stakeholders (Hisschemöller et al. 2001). However, the assessment community for the most part does not integrate information from the pre-instrumental period. In this section we present a few examples of information from the past record which should be seen as highly relevant to integrated impact assessment. There are numerous examples of natural environmental variability, and rapid climate change, with regional impacts far more severe than those contained in future scenarios based on greenhouse gas forcing. We can be very confident indeed that such variability will occur in the future, since we know it occurred in the past. To the extent that the integrated impact assessment community may one day provide information which is applicable to national and international political processes, thereby transcending the ubiquitous box and arrow diagrams, it certainly needs to do a much better job of including paleoclimate information in its prognostications. In this section we present some examples of paleoresearch with direct bearing on assessing the social and economic risks of future climate change. The full range of applicable paleoresearch is of course much wider than these individual examples.

9.4.1
Abrupt Climate Change: Dealing with the Unpredictable

Fig. 9.9 shows a well-known example of abrupt climate change. The end of the Younger Dryas, around 11 500 years ago, was a widespread abrupt climatic event marking the final transition to the Holocene warm period. Ice cores from central

Greenland indicate that a doubling of snow accumulation rate and a staggering warming of order 10 °C occurred within a decade or less (Alley 2000). Yet the generally accepted insolation forcing for the termination of the last glacial period occurred gradually over many thousands of years. It seems that this gradual forcing pushed the Earth system over a nonlinear threshold beyond which internal dynamics generated extremely rapid climatic change through a range of, as yet poorly understood, feedback mechanisms. There is no reason to suppose that similar nonlinear thresholds do not exist today.

Growing attention has been paid to the possible ramifications of potential future abrupt shifts beyond the range of variability upon which planning and construction schemes are based and even outside the envelope of scenario projections generated by climate models (Pfaff and Peteet 2001). By the very nature of nonlinear systems, which are marked by sensitive dependence on initial conditions, such abrupt changes will remain largely unpredictable, no matter how sophisticated coupled, dynamical climate models may become. Thus, the best way to account for them is to develop scenarios based on past events.

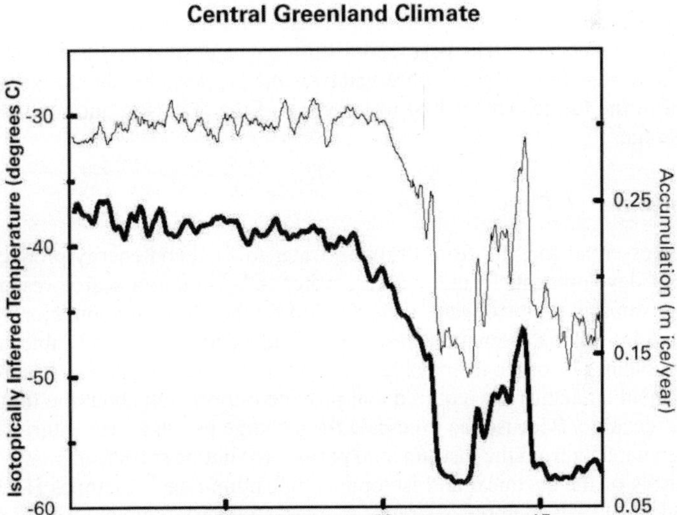

Fig. 9.9. The termination of the Younger Dryas cold event 11.6 thousand years ago was an abrupt climate event. In this record, from central Greenland, it is manifested as a warming of order 10 ºC accompanied by a doubling in annual precipitation, and occurred in less than a decade. Shown here are accumulation and oxygen isotope (interpreted as temperature) records from the GISP2 ice core (Alley et al. 1993)

9.4.2 Preparing for the 1000 Year Drought

Examples of abrupt climate events in the paleorecord are not limited to polar regions or glacial periods. Paleoclimate records also reveal sudden changes in regional, low latitude hydrological balance during the warmth of the recent Holocene, sometimes persistent for decades to centuries. Such variability lies well outside the range of instrumental records. Concepts of future environmental sustainability, water supply and food security are limited and potentially dangerously short sighted if they fail to accommodate such evidence from the longer term record.

Fig. 9.10 shows a transect of lake level records from north Africa (Gasse 2000). Hydrological variability in the region has been abrupt and of extremely large magnitude. Lake level variations are as much as 100 meters or more and are widespread across this transect. They are abrupt. Periods of extreme drought are sustained for decades to centuries. All of this in a region which, today, experiences widespread catastrophic human casualties following on just a few consecutive years of drought. Records indicating similar variability are found all over the world. Nowhere in the world though, has societal infrastructure been developed to withstand such changes. The past record strongly suggests that widespread natural droughts more severe than any observed over the past century should be expected to occur in the future. This is true irrespective of the effects of anthropogenic climate change.

Groundwater: A Paleo Resource

Water is essential for life, from drinking water to food and energy production to industrial development. In many areas, water is becoming a scarce resource. In arid and semiarid regions, water consumption by the growing population is being supported by mining the natural reservoirs, leading to lower water tables and related problems. In order to manage these ground water resources, research and management strategies are required that provide information about the timing and rates of recharge. Because the timescale for recharge processes is centuries to tens of millennia, data from the instrumental period will not be sufficient.

Analysis of the chemical and isotopic composition (stable isotopes H, ^{2}H, ^{16}O, ^{18}O) and the dissolved noble gas ratios (e.g. argon/neon) in ground waters provide information about climatic conditions when the water entered into the ground. For example, groundwater archives provide a robust means of estimating continental temperatures during the last glacial period (Stute et al. 1995). Dating of these waters, for example with radiocarbon methods, suggests that many are relict bodies, formed under wetter, cooler climate conditions in the distant past (8ky – 100ky). In such areas there is little or no actual recharge, therefore populations and economies are fully dependent on fossil water resources. Continuing lowering of these watertables due to continued overuse will eventually lead to, and in some cases already has lead to, the loss of these resources.

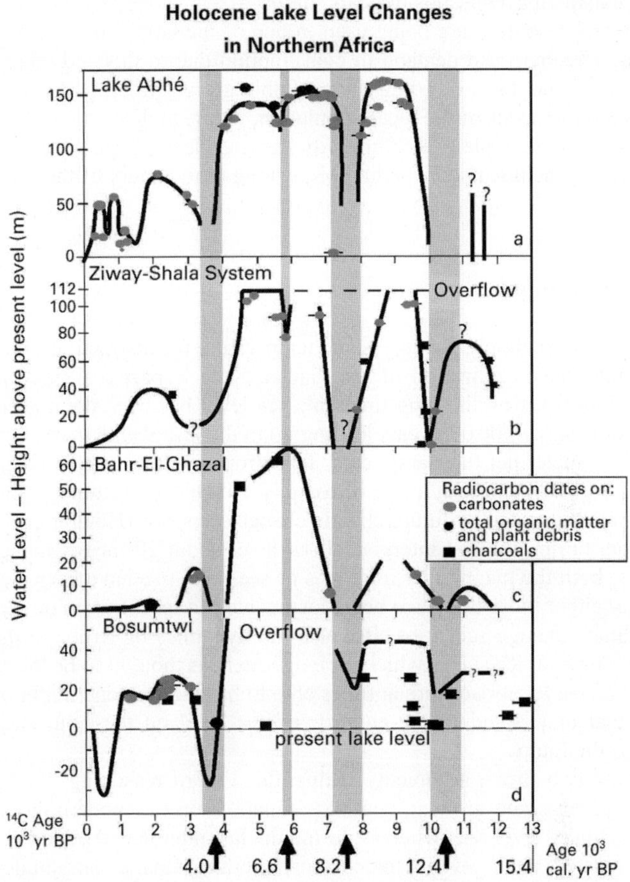

Fig. 9.10. Changes in lake level during the Holocene in an east to west transect of lakes in the northern monsoon domain of Africa. During this period, lake level variations are coherent across the region and have exceeded 100 meters in some cases, indicating widespread changes in regional hydrological balance. Such changes, should they occur in the future, would have enormous societal impact (Gasse 2000)

In the north African Sahara and Sahel, precipitation is low (< 500 mm/yr) and highly variable. Groundwater is found in sedimentary basins of considerable depth, and has been historically used by the local population at oases and artesian wells. More recently, deep drilling and mechanical pumping is lowering the water table at an accelerating rate. These reservoirs were formed during cool humid climate conditions when large parts of the Sahara were covered by Savannas, allowing a landscape with lakes, rivers and diverse Fauna and Flora. Such conditions have not existed for thousands of years. In the Atacama desert of north Chile, non-renewal water resources are being used in one of the largest mining operations in the world. The increased demand in consumption due to this and other economic development is largely exceeding the available groundwater, and leading to conflict between the needs of the local population, nature and economic development. Strategies for sustainable use of groundwater are clearly urgently required. Such strategies must include understanding of the long term history of the resource.

9.4.2
Sea Level Change

In addition to greenhouse gas levels and temperature reconstructions, one tangible, globally integrated realization of the glacial cycle is past sea level history. As shown in Fig. 9.11, on the long timescale sea level has been dominated by a decrease over roughly 100 000 years leading up to the last glacial maximum, at a rate of roughly 1 meter per thousand years. This drop is followed by a rise, occurring in less than 10 000 years, at a rate of roughly 1 meter per century, at times much faster than is suggested by future climate change scenarios (Houghton et al. 2001). In addition, during the last interglacial sea level about 20 meters higher than today. Thus, both the magnitudes and rates of sea level rise have been significantly larger than either those that have been documented in recent years or predicted for future climate change scenarios. Because the dynamics of large, northern hemisphere, continental ice sheets which no longer exist is thought to be the reason that past rates of sea level rise were at times able to be so large, one might be tempted to think that understanding the controls of sea level on these timescales is not relevant to the future.

There are two strong arguments against this line of reasoning. First, and most importantly, large and growing coastal populations in the modern day are highly vulnerable to sea level rise, whereas during the last interglacial period no such settlements existed. Thus, even a much more gradual, smaller amplitude sea level rise occurring today will have much greater human impact than those of the past. Furthermore, extant ice sheets on Greenland and Antarctica represent fresh water storage equivalent to many meters and many tens of meters of sea level respectively. The possibility of rapid destabilization of even a small part of these ice sheets in the future, although less likely than many other climate change predictions, cannot be excluded from consideration. Thus, an understanding of the dynamics of large ice sheets, gained through investigation of their behavior in the past, may be important for our future.

Given the growing, and increasingly urbanized, coastal population and the great potential for nonlinear 'threshold crossing' response of coastal zones and the ecosystems they support to sea level rise (Anderson et al. in press), there is an urgent need to understand the processes driving and responding to sea level rise on timescales from decades to millennia. Many of these processes, such as post-glacial rebound of continental plates, continue to effect sea level change today. The past record of sea level change is thus highly relevant to its likely future course.

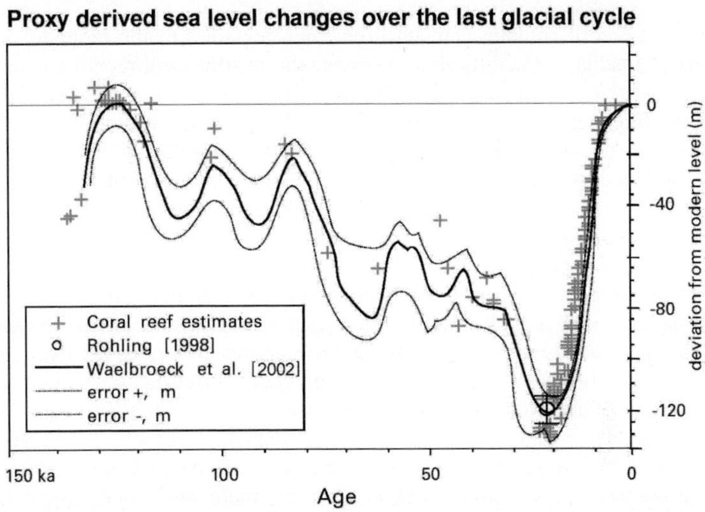

Fig. 9.11. An estimate of sea-level change over the last glacial cycle. According to this reconstruction both the magnitude and the rate of sea level rise in the past have been higher than those predicted in future global warming scenarios. However, the societal and economic impacts of even much smaller changes than those that occurred in the past will be large as a result of the increasingly populated and urbanized coastal zone.

9.5 How is the Past Significant for the Future?

Understanding past climatic and environmental change clearly is relevant to predicting the future. This contention is not new (e.g. Alverson and Oldfield 2000). In this chapter we have provided numerous examples as to why this is the case. Synthesizing these examples leads to a number of key points.

The paleorecord extends instrumental climate records allowing the statistical significance of any potentially detected signs of anthropogenic change to be more robust. Especially in terms of greenhouse gas changes, hemispheric average temperature, and high altitudes changes paleorecords play a vital role in detecting ongoing anthropogenic perturbation to the Earth System

It is clear that there are numerous long term processes operating in the earth system on timescales of decades and longer. In fact, the frequency of earth system behavior, although containing peaks on various frequencies, is basically red, with the most energy at the lowest frequencies. These slow, high energy processes are still occurring and will influence the present and the future as they have in the past.

The past record contains a much fuller representation of the range of climatic and environmental variability than do either the instrumental record or numerical models. Large and abrupt climate changes associated with nonlinearity abound. Some nonlinear climate feedbacks, such as those caused albedo difference between water and ice, and the freezing point of water as a threshold value, are relatively well understood. Others, are seen in the paleorecord, but not understood. Such rapid climate changes, and their environmental impacts, are absent in the instrumental record and the nature of nonlinear systems is that they are not predictable. Thus, developing scenarios based on analogues from the past record is perhaps the best way to deal with the potential problem of future climate surprises.

The paleorecord includes numerous examples of the interaction between climate change ecosystems, and even societies. Studying processes involved in these interactions as they are recorded in the paleorecord provides an excellent way to understand possible interaction of ecosystems with future climate change.

Of course, the paleorecord does not tell us everything. Much of the rest of this book is dedicated to that expression of global climate which has been expressed in the instrumental period. In this recent period, there are surely many effects, including for example those associated with systemic direct human influences on ecosystems worldwide, for which there is no analogue in the past. Although the past record is rich, it certainly does not contain all possible futures.

Acknowledgements

Lonnie Thompson is greatly acknowledged for the data in Fig. 9.1. Thanks also to Ray Bradley, Keith Briffa, Julia Cole, Malcolm Hughes, Isabelle Larocque, Tom Pedersen, Lonnie Thompson and Sandy Tudhope for their input into the call for a Global Paleoclimate Observing System described in Sect. 9.1.1. The examples of developing process based understanding from paleorecords in Sect. 9.3.1 and Sect. 9.3.2 are drawn from published and ongoing collaborative work with Kent Moore and Pascal Le Grand respectively. The ideas on integrated risk assessment in Sect. 9.4 are largely due to discussions with Frank Oldfield.

References

Alley, R. B. (2000). "The Younger Dryas cold interval as viewed from central Greenland." Quaternary Science Reviews 19: 213-226.

Alley, R. B., D. A. Meese, C. A. Shuman, A. J. Gow, K. C. Taylor, P. M. Grootes, J. W. C. White, M. Ram, E. D. Waddington, P. A. Mayewski and G. A. Zielinski (1993). "Abrupt increase in Greenland snow accumulation at the end of the Younger Dryas event." Nature 362: 527-529.

Alverson, K., R. Bradley, K. Briffa, J. Cole, M. Hughes, I. Larocque, T. Pedersen, L. Thompson and S. Tudhope (2001). "A Global Paleoclimate Observing System." Science 293(5527): 47-48.

Alverson, K., R. Bradley, K. R. Briffa, J. Cole, M. K. Hughes, I. Larocque, T. Pedersen, L. G. Thompson and A. W. Tudhope (2001). "Disappearing Evidence: the Need for a Global Paleoclimate Observing System." Global Change Newsletter 46: 2-6.

Alverson, K., R. Bradley and T. Pedersen (2001). Environmental Variability and Climate Change. Stockholm, IGBP Secretariat.

Alverson, K. and M. Eakin (2001). "Making sure that the world's paleodata do not get buried." Nature 412: 269.

Alverson, K. and F. Oldfield (2000). "PAGES - Past Global Changes and their Significance for the future: an Introduction." Quaternary Science Reviews 19: 3-7.

Alverson, K., F. Oldfield and R. Bradley, Eds. (2000). Past Global Changes and Their Significance for the Future. Quaternary Science Reviews, Elsevier Science Ltd.

Amman, B. and F. Oldfield, Eds. (2000). Biotic Respnses to Rapid Climatic Chnages Around the Younger Dryas. Paleo Geography Climatology Ecology, Elsevier.

Anderson, J., D. Belknap, B. Douglas, D. Fitzgerald, C. Fletcher, R. Holman, S. Leatherman, B. Richmond, S. Riggs, A. Rodriguez, S. Tebbens, T. Tornqvist and O. v. d. Plassche (in press). "Forecasting Coastal Change." Eos Trans. AGU.

Archer, D., A. Winguth, D. Lea and N. Mahowald (2000). "What caused the glacial/interglacial atmospheric pCO_2 cycles?" Reviews of Geophysics 38,2: 159-189.

Bacastow, R. B. (1996). "The effect of temperature change of the warm surface waters of the oceans on atmospheric CO_2." Global Biogeochemical Cycles 10(2): 319-333.

Briffa, K. R. (2000). "Annual climate variability in the Holocene: interpreting the message of ancient trees." Quaternary Science Reviews 19: 87-105.

Briffa, K. R., F. H. Schweingruber, P. D. Jones, T. J. Osborn, S. G. Shiyatov and E. A. Vaganov (1998). "Reduced sensitivity of recent tree-growth to temperature at high northern latitudes." Nature 391: 678-682.

Colinvaux, P. A., P. E. D. Oliveira and M. B. Bush (2000). "Amazonian and neotropical plant communities on glacial time-scales: The failure of the aridity and refuge hypotheses." Quaternary Science Reviews 19: 141-169.

Crowley, T. J. (2000). "Causes of climate change over the past 1000 years." Science 289: 270-277.

Cuffey, K. M. and F. Vimeaux (2001). "Covariation of carbon dioxide and temperature from the Vostock ice core after deuterium-excess correction." Nature 412: 523-227.

D'Arrigo, R. (1998). "The Southeaset Asian Dendro Workshop 1998." PAGES News 6(2): 14-15.

Elderfield, H. and R. E. M. Rickaby (2000). "Oceanic Cd/P ratio and nutrient utilization in the glacial Southern Ocean." Nature 405: 305-310.

Falkowski, P., R. J. Scholes, E. Boyle, J. Canadell, D. Canfield, J. Elser, N. Gruber, K. Hibbard, P. Högberg, S. Linder, F. T. Mackenzie, B. M. III, T. Pedersen, Y. Rosenthal, S. Seitzinger, V. Smetacek and W. Steffen (2000). "The global carbon cycle: a test of our knowledge of earth as a system." Science 290: 291-296.

Francois, R., M. Altabet, E.-F. Yu, D. M. Sigman, M. P. Bacon, M. Frankl, G. Bohrmann, G. Bareille and L. D. Labeyrie (1997). "Contribution of Southern Ocean surface-water stratification to low atmospheric CO_2 concentrations during the last glacial period." Nature 389: 929-935.

Fyfe, J. C. and G. M. Flato (1999). "Enhanced Climate Change and its Detection over the Rocky Mountains." Journal of Climate 12: 230-243.

Gagan, M. K., L. K. Ayliffe, J. W. Beck, J. E. Cole, E. R. M. Druffel, R. B. Dunbar and D. P. Schrag (2000). "New views of tropical paleoclimates from corals." Quaternary Science Reviews 19: 45-64.

Gasse, F. (2000). "Hydrological changes in the African tropics since the Last Glacial Maximum." Quaternary Science Reviews 19: 189-211.

Ginot, P., C. Kull, M. Schwikowski, U. Schotterer and H. W. Gäggeler (in press). "Effects of post-depositional processes on snow composition of a subtropical glacier (Cerro Tapado, Chilean Andes)." Journal of Geophysical Research.

Henderson, K. A., L. G. Thompson and P. Lin (1999). "Recording of El Niño in ice core d18O records from Nevado Huascaran, Peru." Journal of Geophysical Research 104: 31053-31065.

Hisschemöller, M., R. Tol and P. Vellinga (2001). "The relevance of participatory approaches in integrated environmental assessment." Integrated Assessment 2: 57-72.

Hoegh-Guldberg, O. (1999). "Climate Change, coral bleaching and the future of the world's coral reefs." Marine and Fresh Water Research 50(8): 839-866.

Hormes, A., B. U. Müller and C. Schlüchter (2001). "The Alps with little ice: evidence for eight Holocene phases of reduced glacier extent in the Central Swiss Alps." The Holocene 11(3): 255-65.

Houghton, J. T., Y. Ding, D. G. Griggs, M. Noguer, P. J. v. d. Linden, X. Dai, K. Maskell and C. A. Johnson, Eds. (2001). Climate Change 2001: The Scientific Basis. Contribution of WOrking Group I to the Third Assessment Report of the IPCC, 2001, Cambridge University Press.

Kaplan, A., M. Cane, Y. Kushnir, C. A., M. Blumenthal and B. Rajagopalan (1998). "Analyses of global sea surface temperature 1856-1991." Journal of Geophysical Research 103: 18,567-18,589.

Keir, R. S. (1993). "Cold Surface Ocean Ventilation and Its Effect on Atmospheric CO_2." Journal of Geophysical Research 98(C1): 849-856.

Körner, C. (2001). Warming Treelines and the Alpine. Climate Change at High Elevation Sites: Emerging Impacts: Highest 2, Davos, Switzerland.

Kull, C. and M. Grosjean (2000). "Late Pleistocene climate conditions in the north Chilean Andes drawn from a climate-glacier model." Journal of Claciology 46(155): 622-632.

Kull, C., M. Grosjean and H. Veit (in press). "Modeling modern and late Pleistocene glacio-climatological conditions in the north Chilean Andes (20 °S)." Climatic Chante.

LeGrand, P. and K. Alverson (in press). "Variations in Atmospheric CO_2 During Glacial Cycles from an Inverse Ocean Modeling Perspective." Paleoceanography.

LeGrand, P. and C. Wunsch (1995). "Constraints from paleotracer data on the North Atlantic circulation during the last glacial maximum." Paleoceanography 10(6): 1011-1045.

Liniger, H. P., R. Weingartner, M. Grosjean, C. Kull, L. MacMillan, B. Messerli, A. Bisaz and U. Lutz (1998). Mountains of the World: Water Towers for the 21st Century, University of Bern, Switzerland: 32.

Mann, M. E., R. S. Bradley and M. K. Hughes (1999). "Northern Hemisphere Temperatures During the Past Millennium: Inferences, Uncertainties, and Limitations." Geophysical Research Letters 26(6): 759-762.

Meybeck, M., P. Green and C. Vörösmarty (2001). "A New Typology for Mountains and Other Relief Classes." Mountain Research and Development 21(1): 34-45.

Moore, G. W. K., K. Alverson and G. Holdsworth (Submitted). "Variability in the Climate of the Pacific Ocean and North America as Expressed in an Ice Core from Mount Logan." Annals of Glaciology.

Moore, G. W. K., G. Holdsworth and K. Alverson (2001). "Extra-Tropical Response to ENSO 1736-1985 As Expressed In An Ice Core From The Saint Elias Mountain Range In Northwestern North America." Geophysical Research Letters 28(18): 3457-3461.

Pedersen, T. F. and P. Bertrand (2000). "Influences of oceanic rheostats and amplifiers on atmospheric CO_2 content during the Late Quaternary." Quaternary Science Reviews 19: 273-283.

Pfaff, A. and D. Peteet (2001). "Generating Probabilities in Support of Societal Decision Making." Eos Trans. AGU 82(20): 222-225.

Porter, S. C., A. Singhvi, Z. An and Z. Lai (2001). "Luminescence Age and Palaeoenvironmental Implications of a Late Pleistocene Ground Wedbe on the Northeastern Tibetan Plateau." Permafrost and Periglacial Processes 12: 203-210.

Sarmiento, J. L. and R. Toggweiler (1984). "A new model for the role of the oceans in determining atmospheric pCO_2." Nature 308: 621-624.
Shiyatov, S., P. Moiseev and O. Tchekhlov (2001). The impact of climate changes on forest-tundra vegetation in the Ural Mountain highlands during the 20th century. Climate Change at High Elevation Sites: Emerging Impacts: Highest 2, Davos, Switzerland.
Siegenthaler, U. and T. Wenk (1984). "Rapid atmospheric CO_2 variations and ocean circulation." Nature 308: 624-626.
Sigman, D. M. and E. A. Boyle (2000). "Glacial/interglacial variations in atmospheric carbon dioxide." Nature 407: 859-869.
Stephens, B. B. and R. F. Keeling (2000). "The influence of Antarctic sea ice on glacial-interglacial CO_2 variations." Nature 404: 171-174.
Stott, P. A., S. F. B. Tett, G. S. Jones, M. R. Allen, J. F. B. Mitchell and G. J. Jenkins (2000). "External Control of 20th Century Temperature by Natural and Anthropogenic Forcings." Science 290: 2133-2137.
Stute, M., M. Forster, H. Frischkorn and A. Serejo (1995). "Cooling of tropical Brazil (5° C) during the last glacial maximum." Science 269: 379-383.
Thompson, L. G. (2000). "Ice core evidence for climate change in the Tropics: implications for our future." Quaternary Science Reviews 19: 19-35.
Thompson, L. G. (2001). Stable Isotopes and their Relationship to Temperature as Recorded in Low-Latitude Ice Cores. Geological Perspectives of Global Climate Change. L. C. Gerhard, W. E. Harrison and B. M. Hanson: 99-119.
Thompson, L. G., Davies, M.E., Mosley-Thompson, E., Sowers, T.A., Henderson, K.A., Zagoronov, V.S., Lin, P.-N., Mikhalenko, V.N., Campen, R.K., Bolzan, J.F., Cole-Dai, J., Francou, B. (1998). "A 25,000-year tropical climate history from Bolivian ice cores." Science 282: 1858-1864.
Thompson, L. G., T. Yao, E. Mosley-Thompson, M. E. Davis, K. A. Henderson and P.-N. Lin (2000). "A High-Resolution Millennial Record of the South Asian Monsoon from Himalayan Ice Cores." Science 289(5486): 1916-1919.
Toggweiler, J. R. (1999). "Variation of atmospheric CO_2 by ventilation of the ocean's deepest water." Paleoceanography 14(5): 571-588.
Tudhope, A. W., C. P. Chilcott, M. T. McCulloch, E. R. Cook, J. Chappell, R. M. Ellam, D. W. Lea, J. M. Lough and G. B. Shimmield (2001). "Variability in the El Nino-Southern Oscillation Through a Glacial-Interglacial Cycle." Science 291(5508): 1511-1517.
Urban, F. E., J. E. Cole and J. T. Overpeck (2000). "Influence of mean climate change on climate variability from a 155-year tropical Pacific coral record." Nature 407: 989-993.
Vaganov, E. A., M. K. Hughes, A. V. Kidyanov, F. H. Schweingruber and P. P. Silkin (1999). "Influence of snowfall and melt timing on tree growth in subarctic Eurasia." Nature 400: 149-151.
Vandenberghe, J. and J. Lowe (2001). Climatic and environmental variability of Mid-Latitude Europe during the last interglacial-glacial cycle. Past Climate Variaiblity Through Europe and Africa, Aix-en-Provence, PAGES.
Villaba, R., R. D. D'Arrigo, E. R. Cook, G. C. Jacoby and G. Wiles (2001). Decadal-scale climatic variability along the extra-tropical western coast of the Americas: Evidence from tree-ring records. Interhemispheric Climate Linkages. V. Markgraf, Academic Press: 155-172.
Wunsch, C. (1996). The Ocean Circulation Inverse Problem, Cambridge University Press.
Wunsch, C. (1999). "The Interpretation of Short Climate Records, with Comments on the North Atlantic and Southern Oscillations." Bulletin of the American Meteorological Society 80(2): 245-255.

10 The Indian Monsoon and its Relation to Global Climate Variability

V. Krishnamurthy[1] and James L. Kinter III[2]

Center for Ocean-Land-Atmosphere Studies, Institute of Global Environment and Society, Inc. 4041 Powder Mill Road, Suite 302, Carlverton, MD 20705-3106, USA
[1]krishna@cola.iges.org
[2]kinter@cola.iges.org

10.1 Introduction.

Monsoon, meaning season, is the wind system over India and adjoining oceanic regions that blows from the southwest half the year and from the northeast during the other half. The seasonal reversal of the wind direction occurring in May brings copious moisture from the warm waters of the tropical ocean to the Indian continent through southwesterlies. Most of the annual rainfall in India occurs from June to September during what is referred to as the summer monsoon or southwest monsoon. The winter monsoon or the northeast monsoon brings rainfall to the southeastern part of India through northeasterlies during October to December and contributes a small percentage to the annual Indian rainfall.

Two remarkable features of the summer monsoon are its regular occurrence every year from June to September and the irregular variation in the amount of seasonal mean rainfall that it brings to India from one year to the other. There are many instances of years with flood (strong monsoon) or drought (weak monsoon) during which India as a whole receives excess or deficient seasonal rainfall, respectively. Even within a season, there is considerable variation, both in space and time, in the rainfall over India. The intraseasonal variation is characterized by "active" periods of high rainfall and "break" periods with weak or no rainfall over central India and the west coast, each phase lasting for a few days. The intraseasonal and interannual variability of the summer monsoon has a tremendous socioeconomic impact on India, especially in the fields of agriculture and health.

The seasonal wind reversal is associated with the surface temperature contrast between the Indian continent and the Indian Ocean, caused by the different responses of land and sea to solar heating, during April and May. The onset of the monsoon over India involves the establishment of a low pressure region called the monsoon trough. A southeasterly wind from the southern Indian Ocean crosses the equator and gets deflected by the Earth's rotation to become a southwesterly wind flowing over India. Cloudiness and rainfall are usually associated with low pressure regions such as the monsoon trough. The day-to-day rainfall in India is caused by synoptic scale disturbances that vary in intensity, increasing from a low to a depression to a cyclone. The active and break periods are thought to be related to where the monsoon trough is over India and to the genesis and growth of the monsoon disturbances within the trough.

The study of the summer monsoon, particularly the rainfall in India, has a long history with long range forecasts of seasonal rainfall being issued by the India Meteorological Department (IMD) since the late 19th century. A large network of rain gauge stations established after the devastating droughts of 1877 and 1899 has continued to provide a long record of rainfall over India. After the 1899 drought, Sir Gilbert Walker became the Director-General of Observatories in India and conducted systematic studies to provide advance warning of floods and droughts by searching for predictors of seasonal Indian rainfall. While establishing the correlation of the Indian rainfall with variables observed at various global locations, Walker (1923, 1924) discovered the Southern Oscillation, the North Atlantic Oscillation and the North Pacific Oscillation.

Walker's view that the Indian monsoon is a global phenomenon has been reinforced with the understanding of various phenomena, such as the El Niño, and their contribution to global climate variability. For example, the Indian monsoon is known to have a strong association with El Niño and the Southern Oscillation (ENSO) events through ocean•atmosphere interactions. Whether the monsoon influence ENSO or ENSO influences monsoon is also of considerable research interest. With similar relations with the Indian Ocean, Eurasian snow and the climate of other parts of the globe, the Indian monsoon is now understood to be an integral part of the global climate system involving coupled atmosphere-land-ocean interactions (Webster et al. 1998). It is also known that the monsoon exhibits its variability even on interdecadal time scales in association with other global climate variables.

The prediction of the monsoon rainfall and circulation is crucial for India and definitely important for other parts of the globe because of the monsoon's relation with such components. With advances made in dynamical prediction of weather and seasonal climate using general circulation models (GCMs), the simulation of monsoon by GCMs is an active area of research. According to a hypothesis by Charney and Shukla (1981), the seasonal monsoon rainfall over India has potential predictability, because it is forced by the slowly varying boundary conditions such as sea surface temperature (SST), soil moisture, sea ice and snow. However, the predictability of the Indian monsoon may depend on the relative contributions from the internal dynamics of the monsoon system and the influences of external forcings. This review is focused on the summer monsoon over India, and discusses the mean monsoon (Sect. 10.2), the variability of monsoon on intraseasonal, interannual and interdecadal time scales and their mechanisms (Sect. 10.3) and modeling the monsoon for prediction (Sect. 10.4).

10.2 Climatology.

An understanding of the mean features of the monsoon is necessary before discussing how monsoon variability occurs through strengthening or weakening of the mean conditions on different time scales. The important features of the mean monsoon are the climatological mean rainfall and circulation over India during the summer monsoon season. However, the annual cycles of other climate variables such as SST, circulation and rainfall over a large-scale region including the Indian and Pacific Oceans also seem to be important in determining the mean monsoon.

10.2.1
Annual cycle.

The annual cycle of the Indian monsoon is associated with that of a larger scale climate system that covers India and the tropical Indian and Pacific Oceans and moves from the Northern Hemisphere (NH) to the Southern Hemisphere (SH) and back following the movement of the vertical ray of sun across the equator. The climatological mean rainfall for the June-July-August (JJA) and December-January-February (DJF) seasons of the annual cycle are shown in Fig. 10.1. During the NH summer season (JJA), the rainfall is confined mostly to the north of equator with intense rainfall in India and the surrounding parts of the Arabian Sea and the Bay of Bengal. The rainfall zone extends over the Maritime Continent and across the Pacific Ocean between the equator and 10°N (Fig. 10.1a). The rainfall over the region south of the equator in the Indian and western Pacific Oceans is weak during this time. All of India and the surrounding oceanic region become dry during the NH winter (DJF) while the intense rainfall zone extends from the equator to about 15°S over the Indian and western Pacific Oceans (Fig. 10.1b). The rainfall is less intense over the central tropical Pacific Ocean north of the equator and mostly absent over the eastern Pacific.

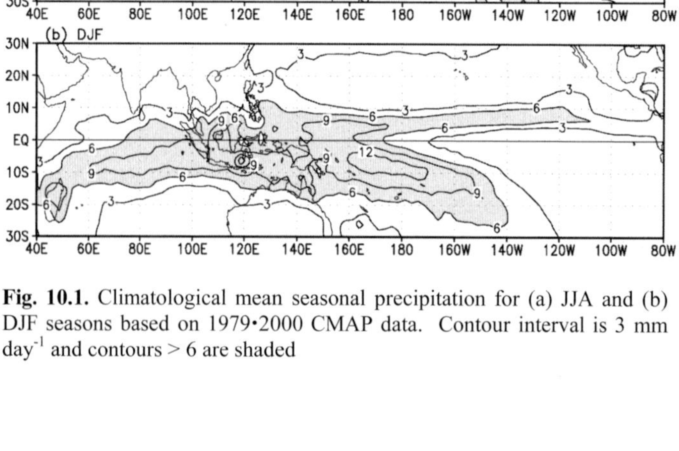

Fig. 10.1. Climatological mean seasonal precipitation for (a) JJA and (b) DJF seasons based on 1979•2000 CMAP data. Contour interval is 3 mm day^{-1} and contours > 6 are shaded

The seasonal movement of the rainfall coincides with the movement of the convective zones observed in the outgoing longwave radiation (OLR) and also indicated by the satellite imagery of reflectivity (Meehl 1987, Philander 1990). The intense convective zones are associated with the east-west intertropical convergence zone (ITCZ) over the tropical Pacific Ocean north of the equator and with the South Pacific convergence zone (SPCZ) that slopes in the northwest-southeast direction over the western tropical Pacific Ocean. By analyzing satellite imagery and the location of the monsoon trough, Sikka and Gadgil (1980) suggested that the summer monsoon rainfall system in the Indian region consists of intense low level convergence and deep moist convection and has dynamical and thermodynamical characteristics similar to the ITCZ. The ITCZ is also strong over the tropical Pacific Ocean during the NH summer. As the season changes from JJA to DJF, the intense convective activity moves toward the southeast from the Indian region to the south Pacific Ocean through the Maritime Continent. The SPCZ becomes strong during DJF while the ITCZ over the northern tropical Pacific becomes weaker. The intense convective zone moves toward the northwest by April when the ITCZ begins to strengthen.

High rainfall is well correlated with low surface pressure and ascending moist air. Strong surface winds therefore flow toward the convective areas from high pressure regions with descending dry air. The most striking feature in the Indian region is the strong southwest monsoon wind during the NH summer as seen in the JJA seasonal surface wind in Fig. 10.2a. The surface wind starts in the southern Indian Ocean as a southeasterly flow, crosses the equator and becomes a southwesterly flow in the northern tropical Indian Ocean. In fact, the monsoon wind is southwesterly from May to September and reverses direction to become northeasterly over the northern Indian Ocean, as shown in Fig. 10.2b for the DJF season, from November to March. Over the tropical Pacific Ocean, the surface wind flowing westward and equatorward is the trade wind that converges into the ITCZ. The northeasterly trade wind is more intense around 10°N and the strong southeasterly trade wind is closer to the equator during JJA (Fig. 10.2a). During DJF, the region of strong northeasterly trade wind has moved toward the equator while the convergence in the SPCZ is also quite evident (Fig. 10.2b).

A conceptual understanding of the large-scale circulation system over the tropical Indian and Pacific Ocean regions can be gained by considering the zonal and meridional components of the wind separately. For example, the trade wind over the Pacific generates an easterly surface zonal wind that forms a link between the ascending motion over the western Pacific and the descending motion over the eastern Pacific. The upper level westerly flow completes the circuit to form a cell that has come to be known as the Walker circulation. Similar Walker circulation cells can be imagined over other parts of the tropics. Similarly, the meridional component of the tropical circulation consists of ascending motion over convective regions such as the ITCZ and descending motion to the north and south. These branches can be thought to be connected by low level equatorward flow and upper level poleward flow to form a cell that is part of the zonal mean meridional circulation. This cell is sometimes referred to as the "regional" or "local" Hadley circulation. The annual cycles of rainfall and wind bring about changes in the location and intensity of the Walker and Hadley circulations.

Because of the interaction between atmosphere and ocean, the influence of the annual cycle of the SST is important in the seasonal movement of the convective zones and the rainfall and wind patterns. The seasonal mean SST patterns depicted in Fig. 10.3 show that the warm waters are co-located with strong precipitation areas over the ocean (Fig. 10.1). Warm SST occurs mostly in the NH during JJA season and mostly in the SH during DJF season. During JJA, SST is greater than 28°C over the western Pacific Ocean (where the ITCZ is located), Bay of Bengal and central Indian Ocean. The warm SST zone moves south in the following months with warmest water (SST > 28°C) occurring in the SPCZ in DJF. Based on satellite data and observed OLR, Gadgil et al. (1984) and Graham and Barnett (1987) have hypothesized that organized atmospheric convection is inhibited below a threshold SST of 28°C. When the SST exceeds the threshold, the convection is associated more with the convergence of moist air rather than evaporation. In the Indian Ocean, the region of SST greater than 28°C occurs south of the equator in January and expands to a large region (20°S-20°N) during February•April (Krishnamurthy and Kirtman 2001). The SST is warmest over most of the Indian Ocean north of 10°S when the southwest monsoon begin in India at the end of May. However, the Arabian Sea cools rapidly during JJA (Fig. 10.3), soon after the monsoon is established over India. Since the SST above the threshold is confined to the Bay of Bengal and to the south of India during the monsoon season, convection may be sensitive to both atmospheric dynamics and SST. The warm SST shifts southward during October-December passing over the region around Indonesia.

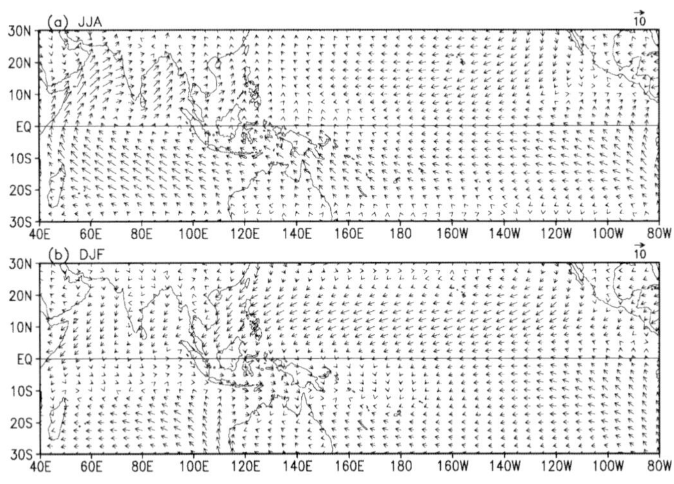

Fig. 10.2. Climatological mean seasonal surface horizontal wind for (a) JJA and (b) DJF seasons based on 1948-2000 NCEP-NCAR reanalysis data. Unit vector is 10 m s^{-1}.

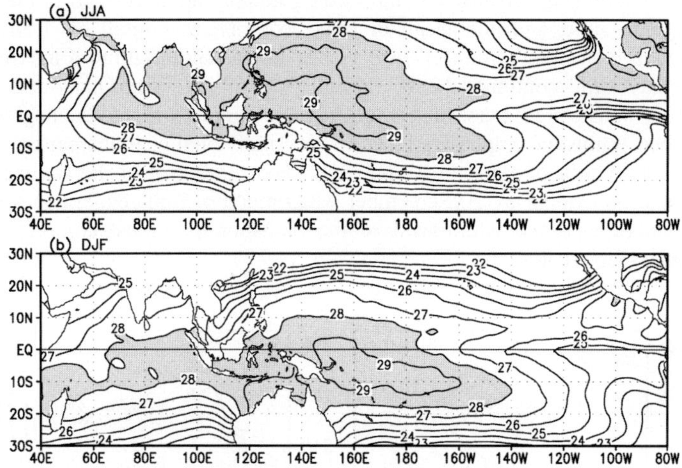

Fig. 10.3. Climatological mean seasonal SST for (a) JJA and (b) DJF seasons based on 1870-1998 period. Contour interval is 1EC and contours > 28 are shaded. The SST data set (HADISST1.1) was provided by the Hadley Center, Meteorological Office, U.K.

10.2.2 Monsoon rainfall over India.

It is well-known that most of the rainfall in India occurs during the June-July-August-September (JJAS) season. In addition to year-to-year variability, there is also large spatial and intraseasonal variability of the summer monsoon rainfall in India. The primary reason for this variability is that the rainfall is associated either with the intensification and/or displacement of the monsoon trough (or ITCZ) over northern India or with the monsoon depressions that form over the adjoining seas and move over land. A small fraction of the annual rainfall occurs over the south-eastern region during October-December winter monsoon.

The summer monsoon rainfall first arrives in the southwestern state of Kerala between late May and early June every year. The climatological mean date of onset of the monsoon over Kerala is 2 June with a standard deviation of 8 days (Mooley and Shukla 1987). Using daily rainfall data for the period 1901-70, Krishnamurthy and Shukla (2000) studied the climatology of the onset, advance, persistence and withdrawal of the summer monsoon in India by constructing daily rainfall climatology composites normalized to the climatological date of onset over Kerala. The daily rainfall climatology is shown in Fig. 10.4 for three selected dates. After the onset on 2 June, the monsoon covers the Western Ghats with heavy rainfall and with at least 3 mm day^{-1} over the southern and eastern parts of

India by 5 June (Fig. 10.4a). The monsoon gradually advances across the rest of India during June, and almost all of India experiences rainfall by 1 July (Fig. 10.4b) with considerable rainfall over central India ranging up to a maximum of 12 mm day^{-1} and continued heavy rainfall over the Western Ghats and eastern hilly areas. This pattern persists through July and August and into the beginning of September. It must be emphasized that, in each individual year, the pattern of rainfall persists with active and break periods. The withdrawal of the summer monsoon becomes evident by 20 September when much of India experiences less intense rainfall, and the withdrawal of the monsoon is almost complete by 28 September (Fig. 10.4c).

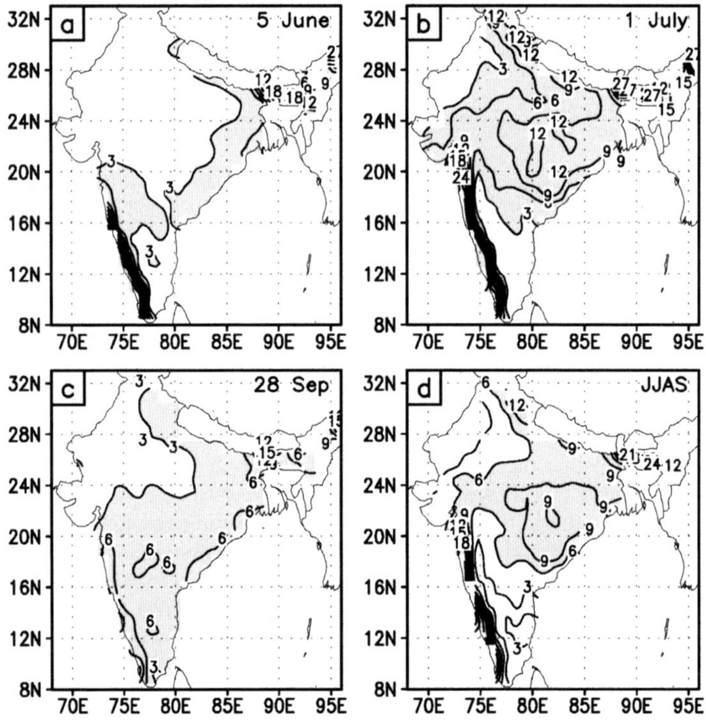

Fig. 4. Climatological mean daily rainfall for selected days showing the (a) onset (5 June), (b) persistence (1 July) and (c) withdrawal (28 September) phases of the Indian summer monsoon. The daily climatology is based on onset composites formed by shifting the onset date of each year to the climatological mean onset date (2 June) of the monsoon in Kerala. (d) JJAS climatological mean seasonal rainfall. Contour interval is 3 mm day^{-1}. Contours > 2 are shaded in (a), (b) and (c), and contours > 6 are shaded in (d). The climatologies are based on IMD rainfall data for 1901-70 period. The rainfall data were provided by Marc Michelsen

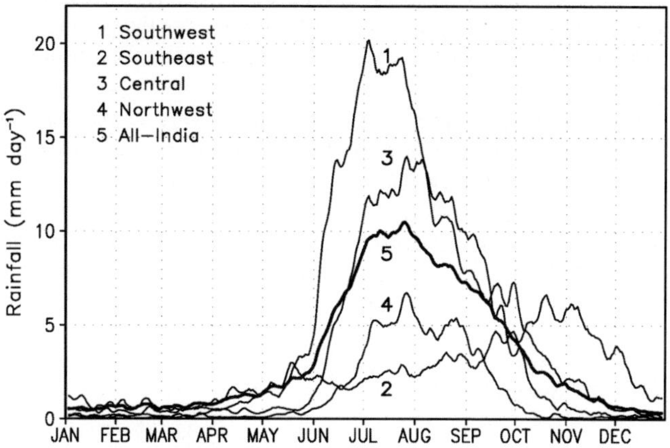

Fig. 10.5. Climatological mean daily rainfall for 5-day running mean, area averages of the (1) Southwest, (2) Southeast, (3) Central, (4) Northwest India and (5) all India regions. See the text for the definitions of the regions. The daily climatology is based on rainfall for calendar days of 1901-1970 period (reproduced from Krishnamurthy and Shukla 2000.)

The climatological mean rainfall for the entire JJAS season shown in Fig. 10.4d exhibits certain features that are similar to the daily rainfall climatology. The northeastern hilly regions and the Western Ghats are the areas of maximum rainfall in excess of 24 mm day^{-1}. The bulk of the seasonal rainfall in the rest of India occurs in central India which receives up to about 9 mm day^{-1}. The northwest and southeast regions however receive about 3 mm day^{-1} or less. The area average of seasonal rainfall shown in Fig. 10.4d over all of India including the northeast region is 7.57 mm day^{-1} (or a total of 923 mm for the JJAS season). A noteworthy feature of the seasonal climatology of rainfall is the appearance of the local maximum in central India (Fig. 10.4d).

To show that the climatological mean rainfall has considerable spatial variability, the annual cycle of the rainfall over different regions of India is presented in Fig. 10.5. Based on the geographical uniformity in the rainfall, the daily climatological rainfall has been averaged over four regions: southwest (73°-76°E, 11°-21°N), southeast (77°-80°E, 8°-16°N), central (79°-85°E, 20°-26°N), and northwest (71°-76°E, 23°-31°N) (Krishnamurthy and Shukla 2000). The all-India average is also shown. The monsoon rainfall season lasts from June to September for all the regions except the southeast. The onset of the monsoon is rapid in the southwest region (Western Ghat region) which receives the maximum amount of rainfall throughout the JJAS season, with a peak value around 20 mm day^{-1}. The central region receives about 10-15 mm day^{-1} during July and August and closely follows the all-India average. Not surprisingly, the desert region in the northwest receives very little rainfall throughout the JJAS season, with peak values only around 5 mm day^{-1}. The decline in the amount of rainfall received in the above regions is also rapid during the withdrawal of the summer monsoon in September. The southeast region receives the least amount of rainfall (less than 3 mm day^{-1}) during JJAS but receives a higher amount of rainfall from the winter monsoon during October-December.

Figure 10.1 clearly shows heavy rainfall over the Bay of Bengal and over the Arabian Sea along the west coast of India during the summer monsoon season. The rainfall over a larger region that includes India and the adjoining oceanic region was shown to vary coherently on seasonal and interannual time scales by Goswami et al. (1999). They defined a broad scale index, called the Extended Indian monsoon rainfall (EIMR), by averaging the rainfall over (70°E-110°E, 10°N-30°N). The EIMR index is expected to better represent the fluctuations of convective heating associated with the Indian monsoon. The annual cycle of the EIMR index is similar to that of all-India rainfall shown in Fig. 10.5 but with higher values during JJAS (Goswami et al. 1999).

10.3 Monsoon variability.

The Indian monsoon exhibits large variations on intraseasonal to interannual and interdecadal time scales. These are manifested in many features of the monsoon including rainfall and circulation variables. The monsoon variability on different time scales and the factors responsible for such variations are discussed in this section. The mechanisms include internal dynamics, the influences of land and ocean variability and teleconnections to climate variability in other regions. The relation between the variability on different time scales is relevant to the predictability of monsoon.

10.3.1 Intraseasonal variability.

The most remarkable character of the subseasonal variability of rainfall over India is the occurrence of active periods with high rainfall over central India and break periods with weak or no rainfall over central India and high rainfall over northern and southeastern India, each phase lasting for a few days. Ramamurthy (1969) conducted an exhaustive analysis of the daily rainfall over India and related the active periods to occur when the monsoon trough is over the northern plains and the break periods to when it is over the foothills of the Himalayas. Sikka and Gadgil (1980) have suggested that the monsoon trough waxes, wanes and fluctuates in the region north of 15°N during the peak monsoon months.

Active and break phases

Although the onset of the southwest monsoon over Kerala has a standard deviation of 8 days with extremes being 38 days apart (Mooley and Shukla 1987), the onset date is found to have no significant correlation with the subsequent seasonal rainfall over India (Bansod et al. 1991). The nature of the intraseasonal variability of the rainfall is not different during the years of strong and weak monsoons. Fig. 10.6 shows the time series of rainfall averaged over all of India for each day during JJAS months of 1942 and 1965, the years during which the seasonal mean rainfall was excess and deficient, respectively, over India. Active and break periods of rainfall are clearly evident during both years with no noticeable difference in the frequency of occurrence. In 1942, the daily rainfall is generally much higher than the long-term daily mean rainfall during the active periods, and lower than the mean during break periods (Fig. 10.6a). However, the daily rainfall in 1965 is generally below the daily climatological means with only a few days of above-normal rainfall during the active periods (Fig. 10.6b). A similar character has been revealed in the analysis of daily rainfall during 1901-70 (Krishnamurthy and Shukla 2000).

The variance of daily rainfall also has considerable spatial variation over India (Krishnamurthy and Shukla 2000). The standard deviation of daily rainfall anomalies for the JJAS season of the period 1901-70 presented in Fig. 10.7a shows large variability in central India and along the Western Ghats. The spatial character of rainfall is clearly seen in two examples of daily rainfall anomaly maps shown, one for 1 July 1942 (Fig. 10.7b) and the other for 5 August 1965 (Fig. 10.7c). On 1 July 1942, the rainfall anomalies are positive in most of India, except for the southeastern rain shadow region, with anomalies greater than 60 mm day^{-1} along the Western Ghats and about 20-40 mm day^{-1} in central India. The situation is reversed on 5 August 1965 with negative anomalies in most of India. During a typical day in a normal monsoon year or a normal day in strong or weak year, the spatial distribution of the rainfall anomalies is more varied with both positive and negative anomalies in different regions.

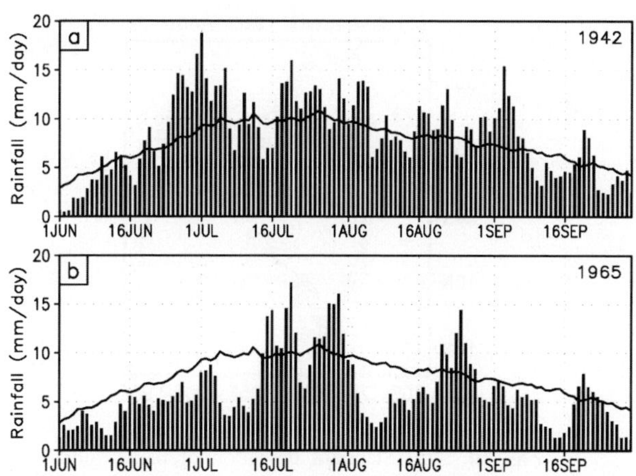

Fig. 10.6. Daily mean rainfall (mm day^{-1}) (bars) during 1 June to 30 September for (a) 1942 and (b) 1965. The daily climatological mean rainfall for 1901-70 is also plotted (solid line).

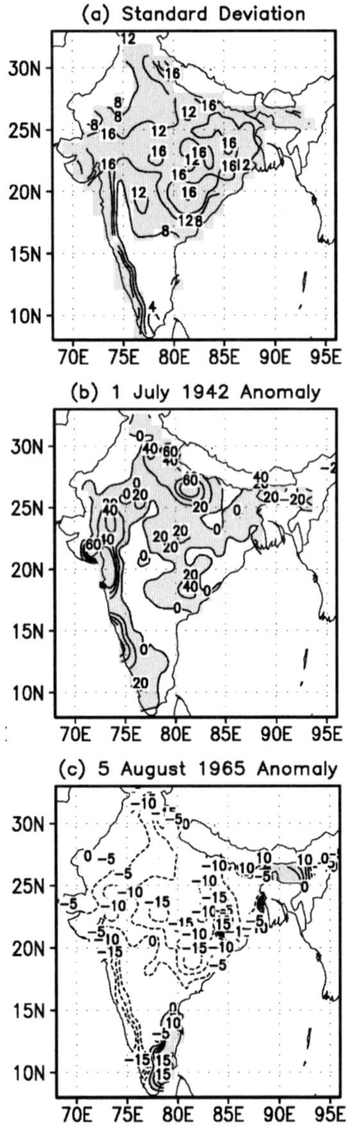

Fig. 10.7. (a) Standard deviation of daily rainfall anomalies of JJAS 1901-1970. Daily rainfall anomalies for (b) 1 July 1942 and (c) 5 August 1965. Contour intervals are 4, 20 and 5 mm day^{-1} in (a), (b) and (c), respectively. Positive contours are shaded.

The dominant spatial structure of the rainfall and wind anomalies during the active and break periods was determined by Krishnamurthy and Shukla (2000). The difference map between the active phase composite and break phase composite of daily rainfall anomalies (Fig. 10.8a) reveals that positive (negative) anomalies exist in the Western Ghats and in Central India and negative (positive) anomalies are in the foothills of the Himalayas and in southeastern India during active (break) periods. The composites in Fig. 10.8a were based on defining the active (break) periods to have rainfall anomalies beyond a certain positive (negative) threshold for 5 or more consecutive days. Since these composites include all the years during 1901-70 period, it implies that the nature of the fluctuating intraseasonal component of the rainfall remains the same during normal, weak and strong monsoon years.

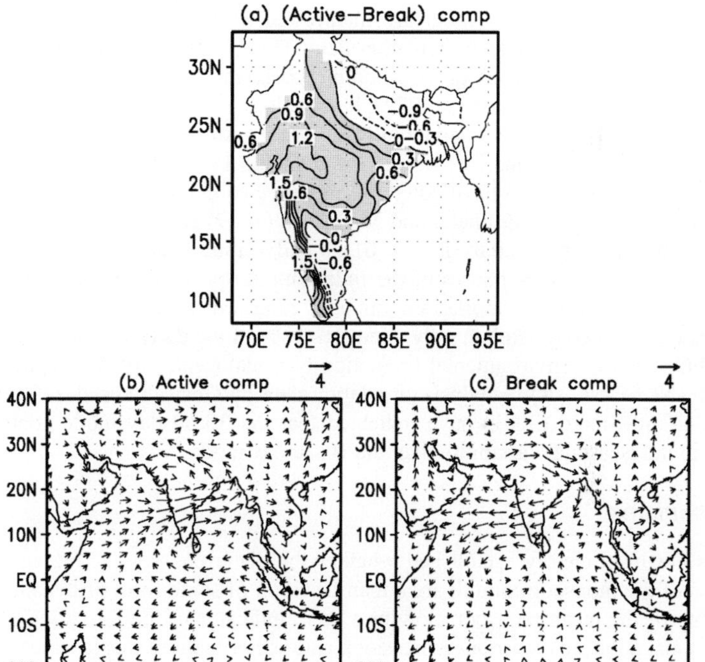

Fig. 10.8. (a) Difference between active phase composite and break phase composite of standardized daily rainfall anomalies during JJAS 1901-1970. Contour interval is 0.3 standard deviation and positive contours are shaded. (b) Active phase and (c) break phase composites of daily horizontal wind anomalies at 850 hPa during JJAS 1948-70. Unit vector is 4 m s^{-1}. (Partly reproduced from Krishnamurthy and Shukla 2000).

Based on a similarly-constructed composite of low-level horizontal circulation, Krishnamurthy and Shukla (2000) have shown that the character of the mean monsoon flow (which is similar to that in Fig. 10.2 over India and Indian Ocean region) does not change during active and break periods. The composites of horizontal surface wind anomalies in Figs. 10.8b and 10.8c show that the entire monsoon flow gets strengthened or weakened during active and break periods, respectively, of the monsoon in India. During active (break) phases, the westerlies to the north of the equator and the easterlies to the south of the equator get stronger (weaker).

Intraseasonal modes

The intraseasonal variation of the Indian monsoon is found to be dominated by fluctuations on time scales of 10-20 days and 30-60 days. These intraseasonal fluctuations are not periodic but are manifested as two dominant bands in the spectra of rainfall, OLR and circulation data. The 10-20 day fluctuation was observed in various parameters of monsoon by Krishnamurti and Bhalme (1976) and in cloudiness data by Yasunari (1979). It was also detected in a principal oscillation pattern analysis of OLR and circulation data by Annamalai and Slingo (2001).

Several studies (Yasunari 1979; Sikka and Gadgil 1980; Gadgil and Asha 1992; Annamalai, Slingo 2001) have noted the presence of the 30-60 day oscillation in convection data while Hartmann and Michelsen (1989) confirmed a spectral peak on this time scale in the analysis of 1901-70 daily rainfall data. In several of these studies, the active-break phases of the Indian monsoon rainfall have been associated with intraseasonal oscillations on both time scales. The two intraseasonal modes were shown to explain only about 10-25 % of the daily variance in the National Centers for Environmental Prediction-National Center for Atmospheric Research (NCEP-NCAR) reanalysis circulation data by Goswami et al. (1998). Annamalai and Slingo (2001) have indicated that the 30-60 day mode contributes more to intraseasonal variability than the 10-20 day mode.

Mechanisms

It has not been clearly established whether the occurrences of active and break phases of the monsoon rainfall are a manifestation of some form of dynamical instability of the mean monsoon flow or a mere indicator of the formation, growth and propagation of monsoon depressions and/or north-south displacement of the monsoon trough. The rainfall over India on a particular day is related to the synoptic scale disturbances, convergence zones and the interaction of the monsoon flow with orography. The frequency, intensity, life cycle and propagation characteristics of the monsoon disturbances determine the regional distribution of rainfall (Shukla 1987). During break periods, there is hardly any generation of monsoon disturbances. In intensity, the monsoon disturbances range from a low to a depression to a cyclone. Sikka (1980) found that there was no variation in the number of monsoon depressions between strong and weak monsoon years. However, he found more monsoon lows during strong years than in weak years and concluded that a higher number of lows means greater instability of the monsoon trough.

The spatial and temporal variations of the monsoon trough, or the tropical convergence zone (TCZ) as it is sometimes called, are related to the genesis, growth and propagation of the monsoon disturbances embedded in it. The break period in the rainfall is associated with the monsoon trough moving north and with the strengthening of the convergence zone over the south of India in the equatorial region. Two possible explanations have been provided for the transition to an active period of rainfall to occur. The revival may happen due to the formation of disturbances over the Bay of Bengal and subsequent westward movement over India. The other possibility is the northward movement of the convergence zone formed over the equatorial Indian Ocean.

Sikka and Gadgil (1980) have presented observational evidence using satellite data of cloudiness to suggest that the active phase is reestablished with the northward propagation of the oceanic convergence zone onto the Indian continent. The active-break cycles and their association with the fluctuations in the strengths of the continental and oceanic TCZs have been linked to 30-60 day intraseasonal oscillation (Yasunari 1979, Sikka and Gadgil 1980, Gadgil and Asha 1992). Some studies (Yasunari 1979, Lau and Chan 1986, Singh et al. 1992, Annamalai and Slingo 2001) have also noted eastward propagation of convection associated with the northward propagation of TCZ in the 30-60 day time-scale, implying that the Madden-Julian Oscillation (MJO, Madden and Julian 1972) has an influence on the active-break cycles of monsoon rainfall. Annamalai and Slingo (2001) found that the 30-60 mode is active during the onset phase of the monsoon but shows less coherent propagation during the subsequent summer months. The zonal propagation is found to be more robust than the meridional propagation (Goswami et al. 1998).

The 10-20 day mode is found to contribute less than the 30-60 day mode to the total intraseasonal variability as revealed in their relative contributions to the convective heating (Annamalai and Slingo 2001). The 10-20 day mode has been associated with westward propagating disturbances originating in the west Pacific (Krishnamurti and Ardunay 1980). These westward propagating events are also seen to be responsible for the reestablishment of the active monsoon phase after a break phase by the genesis of the disturbances within the monsoon zone. Annamalai and Slingo (2001) have suggested that the 10-20 day mode is more regional in character with influence over the local Hadley circulation.

10.3.2 Interannual variability.

Although the summer monsoon over India occurs regularly during JJAS, the year-to-year variation of the seasonal mean monsoon is quite considerable and has a major impact on India. The interannual variability of the seasonal monsoon is nonperiodic, and may result from the inherent atmospheric dynamics that is nonlinear. The variability of a nonlinear system can be nonperiodic even if the forcing is constant. The internal atmospheric dynamics includes instabilities in the form of synoptic scale disturbances, nonlinear interactions among various scales of motion, topographic and thermal forcings. Shukla (1987) has provided a discussion of some these processes involved in the monsoon variability.

The interannual variability of the monsoon can be further influenced by the slowly varying forcings such as SST, soil moisture, sea ice and snow at the surface (Charney and Shukla 1981). These global boundary forcings can modify the location and intensity of heat sources and circulation such as Hadley and Walker circulations in the tropics. Understanding the influences of these boundary forcings is important to establish the relation between the monsoon variability and other global climate variability. It is also known that the monsoon may have teleconnections with the climate of remote locations such as Africa and the Atlantic Ocean. The strength of the seasonal monsoon in a particular year may depend on the relative contributions from the internal dynamics and external forcings.

Variability of summer monsoon

The most widely used measure of the intensity of the planetary scale monsoon over India is the seasonal mean rainfall spatially averaged over India. The Indian monsoon rainfall (IMR) index is defined as the area-weighted average of rainfall observed at well-distributed rain gauge stations all over India. A long record of the monthly mean and seasonal mean IMR index has been prepared by Parthasarathy et al. (1994, 1995) on the basis of observations at a network of 306 homogeneously distributed stations. Although about 2000-5000 rain gauge stations have existed all over India since the late 19th century, Parthasarathy et al. (1994) carefully selected the 306 stations covering each district (a small administrative area) in the plains regions of India. Appropriate area weights have been assigned to the rainfall at each station.

The time series of JJAS seasonal mean IMR index for the period 1871-1999 is presented in Fig. 10.9. The long-term mean of the JJAS IMR index for the period 1871-1999 is 852 mm (about 7 mm day^{-1}), and the standard deviation for this period is 83 mm (about 0.7 mm day^{-1}) or about 10% of the long-term mean.

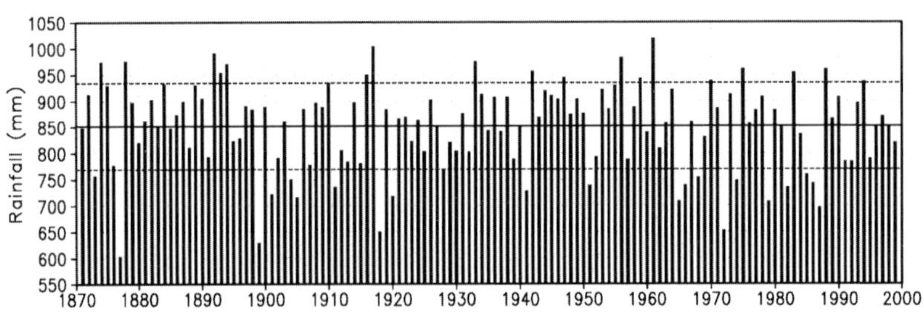

Fig. 10.9. Time series of JJAS seasonal rainfall (mm) area averaged over all-India for 1871-1999 (bars). The horizontal line represents the mean (852 mm) for 1871-1999 and the dashed lines represent the mean standard deviation (83 mm). The rainfall data were obtained from the Indian Institute of Tropical Meteorology, Pune, India.

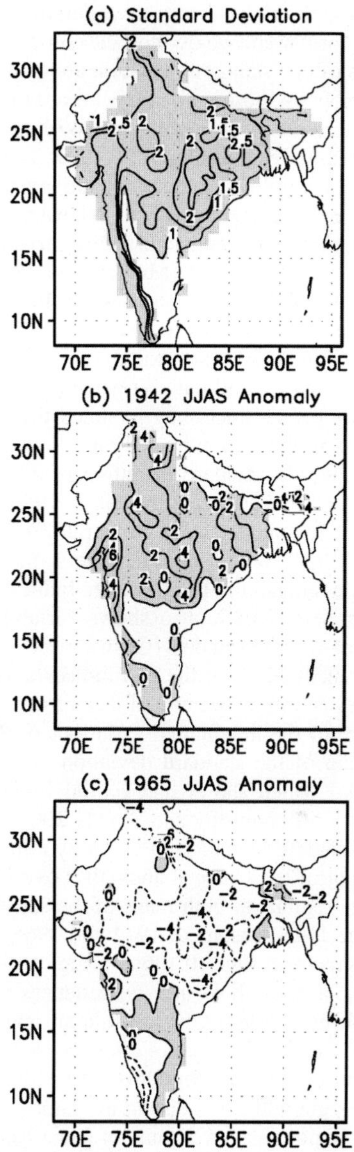

Fig. 10.10. (a) Standard deviation of JJAS seasonal rainfall anomalies for 1901-1970, and JJAS seasonal rainfall anomalies for (b) 1942 and (c) 1965. Contour intervals are 0.5, 2 and 2 mm day^{-1} in (a), (b) and (c) respectively. Positive contours are shaded.

The IMR shows considerable year-to-year variation with quite a few years having substantial above-normal and below-normal rainfall. Usually, years with seasonal rainfall in excess of one standard deviation above the long-term mean are referred to as flood years while those with rainfall more than one standard deviation below the mean are referred to as drought years. With this criterion, there were 18 flood years and 21 drought years during 1871-1999. There are a few instances of drought year or flood year following the other, the most notable being 1987 and 1988 which also happen to coincide with El Niño and La Niña events, respectively. There is a strong biennial component in the variability of IMR with a broad peak around 2 years in its spectrum (Mooley and Parthasarathy 1984, Meehl 1987). The biennial variability and the connection between the Indian monsoon and ENSO will be discussed in later sections.

Krishnamurthy and Shukla (2000) prepared a time series of IMR for the period 1901-70 based on a gridded rainfall data set generated from observations at more than 3700 stations over India. Since they included the rainfall over the hilly regions of northeast India in the area averaged rainfall, the long term mean of their JJAS IMR is 923 mm with a standard deviation of 87 mm, both values slightly higher than those obtained by Parthasarathy et al. (1994). However, the time series of IMR prepared by Krishnamurthy and Shukla (2000) closely resembles the time series in Fig. 10.9 for 1901-70 and identifies the same drought and flood years. The EIMR index that averages the rainfall over India and adjoining oceanic regions, as discussed in Sect. 10.2.2, also shows variability of its JJAS seasonal mean similar to that of the IMR index (Goswami et al. 1999). However, there were noticeable differences between the two indices during 1986-90. The length of the EIMR time series is limited because of lack of rainfall data over the ocean.

The spatial structure of the interannual variations of the Indian rainfall can be summarized by the map of the standard deviation of the JJAS seasonal rainfall anomalies (Fig. 10.10a) which shows that central India and the Western Ghats have large variability (Krishnamurthy and Shukla 2000). A comparison with Fig. 10.7a shows that the standard deviation of the daily rainfall anomalies is about 7-10 times larger than that of the seasonal anomalies over most of India. Large values appear in the same regions in both maps of the standard deviations. The seasonal anomalies for the JJAS seasons of 1942 and 1965, two years with contrasting IMR values, are shown in Fig. 10.10b and Fig. 10.10c, respectively. During the flood year of 1942, almost all of India experiences excess rainfall, especially over central India and the Western Ghats, while an opposite structure with deficient rainfall is seen in the drought year of 1965. Such spatial coherence may not, however, be present during normal years. It is not uncommon to have different regions experiencing drought and flood during the same year. The variability of the rainfall over different regions (subdivisions) of India has been discussed in detail by Shukla (1987). The dominant pattern of the interannual variability is obtained by examining the composites of JJAS seasonal rainfall anomalies for strong (flood) and weak (drought) monsoon years (Krishnamurthy and Shukla 2000). The difference between the strong and weak composites of rainfall anomalies presented in Fig. 10.11a shows that positive (negative) anomalies cover all of India during flood (drought) years with fairly uniform values. The corresponding strong and weak monsoon composites of the horizontal surface wind anomalies for the JJAS season are shown in Fig. 10.11b and Fig. 10.11c, respectively. There is not much difference in the mean surface winds between strong and weak monsoon years (not shown); however, as seen in Fig. 10.11b and Fig. 10.11c, the anomalies strengthen (weaken) the seasonal mean monsoon flow during flood (drought) years.

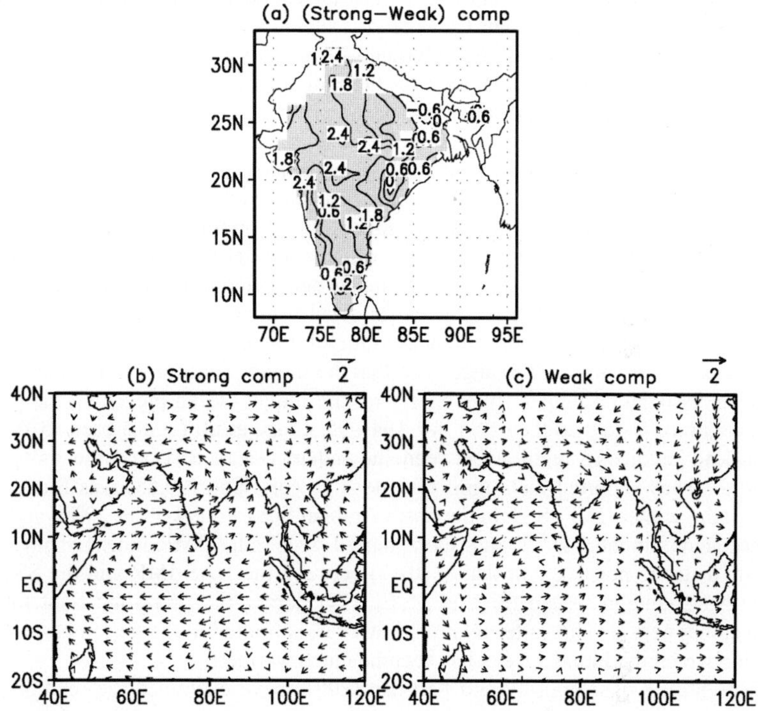

Fig. 10.11. (a) Difference between composites of strong years and weak years of standardized JJAS seasonal rainfall anomalies during JJAS 1901-1970. Contour interval is 0.6 standard deviation and positive contours are shaded. (b) Strong years and (c) weak years composites of JJAS seasonal horizontal wind anomalies at 850 hPa during JJAS 1948-70. The unit vector is 2 m s^{-1}.

Influence of SST

Because of the interaction between atmosphere and ocean, the SST of the Indian and Pacific Oceans may influence the variability of the Indian monsoon, and in turn, the monsoon winds and rainfall may affect the variability of SST of the oceans. This mutual interaction introduces the possibility that the monsoon and the oceans form a coupled climate system (Webster and Yang 1992).

Indian Ocean. Even when the available SST data were sparse, earlier observational studies (Ghosh et al. 1978, Cadet and Reverdin 1981) examined the role of the Indian Ocean and estimated the relative magnitudes of the local evaporation in the Arabian Sea and the cross-equatorial moisture flux in contributing to the moisture flux across the west coast of India during the monsoon. Studies disagreed in the relative estimates. In a sensitivity study with a GCM, Shukla (1975) found that the monsoon rainfall over India was significantly reduced with cold SST over the Arabian Sea. In a more recent GCM study, Shukla and Fennessy (1994) showed that the annual cycle of SST in the Indian Ocean plays an important role in establishing the rainfall and circulation over the Indian monsoon region. Using observations along a ship track, Shukla and Misra (1977) found weak positive correlation between the SST over the Arabian Sea and the seasonal mean rainfall over the west coast region of India. A later analysis by Shukla (1987) using a longer SST record over a larger region of the Arabian Sea showed that above-normal (below-normal) monsoon rainfall followed warm (cold) SST anomalies but cautioned that the predictive value of this relationship was limited because of the small magnitude of SST anomalies. However, his analysis obtained a more robust relation that excess (deficient) monsoon rainfall is followed by large negative (positive) SST anomalies in the Arabian Sea with peak values during October-November.

Rao and Goswami (1988) suggested that the Indian monsoon rainfall has a significant association with warmer SST over the Indian Ocean preceding the monsoon season. Using gridded SST for 1945-95, Clark et al. (2000) tried to find a long-lead predictive relation between the Indian Ocean SST and the monsoon rainfall. They reported that significant positive correlation (0.53) was found between the winter (DJF) SST in the north Arabian Sea (around 66°E, 20°N) and the subsequent summer (JJAS) monsoon rainfall over India (IMR index). They also found a strong positive correlation (0.87) between the JJAS IMR and the SST in the central Indian Ocean (around 86°E, 4°N) during the preceding September-November period. This latter relation, however, is limited to the period after 1976 during which the Indian Ocean has become much warmer.

Recently, Krishnamurthy and Kirtman (2001) have examined the relation between the monsoon and the variability of the Indian Ocean using long records of SST, IMR and circulation data. Based on their analysis, the point correlation between the JJAS seasonal anomaly of the IMR index and the SST anomalies in the Indian and Pacific Oceans during four successive seasons are presented in Fig. 10.12. The correlations with the Pacific Ocean will be discussed in the next subsection. During the March-April-May (MAM) season preceding the summer monsoon, much of the significant but weak positive correlations in the Indian Ocean are found in the SH (Fig. 10.12a). There is no significant correlation of IMR with the JJA SST (Fig. 10.12b) although a faint signature of cooling associated with positive rainfall anomalies is seen in the Arabian Sea. The negative correlation of IMR with SST in most of the Indian Ocean west of 90°E is significant during September-October-November (SON) season following the monsoon, while there is also a positive correlation to the east of 90°E in the Southern Indian Ocean. IMR and the entire Indian Ocean are negatively correlated during the subsequent DJF season. If the Indian Ocean is considered in isolation, the contrasting east-west correlation during the SON season may be related to the so-called "dipole mode" of the SST (e.g., Saji et al. 1999, Webster et al. 1999). Krishnamurthy and Kirtman (2001) have shown that the evolution of the monsoon wind plays an important role in bringing the changes in the Indian Ocean SST through local air-sea interactions and that the evolution of the SST in the Indian Ocean may be related to that of the Pacific Ocean. An analysis of the satellite data over the Indian Ocean and the monsoon region by Gautier et al. (1998) has shown that the monsoon strongly affects the air-sea interactions over the Indian Ocean by inducing a strong exchange of heat between air and sea.

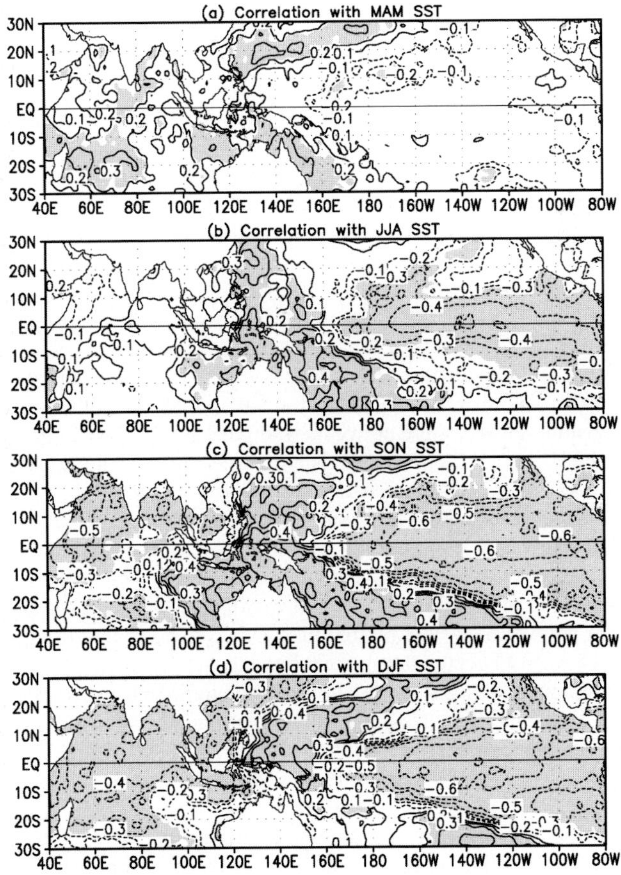

Fig. 10.12. Point correlations of JJAS IMR seasonal anomaly with SST seasonal anomalies of (a) preceding MAM, (b) simultaneous JJA seasons, (c) succeeding SON and (d) succeeding DJF seasons for 1871-1998. Shaded areas represent correlation above 99% confidence level.

Pacific Ocean. The correlation between the Indian monsoon rainfall and the Southern Oscillation, discovered by Walker (1923, 1924), long ago established the link between the monsoon and the climate of the Pacific Ocean region. However, much later than it was established that El Niño and Southern Oscillation are the oceanic and atmospheric manifestations of the same phenomenon (e.g., Bjerknes 1969), the tropical Pacific Ocean SST became known to be related to the Indian monsoon. In fact, the Indian monsoon rainfall is understood to have a stronger relation with the Pacific Ocean SST than with the Indian Ocean SST. Using observed data for the period 1875-1975, Sikka (1980) showed that there is a tendency for the failure of the summer monsoon rainfall to occur during El Niño years that were mainly identified from Line Island precipitation index. By identifying the El Niño years from observed Pacific SST data, Rasmusson and Carpenter (1983) also showed a similar relation with about 66% of the El Niño years accompanying drought over India during 1875-1979.

A significant correlation (-0.62) between the summer monsoon rainfall and the Pacific SST anomalies of the following December-January period was obtained by Angell (1981). It is now well-known that a significant number of La Niña (cold events in the east Pacific) years are associated with strong monsoon or flood years. The association between the tropical Pacific events and the Indian monsoon rainfall are clearly seen in Fig. 10.13 in which the JJAS seasonal anomaly IMR index is plotted along with the Niño-3 index (SST anomalies averaged over 150°W-90°W, 5°S-5°N) of the following DJF season. The Niño-3 index is a good indicator of the El Niño and La Niña events. The correlation between IMR and Niño-3 indices is -0.58 for the period 1871-1998. During this period, while there are a significant number of El Niño (La Niña) events coinciding with weak (strong) monsoon rainfall, there are years when droughts (floods) occur without El Niño (La Niña). Some El Niño (La Niña) years are also not accompanied by droughts (floods) over India (Fig. 10.13).

The reason for choosing the DJF instead of JJAS season for plotting the Niño-3 index in Fig. 10.13 is to point out the varying nature of the relation between IMR and Niño-3 indices with season, as shown, for example, by Goswami et al. (1999). The lag correlation of monthly Niño-3 index with the JJAS seasonal anomaly IMR index for the period 1871-1998 plotted in Fig. 10.14 shows that the strongest correlation (-0.58 to -0.6) occurs during October-January following the monsoon season. The simultaneous (lag 0) correlation is also significant but smaller in magnitude (-0.48). However, there is almost no correlation between IMR and the Niño-3 of preceding seasons. Using the Niño-3.4 index (SST anomalies averaged over 170°W-120°W, 5°S-5°N), covering a region slightly to the west of the Niño-3 region, also gives the same kind of lagged relation with the IMR (Fig. 10.14). A similar behavior is also seen in the lag correlation of the Southern Oscillation Index (SOI, i.e., normalized Tahiti-Darwin sea level pressure) with the IMR but with sign opposite to that with Niño-3 index (Fig. 10.14). Shukla and Paolino (1983) obtained a similar lagged relation between IMR and sea level pressure (SLP) at Darwin.

The spatial structure of the point correlation of the JJAS IMR time series with the tropical Pacific SST for different seasonal lags is provided in Fig. 10.12. The

MAM SST anomalies over the entire tropical Pacific have almost no correlation with the subsequent season's monsoon rainfall (Fig. 10.12a). Moderate negative values in the central and eastern tropical Pacific and weak positive values in the western Pacific are present in the simultaneous correlation (Fig. 10.12b). The SON SST (following the monsoon season) is significantly correlated with the IMR with strong negative correlation in the central and eastern Pacific and western Indian Ocean (Fig. 10.12c). Positive correlation, in a horse shoe pattern similar to the El Niño SST anomaly pattern, is present in the western Pacific and part of the eastern Indian Ocean. During DJF following the monsoon season, the negative correlation over the central and eastern Pacific and the positive correlation over the western Pacific slightly diminish while the entire Indian Ocean becomes negatively correlated (Fig. 10.12d).

Based on a similar analysis, Krishnamurthy and Kirtman (2001) have suggested that the Indian monsoon plays an important role in the combined evolution of the SST anomalies over the Indian and Pacific Oceans. They showed that the dominant combined variability of the Indian and Pacific Oceans consists of cold (warm) SST anomalies in the in the western Pacific that extends into the eastern Indian Ocean during July-November and warm (cold) SST anomalies to the east (central and eastern Pacific) and west (Indian Ocean). A similar evolution has also emerged in a different kind of analysis by Huang and Kinter (2001).

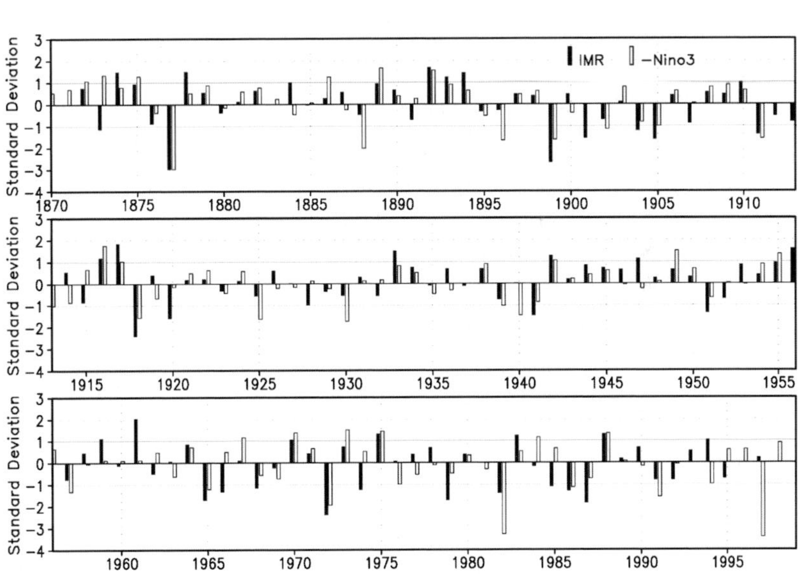

Fig. 10.13. Time series of JJAS seasonal anomaly of IMR and DJF seasonal anomaly of Niño-3 index for 1871•1999. Niño-3 index plotted in negative sign for easy comparison with IMR. Units are in standard deviations.

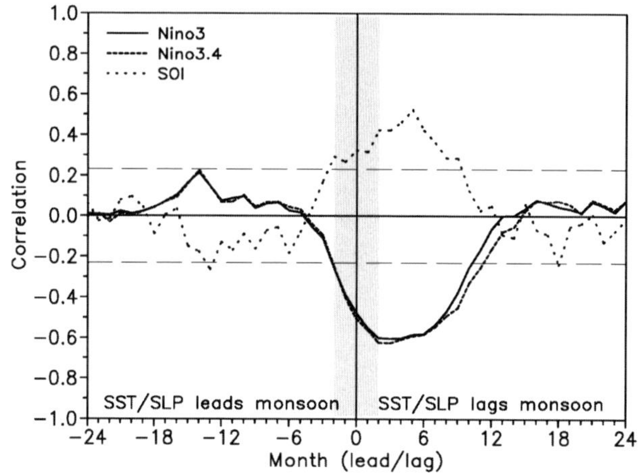

Fig. 10.14. Lag correlation of monthly anomalies of Niño-3 index (solid), Niño-3.4 index (dashed) and SOI (dotted) with the JJAS seasonal anomaly of IMR for the period 1871•1998. The shaded area represents the JJAS season of the monsoon rainfall and the long dashed lines represent the 99% confidence level. The SOI time series was obtained from NCEP.

The spatial patterns of the rainfall over India associated with the eastern tropical Pacific SST at different seasonal lags are shown in Fig. 10.15, which contains the point correlation of the Niño-3 index of three successive seasons with the JJAS seasonal rainfall anomalies over India. The MAM Niño-3 is found to have no significant correlation with the subsequent JJAS rainfall over almost entire India (Fig. 10.15a). The JJAS Niño-3 index shows significant simultaneous negative correlation with the rainfall over the western part of India (Fig. 10.15b). The JJAS rainfall over most of India west of about 82°E has a slightly stronger negative correlation with the following DJF Niño-3, while the eastern part of India is not correlated (Fig. 10.15c).

Studies so far have suggested that an apparent eastward shift in the Walker circulation during El Niño events is associated with reduced rainfall over India. This shift, in turn, supposedly affects the so-called local Hadley circulation in its location over the monsoon region and its effect on the rainfall. Different scenarios involving shifts in the Walker and Hadley circulations have been put forward to explain the occurrence or nonoccurrence of drought during El Niño years (e.g., Ju and Slingo 1995, Goswami 1998, Slingo and Annamalai 2000, Lau and Wu 2001). Goswami (1998) has hypothesized that the eastward shift in the Walker circulation during El Niño episodes results in increased low level convergence and more frequent formation of the TCZ over the equatorial Indian Ocean. The enhanced precipitation over the Indian Ocean is linked to subsidence over India by means of the local Hadley circulation.

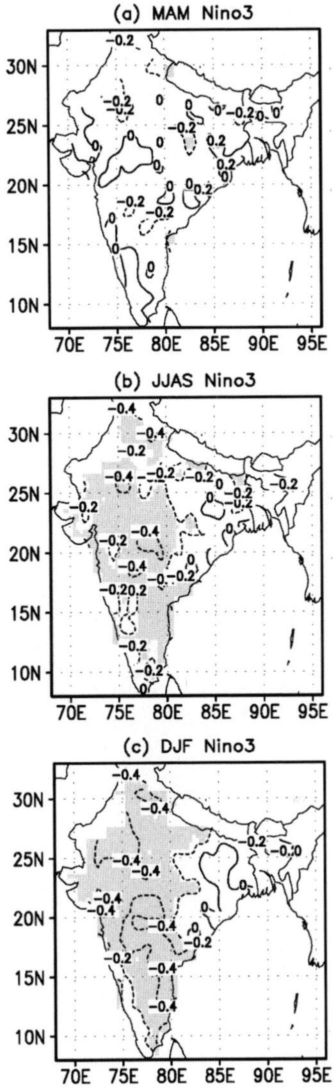

Fig. 10.15. Point correlations of Niño-3 index time series with the JJAS seasonal anomalies of rainfall over India for 1901•1970. The Niño-3 index time series are the anomalies of (a) MAM season preceding the monsoon, (b) JJAS season simultaneous with the monsoon and (c) DJF season succeeding the monsoon. Shaded areas represent correlation above 95% confidence level.

On the other hand, Slingo and Annamalai (2000) argued that the convection was more markedly suppressed over the Maritime Continent and the equatorial Indian Ocean during the 1997 El Niño event and led to the formation of a TCZ over the Indian continent due to the local Hadley circulation. Most of these studies emphasize the role of the local Hadley circulation that may be related to the internal dynamics of the monsoon system and suggest that relative contributions of the effects related to the changes in the Walker circulation (in turn related to El Niño and Pacific SST) and the local Hadley circulation may determine the occurrence or nonoccurrence of droughts during El Niño years. Relative subtle changes in the location and extent, both in zonal and meridional directions, of the local Hadley circulation my influence the extent of rainfall or the lack of it over India.

The possible influence of monsoon on ENSO has also been explained in terms of shifts in tropical wind systems and ocean-atmosphere interaction. Barnett (1983) found a strong interaction between observed monsoon wind and the Pacific trade wind system on the interannual time scale. He suggested that these interactions result from pulsations and an eastward shift of the low level convergence in the region of the Indian and Pacific Oceans and are strongly related to the subsequent changes in the SST anomalies over the central and eastern Pacific Ocean related to El Niño events. Barnett (1984) expanded this view to nearly the entire tropical belt. Using a linear model for diagnostic analysis, Nigam (1994) concluded that the anomalous monsoon rainfall over Asia and the Indian Ocean forces only modest surface westerlies over the tropical Pacific and contributes to the development of ongoing El Niño events.

In a model study, Kirtman and Shukla (2000) reproduced the observed lead-lag relation between the monsoon and ENSO and also found that a strong (weak) monsoon impacts ongoing ENSO events three to six months later either by enhancing (reducing) the magnitude of a cold event or by reducing (enhancing) the magnitude of a warm event. Coupled air-sea interactions in the Pacific were found be crucial in amplifying the effects of the monsoon. They found the uncoupled and contemporaneous effect of the monsoon forcing on the Pacific to be small. In another study, adopting an innovative coupling strategy in an ocean-atmosphere coupled model, Kirtman and Shukla (2001) have concluded that the observed lead-lag relation between the Indian monsoon and ENSO cannot be reproduced without the ocean-atmosphere interactions and that the monsoon variability has an impact on the timing and evolution of ENSO events.

Tropospheric biennial oscillation

The spectral peak in the 2-3 year period found in the Indian monsoon rainfall by Mooley and Parthasarathy (1984) has been linked to an oscillation in the tropics called the tropospheric biennial oscillation (TBO). The existence of TBO has been extensively discussed in the research literature (Trenberth 1975,1976a,b, 1980, Brier 1978, Nicholls 1978, 1979, 1984, Van Loon 1984, Van Loon and Shea 1985, Meehl 1987, 1993, 1994a, 1997, Kiladis and Diaz 1989, Rasmusson et al. 1990, Yasunari 1985, 1989, 1990, 1991, Barnett 1991; Ropelewski et al. 1992, Goswami 1995, Terray 1995, Clarke et al., 1998, Clarke and van Gorder 1999).

Rasmusson et al. (1990), in reviewing the biennial oscillations that had been documented until that time, noted that ENSO has two modes of variability, a low-frequency mode and a biennial mode.

Two physical mechanisms have been proposed for the biennial oscillation in the troposphere and near-surface ocean. Meehl (1987, 1993) proposed that the annual northwest (south Asia) to southeast (Australia) march of the region of strong convection and heavy monsoon rainfall could be associated with a biennial oscillation. Meehl's mechanism proceeds as follows. The air-sea interaction is stronger than normal in the warm pool region of the far western Pacific and eastern Indian Ocean when the SST is warmer than normal. When the strongest convection reaches that region, the strong air-sea interaction reduces the SST to below normal. In the following year, when the region of strong convection again reaches the warm pool region, the lower-than-normal SST is associated with weaker-than-normal air-sea interaction and the convection returns the SST to above-normal values. This leads to an alternation from warm to cold SST and back again every two years.

Nicholls (1978, 1979, 1984) proposed an alternative view based on the fact that the mean winds along the equator north of Australia are westerly in boreal summer and easterly in boreal winter. He assumed that the tendency of anomalous pressure over Darwin, Australia is linearly related to the anomalous warm pool SST (with opposite sign). He further assumed that the tendency of the warm pool SST is linearly related to the Darwin pressure anomaly with positive sign during summer and negative sign during winter. Superimposing a positive wind anomaly on westerly mean flow leads to stronger than normal wind speed and stronger-than-normal ocean mixing, thereby cooling the SST. In contrast, a negative wind anomaly would, by the same reasoning, lead to warmer-than-normal SST. By casting these relationships in a set of simple equations with coefficients determined from a statistical analysis of the observations, Nicholls was able to derive a biennial oscillation.

Influence of snow

Charney and Shukla (1981) suggested that the low frequency variations of the Earth's surface can predispose the Indian monsoon toward a wet or a dry state, thereby providing a hypothetical basis for the predictability of the monsoon. Among the surface boundary conditions that were supposed to have some influence on monsoon predictability, the snow mass on the Eurasian continent has been under consideration for some time since Blanford (1884) formulated the hypothesis that the snow accumulation during the Himalayan winter and spring is inversely related to the subsequent summer monsoon rainfall over India. Hahn and Shukla (1976), with corroboration from Dey and Bhanu Kumar (1982, 1983), Dickson (1984) and Dey et al. (1985), showed that the inverse relationship exists in available observations of snow cover. The underlying physical mechanism was thought to be a reduction (enhancement) of the typical land-sea thermal contrast and latent heat release that drives the monsoon (Webster and Chou 1980. Fein and Stephens 1987. He et al. 1987. Hastenrath 1988) through the albedo and thermal insulation effects of heavier (lighter) than normal snow cover. Greater (less) than normal snow accumulation in winter and spring would also lead to greater (less) than normal early summer snow melt water that would increase (decrease) the soil moisture and evaporation from the surface keeping the surface cooler (warmer) than normal and likewise reducing (increasing) the land-sea thermal contrast. Shukla (1987) also pointed out that a heavier than normal snow accumulation in the winter and spring preceding a monsoon season would delay the development of the monsoonal temperature gradient by reflecting more solar energy (albedo effect) and by using more of the insolation energy to melt the anomalous snow.

These hypotheses were examined in several studies recently. Li and Yanai (1996) described the relationship between the onset of the Indian monsoon, as well as its interannual variability, and land-sea temperature contrast. They found that snow cover over Tibet and the Himalayas has an influence on both the onset and the interannual variability of monsoon rainfall. Matsuyama and Masuda (1998) suggested that there may be a linkage between the soil moisture in the extratropical Eurasia and the Indian monsoon. They found that the enhanced thermal inertia of anomalously high soil moisture in the former Soviet Union could reduce the thermal contrast between the Eurasian land mass and the Indian Ocean and thereby reduce the intensity of the monsoon. Bamzai and Shukla (1999) showed that there is a statistically significant relationship between the snow cover (and probably snow depth) over the northwestern part of Eurasia and the Indian monsoon. This may be related to variations in the climate of the north Atlantic (Chang et al. 2001, see also the comments about teleconnections in this section).

Several GCM studies have been conducted to examine the snow-monsoon hypothesis (Barnett et al. 1989, Vernekar et al. 1995, Douville and Royer 1996, Bamzai and Marx 2000). All the model results indicate that there is a snow effect on the monsoon, but it is subtle, and there is disagreement over whether it is Himalayan or the Northwest Eurasia snow pack that has the most influence. The model results are all vulnerable to the criticism that model error may overwhelm the relatively small signal that is detected.

Influence of land surface

The mean annual cycle of the Indian monsoon has long been held to be associated with the surface temperature contrast between the Indian Ocean and the rapidly warming Asian continent in late spring and summer (e.g., review by Webster 1987). It has also been suggested that the interannual variations of the Indian monsoon may be associated with fluctuations of the surface temperature and other surface conditions, notably soil moisture (e.g., Walker and Rowntree 1977, Shukla and Mintz 1982, Yeh et al. 1984, Delworth and Manabe 1989). As pointed out by Meehl (1994b), the effect of soil moisture variability may be either a positive feedback or a negative one. The positive feedback can occur through the increased (decreased) availability of moisture for convection when the soil is wetter (drier) than normal. The negative feedback can arise as follows. An increased (decreased) soil moisture can increase (decrease) evaporation which can cool (warm) the surface temperature and thereby decrease (increase) the temperature contrast that drives the mean monsoon.

Unfortunately, due to the subtlety of the signal, there are insufficient data to quantitatively distinguish the positive and negative feedbacks to determine which is dominant and under what conditions the effect is comparable to the effect of SST variations. Several modeling experiments have been conducted, most of which have found the soil moisture effect to be detectable but relatively less important than other effects. In extreme, idealized numerical experiments, Shukla and Mintz (1982) and Sud and Smith (1985) reached opposite conclusions, the latter finding that the large-scale moisture flux convergence in the atmosphere completely overwhelmed any signal attributable to surface water balance considerations. Ferranti et al. (1999) likewise found that surface conditions had a lesser impact on monsoon fluctuations than the internal atmospheric dynamics, but they noted that the low frequency behavior could be affected due to the inertia of soil moisture storage. Dirmeyer (1999) and Douville et al. (2001) confirmed the basic result of Sud and Smith (1985) with more sophisticated models and more realistic specification of soil moisture. Several attempts to include the soil moisture effect for seasonal prediction have met with modest success (e.g., Fennessy and Shukla 1999, Becker et al. 2001). Most model experiments suffer from the inadequacies of the models that make it difficult to determine to what extent the results are model-dependent.

Influence of topography

There are two principal effects on the Indian monsoon that may be ascribed to the elevated topography in south Asia. First, the very large massif of the Himalayas presents a major obstacle to the flow of the atmosphere, both the prevailing westerlies of the extratropics north of the massif and the southerly flow that might be associated with outflow from the convective systems in the tropical latitudes. Because the mountains effectively block the low-level and mid-level flow of the atmosphere, the circulation is significantly altered from what it would be with no mountains present. Second, the high elevation of the Tibetan plateau in the middle of the Himalayas is strongly heated in summer by insolation, creating a mid-tropospheric heat source. This elevated sensible heat source and the radiative cooling in the surrounding environment maintain a strong temperature gradient that drives a vigorous ascent in the region with strong descent in the surrounding regions. This local circulation can interact with the convectively-driven circulation in the monsoon region, influencing the onset of the monsoon as well as its development across the subcontinent and its eventual withdrawal.

Hahn and Manabe (1975) showed, through numerical experiments in which they attempted to simulate the mean Indian monsoon with a general circulation model. In one experiment, they included a realistic representation of the mountains and, in another, the model did not have any mountains. They found that the presence of the Himalayan mountains effectively maintains the Asian low pressure system in summer and also allows high temperatures on the Tibetan plateau to extend low pressure toward the southwest. The net effect is to strengthen the

Indian monsoon and produce high rainfall well into the subcontinent through both the mechanical/dynamical and thermodynamical mechanisms. Both these effects were also found in a GCM study by Zheng and Liou (1986) and later verified in an analysis of observations made during the FGGE experiment by He et al. (1987), generalized to tropical climates worldwide by Meehl (1992) and again found in FGGE data by Yanai and Li (1994). Interestingly, geologic evidence strongly suggests that the climate of south Asia changed significantly in the late Cenozoic era at about the time that the Himalayan uplift was occurring (Ruddiman et al. 1989). Atmospheric modeling experiments (Kutzbach et al. 1989, Ruddiman and Kutzbach 1989) confirmed the role of topography, primarily the elevated heating effect, in changing the climate of south Asia from one with little or no monsoon to one in which a vigorous monsoon occurs each summer as is observed today.

Teleconnections to the NAO

As noted before on the relationship between Eurasian snow and the Indian monsoon, there is a hypothesis that anomalous snow cover and/or depth can alter the albedo and surface wetness characteristics of the Asian land mass in the spring season in such a way as to influence the land-sea thermal contrast that drives the Indian monsoon and thereby introduce variability of the monsoon.

Taking one step back from this hypothesis, it is reasonable to ask how Eurasian snow cover and depth anomalies can arise that are sufficiently large-scale and persistent to affect the monsoon in the subsequent summer. It is well known that the climate regimes over western Europe undergo substantial low frequency variability as a result of an oscillation called the North Atlantic Oscillation (NAO; first identified by Walker 1924, defined by Rogers 1984, and reviewed by Lamb and Peppler 1987). In one phase of the NAO (negative), the north Atlantic winter atmospheric circulation is dominated by a large high pressure anomaly near Iceland that represents a weaker than normal Icelandic low and an associated low pressure anomaly (weaker than normal subtropical high) near the Iberian peninsula. In this phase, with a weaker winter meridional pressure gradient, western Eurasia experiences fewer and weaker storms from the north Atlantic whose trajectory is more zonally oriented than normal. This results in less than normal snow depth across western Eurasia. The opposite (positive) phase of the NAO is associated with more and stronger winter storms that enter Eurasia at a higher latitude than normal, producing substantially more snow than normal.

Recently, Liu and Yanai (2001) suggested that Indian monsoon rainfall is strongly modulated by tropospheric temperature over Eurasia which is, in turn, strongly affected by the NAO. Chang et al. (2001) noted that the relationship between surface air temperature over western Eurasia and the Indian monsoon rainfall has become stronger in recent years, over about the same period that the relationship between the monsoon and ENSO has diminished. They suggested that, as the ENSO-monsoon relationship has weakened, the possibility for the NAO to influence the monsoon through the above mechanism has increased.

Teleconnections to Africa and beyond

The Indian monsoon is known to be affected by climatic phenomena in other parts of the world and is also known to have a far-reaching influence on remote regions itself (Hastenrath 1988). Trenberth et al. (2000) have characterized this in terms of a "global monsoon" by which they mean a divergent overturning circulation with meridional (Hadley) and zonal (Walker) components that is thermally direct and transports heat from warmer to cooler regions. In this context, the Indian monsoon is a regional manifestation of the summer season global monsoon. The analysis of Trenberth et al. (2000) shows that the Indian monsoon has connections to other regions, notably northern Africa and the southern Indian Ocean. In particular, it is clear from their and others' analyses that the large convective region of the Indian monsoon drives a divergent circulation in the upper troposphere that reaches to north Africa where the large subsidence strengthens the tendency for desert formation in the Sahara.

Bhatt (1989) showed that, contrary to prevailing wisdom at that time, the rainfall anomalies in the eastern sub-Sahara are more closely related to the Indian monsoon rainfall anomalies and circulation anomalies over the Indian Ocean than to anomalies in the tropical Atlantic regional atmospheric circulation. Her analysis, based on several long records of hydrometeorological data from the 20th century, indicated that the Nile discharge, thought to be associated with an Atlantic moisture source, might be closely related to Indian monsoon anomalies through a large-scale atmospheric circulation mechanism. The link between eastern sub-Saharan rainfall anomalies and Indian monsoon rainfall anomalies was further established by Hoskins and Rodwell (1995) and Camberlin (1997) who suggested that the connection has been robust at seasonal time scales for the past 200 years. He suggested that the mechanism connecting the two regions is a lagged circulation change in which pressure changes over India can strongly influence the winds over the eastern sub-Sahara a few days later causing anomalous moist westerlies over the east African highlands.

It is also known that fluctuations in the Somali jet and the Findlater jet (Findlater 1969) are associated with variations in Indian monsoon rainfall, both in terms of the onset and the seasonal total. The jet originates over Kenya and crosses the equator, giving it very interesting dynamical importance insofar as it can transport negative potential vorticity from the Southern Hemisphere into the atmosphere overlying the Arabian Sea, possibly inducing a symmetric instability (Rodwell and Hoskins 1995). As in earlier studies, Rodwell and Hoskins (1995) found that the eastern African highlands are essential in supporting the existence of the cross-equatorial jet, but they also noted that the potential vorticity has to be modified along the jet trajectory, and they suggested a frictional and diabatic heating mechanism that could explain the observed behavior.

Relation between intraseasonal and interannual variability

In view of the hypothesis by Charney and Shukla (1981) that the interannual variability of monsoon rainfall is largely determined by changes in the slowly varying boundary conditions, the predictability of the monsoon rainfall depends on knowing whether droughts and floods over India (interannual variability) are a manifestation of large-scale persistent rainfall anomalies, or are due to a change in the nature of the intraseasonal variability. Although the importance of the influence of the boundary forcings such as SST is generally recognized, views differ on the characterization of the year-to-year changes in the nature of the intraseasonal variability and on what factors determine the seasonal mean rainfall over India.

Palmer (1994) has hypothesized that slowly varying boundary conditions change only the nature of the intraseasonal variability. He suggested that the summer monsoon has a regime-like behavior with nonperiodic evolution between active phase and break phase during a monsoon season. The effect of the boundary forcing is to simply alter the probability distribution function (PDF) of rainfall to have a bias toward the active or break phase. For example, during an El Niño (La Niña) year, the PDF is biased toward the break (active) phase, although the evolution of the monsoon system continues to alter between active and break phases. The seasonal mean rainfall is therefore determined by the bimodal PDF of rainfall, depending on the frequency and length of active and break periods. The spatial patterns of the interannual variability of the monsoon rainfall, for example, should correspond to those of intraseasonal active and break periods. The best course for the prediction of the seasonal monsoon rainfall is therefore probabilistic. In an evaluation of NCEP-NCAR reanalysis circulation data, Goswami and Ajaya Mohan (2001) lend support to Palmer's hypothesis by identifying a mode of variability common to both intraseasonal and interannual time scales and an asymmetric bimodal PDF of active and break conditions.

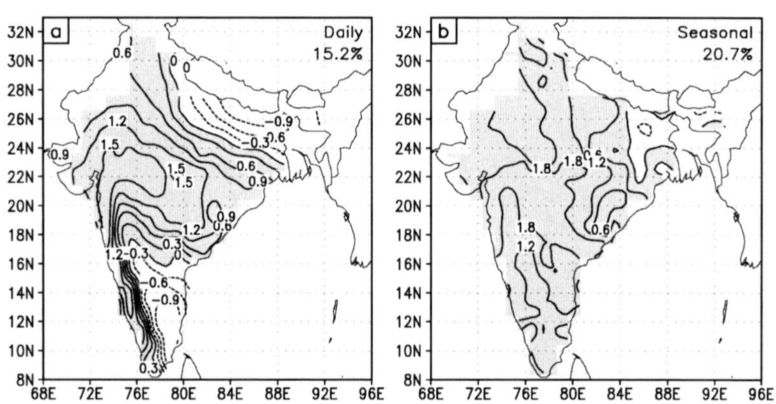

Fig. 10.16. EOF 1 of (a) standardized daily anomalies (5-day running means) and (b) standardized seasonal anomalies of rainfall for JJAS of 1901-70 period. Units are arbitrary and positive contours are shaded. The percentage variance explained by the EOF is given in the top right corner (reproduced from Krishnamurthy and Shukla 2000).

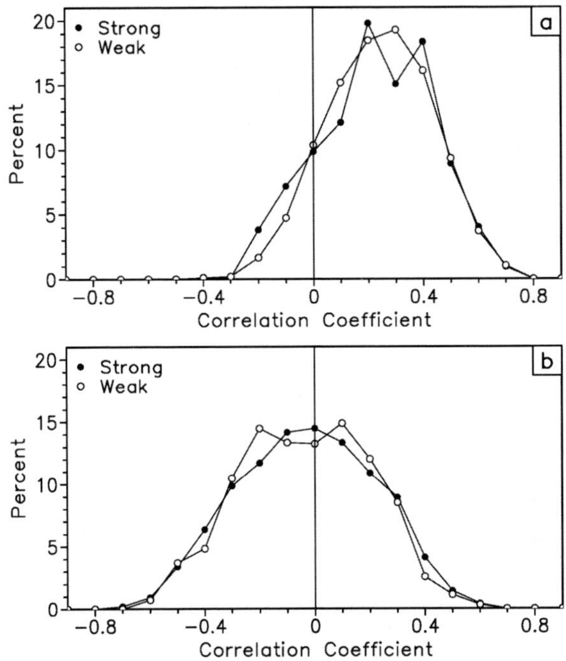

Fig. 10.17. Frequency distribution of (a) spatial pattern correlation between standardized JJAS seasonal anomalies and standardized daily anomalies that include the seasonal anomaly and (b) spatial pattern correlation between standardized JJAS seasonal anomalies and standardized daily anomalies that exclude the seasonal anomaly of rainfall over India for eight strong monsoon years (filled circles) and for eight weak monsoon years (open circles). The daily anomalies are 5-day running means (reproduced from Krishnamurthy and Shukla 2000).

By analyzing a long record of observed high resolution daily rainfall data over India for the 1901-70 period, Krishnamurthy and Shukla (2000) came to a different conclusion that the nature of the intraseasonal variability is not different from year to year, especially during flood and drought years. They found that the dominant modes of intraseasonal and interannual variability are different. These dominant modes, shown in Fig. 10.16, are the first empirical othogonal functions (EOFs) of daily and seasonal rainfall anomalies of JJAS. The seasonal rainfall EOF has the same sign over most of India whereas, in the daily EOF, the foothills of the Himalayas and southeast India have sign opposite to that over central India and the Western Ghats. With its striking resemblance to the (active-!break) composite of daily rainfall anomalies shown earlier in Fig. 10.8a, the leading daily rainfall EOF (Fig. 10.16a) seems to be a good representative of the intraseasonal variability. The principal component (PC) of daily EOF has a very high correlation with the time series of daily rainfall anomalies for all the years (Krishnamurthy and Shukla 2000). The leading seasonal rainfall EOF (Fig. 10.16b) similarly shows close resemblance to the composite of seasonal rainfall anomalies (Fig. 10.11a), and its corresponding PC captures the interannual variability of JJAS rainfall shown in Fig. 10.9 and in Fig. 10.13.

Krishnamurthy and Shukla (2000) showed that there is a signature of the seasonal anomaly pattern throughout the monsoon season. The correlation between the spatial patterns of daily and seasonal rainfall anomalies was determined for all days of the season, and its frequency distribution is shown in Fig. 10.17a for strong and weak monsoon years. The distribution is skewed toward positive correlation, indicating the presence of a persistent signature of the seasonal pattern. By removing the seasonal mean (although of small amplitude) from the daily anomalies, the distribution of the correlation of spatial patterns shows no preference and becomes almost gaussian about zero correlation (Fig. 10.17b). By projecting the daily rainfall anomalies on the daily rainfall EOF shown in Fig. 10.16a, Krishnamurthy and Shukla (2000) showed that there is no bimodality in the projection. Similar behavior has also been found in the analysis of circulation data by Sperber et al. (2000). Krishnamurthy and Shukla (2000) suggested a simple conceptual model of the interannual variability of the Indian monsoon rainfall to consist of a linear combination of a large-scale persistent seasonal mean component and a statistical average of intraseasonal variations. The persistent pattern may be thought of as the low-frequency component of the coupled ocean-land-atmosphere system including the influence of the boundary forcings. Therefore, the ability to predict the seasonal mean rainfall over India depends on the relative contributions of the intraseasonal component and the externally forced component. An analysis of the observed OLR and rainfall data by Lawrence and Webster (2001) estimates that the variance of the seasonal mean rainfall explained by the intraseasonal variability, although modest, is comparable to that explained by ENSO SST for 1975-97.

10.3.3
Interdecadal variability.

The variability of the Indian summer monsoon is also known to have a low-frequency component that alternates between epochs of above-normal and below-normal rainfall, each about three decades long. Several studies have provided evidence for the interdecadal variability in the monsoon rainfall (Parthasarathy et al. 1994, Kripalani and Kulkarni 1997, Kripalani et al. 1997, Krishnamurthy and Goswami 2000) and in circulation parameters (Kripalani et al. 1997). As ENSO is also known to exhibit interdecadal variability (e.g., Zhang et al. 1997), there have been studies (Kripalani and Kulkarni 1997, Mehta and Lau 1997; Krishnamurthy and Goswami 2000) to investigate whether the monsoon-ENSO relation observed in the interannual variability extends to interdecadal time scale.

IMR variability and monsoon-ENSO relation

The time series of low-pass filtered (21-year running mean) IMR and Niño-3 indices based on the period 1871-1998 are shown in Fig. 10.18. Knowing that the JJAS IMR has maximum negative correlation with the Niño-3 index of the subsequent season on the interannual time scale (as discussed in the previous section), the negative DJF Niño-3 index has been plotted in Fig. 10.18, which clearly shows that the two low-pass filtered time series vary together most of the time. Other ENSO parameters, such as the JJAS Niño-3 index and SOI, also show similar covariability with the monsoon rainfall (Krishnamurthy and Goswami 2000). On interdecadal time scales, the monsoon rainfall is in an above-normal phase during 1880-1895 and 1932-1965 and in a below-normal phase during 1896-1931 and 1966-1990 (years refer to the center of the running mean). The interdecadal IMR undergoes major transitions around 1895 and 1935, with close correspondence with similar transitions of the interdecadal Niño-3 index. However, the correspondence between the two interdecadal signals has broken down since 1970. The standard deviations of the low-pass filtered IMR and Niño-3 time series are 0.15 mm day^{-1} and 0.13°K, respectively, and the correlation between them is -0.72.

The interannual variability of IMR and ENSO indices also seem to be modulated on an interdecadal time scale. The variances of the unfiltered IMR and Niño-3 indices in a moving 21-year window are presented in Fig. 10.19a and 10.19b, respectively. For both IMR and Niño-3, the variances are higher during 1890-1920 and 1960-1989 than during 1930-60. The periods of high and low variances roughly correspond to the below-normal and above-normal periods, respectively, of the interdecadal IMR variability (Fig. 10.18). The interannual variability of the monsoon and ENSO appear to go through the same interdecadal modulation.

The correlation between unfiltered IMR and Niño-3 indices in a moving 21-year window is shown in Fig. 10.19c. Although this correlation shows low-frequency variation, it is not as strongly modulated as the variances are on an interdecadal time scale. There are periods of very strong correlation (-0.7 to -0.8), e.g., during 1880-1915, and strong correlation (-0.6 to -0.7), e.g., during 1920-1980, although there are some periods of weak correlation in between. The most noticeable change has taken place after 1980 when the correlation has sharply declined (Goswami et al. 1999, Krishna Kumar et al. 1999, Krishnamurthy and Goswami 2000).

Based on their examination of 50 years of circulation data, Krishnamurthy and Goswami (2000) suggested that the local Hadley circulation associated with the interannual El Niño (La Niña) reinforces (opposes) the prevailing anomalous interdecadal Hadley circulation during the phase when the east Pacific is warm in the interdecadal variation. During the warm interdecadal phase, El Niño events may be strongly related to monsoon droughts whereas La Niña events may not have a significant association with the monsoon. Similarly, during the phase of cold east Pacific in the interdecadal variation, the La Niña events may have strong association with monsoon floods while El Niño events may not have a significant relation with the monsoon.

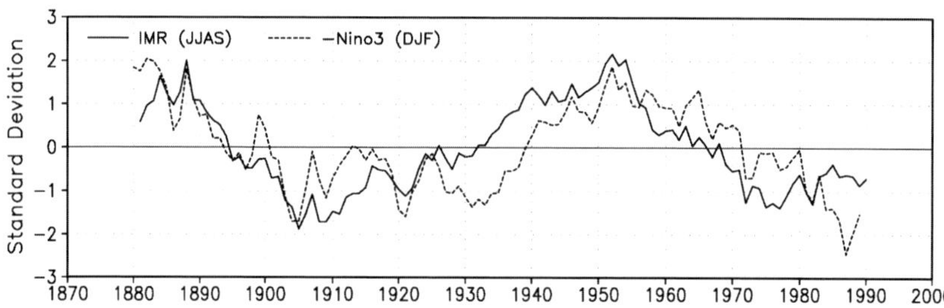

Fig. 10.18. Low-pass filtered (21-year running mean) time series of JJAS seasonal anomalies of IMR (solid) and DJF seasonal anomalies of the Niño-3 index (dashed). The sign of the Niño-3 index is negative for easy comparison. JJAS season of IMR precedes the DJF season of Niño-3.

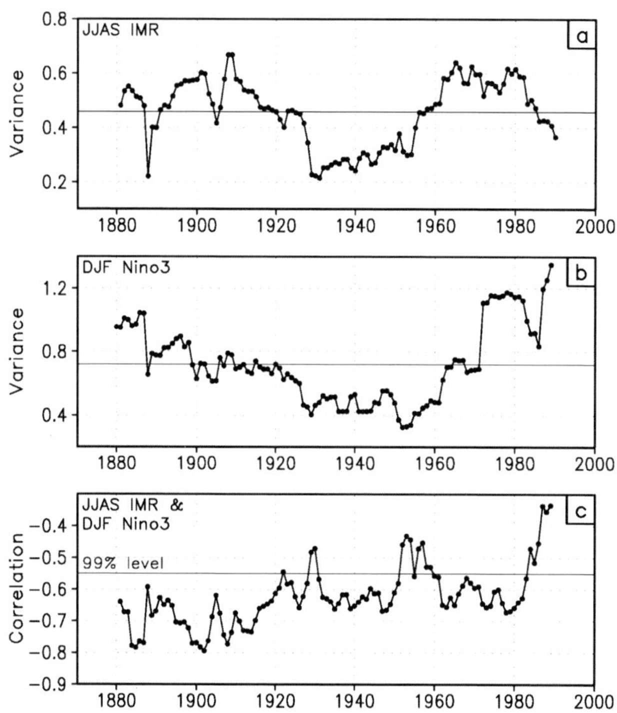

Fig. 10.19. Variance of (a) JJAS seasonal anomaly of IMR [(mm day^{-1})^2] and (b) DJF seasonal anomaly of Niño-3 index (°K^2) in a 21-year moving window. (c) Correlation between JJAS seasonal anomaly of IMR and succeeding DJF seasonal anomalies of Niño-3 index in a 21-year moving window. The years in all panels represent the centers of the window.

Two possible reasons for the weak monsoon-ENSO relation after 1980 have been suggested by Krishna Kumar et al (1999) -an apparent southeastward shift in the Walker circulation that they noticed in the NCEP-NCAR reanalysis circulation data and increased surface temperature over Eurasia in winter and spring. They speculate that global warming may also be responsible for the breakdown.

Changes circa 1976

Trenberth (1990) and Trenberth and Hurrell (1994) derived time series to depict changes in the Northern Hemisphere sea level pressure, including the intensity of the Aleutian low. They found that the pressure averaged over a vast area of the North Pacific was 2 hPa lower after 1976. Wang (1995), Nakamura and Yamagata (1997), Yasuda and Hanawa (1997), Zhang et al. (1997), Torrence and Webster (1999) and Deser et al. (1999) all reported on evidence for a climate transition in 1976-77 with cooling of the extratropical Pacific whose effects may also be found in ecosystem records (Mantua et al. 1997, Minobe 1997). In association with these climatic shifts, Kinter et al. (2002) showed that the relationship between the Asian monsoon and ENSO had likewise shifted. The much lower level of correlation between the Asian monsoon and ENSO had previously been discussed by Krishna Kumar et al. (1999) and Krishnamurthy and Goswami (2000). Kinter et al. (2002) also showed that there were large-scale shifts in the tropical and sub-tropical atmospheric circulation over east Asia and the Pacific circa 1976. They found an anomalous cyclonic pattern, or a weakened anticyclonic pattern, over the subtropical western Pacific that is consistent with both the cooler north Pacific SST and the weakening of the relationship between the monsoon and ENSO.

Pre-instrumental records of monsoon variability

A number of dendroclimatological (tree ring) and coral studies have been conducted in the Indian continental and Indian Ocean regions to determine the longer time scale variability of the Indian monsoon (see review by Borgaonkar and Pant 2001). The tree-ring analysis, begun by Pant (1979, 1983) and continued with collaborators (Pant and Borgaonkar 1984, Borgoankar et al. 1994, Pant et al. 1998), has shown that the alpine conifers in the high altitude regions of the Kashmir and the Himalayas are suitable for reconstructing the temperature and precipitation record in those regions extending back in time about 200 years. They found that there is a strong relationship between the pre-monsoon temperature and tree growth and that the monsoon rainfall is also related to tree growth. Based on the reconstructions made to date, there is little evidence of long-term trends in either pre-monsoon temperature or monsoon rainfall.

The coral analysis, in which an oxygen isotope analysis of samples taken from cores of coral that can be reliably dated to within one month up to 150 years before present, has been undertaken in the western Indian Ocean near the Seychelles Islands. Charles et al. (1997) showed recently that the coral record can be used to determine the strength of the connection between the Asian monsoon and ENSO, and they found that the strength of the correlation between ENSO and the tropical

Indian Ocean SST has remained essentially constant over the last 100 years. They also found decadal modulations of the Indian Ocean SST that they related to similar modulations of the Asian monsoon.

Long proxy records and climate model simulations suggest that the Indian monsoon has undergone a rich spectrum of variability over the past 150 000 years (Prell and Kutzback 1987). These variations are attributed to changes in the glacier boundary conditions and orbitally-produced solar radiation changes.

10.4 Modeling.

As already discussed, the seasonal mean monsoon has potential predictability because of the influence of slowly varying boundary forcings. Long range forecasts (LRF) of the seasonal mean monsoon have been issued by the IMD for over a hundred years using empirical models. These are statistical models based on the correlation of the monsoon rainfall with local and global climate variables found over a long period of time. However, the statistical forecasts lack the ability to predict the spatial distribution of rainfall and on a time scale shorter than a season. A different approach is the dynamical prediction using GCMs for both daily and seasonal forecasts. There have been several organized international efforts to simulate the Indian monsoon by modeling groups since the mid-1980s.

10.4.1 Empirical models.

These empirical models used to predict the monsoon are basically regression models that have undergone modifications, in both technique and predictors used, since the time IMD started issuing LRF (Thapliyal and Kulshrestha 1992; Rajeevan 2001). Reflecting the fact that the seasonal mean monsoon is determined by internal dynamics as well as by the influences of slowly varying boundary conditions, the predictors used are both regional and global climate parameters, including SST, surface temperature and SLP at various locations.

Forecasts of seasonal monsoon rainfall over large regions of India issued by the IMD during 1924-87 were based on multiple linear regression models. Earlier regression models introduced by Walker (1923) used parameters such as the snow accumulation over the Himalayas in May and South American pressure during spring. The performance of these models was found to be correct in 65% of the forecasts during 1924-87 by Thapliyal (1990) who also provides the details of the predictors used during this period.

Since 1988, the IMD has been issuing LRF of seasonal mean monsoon rainfall over India as a whole using parametric and power regression models with 16 predictors (Gowariker et al. 1989, 1991). The parametric model is purely qualitative, with equal weights given to all the parameters, and it provides forecasts of whether the monsoon rainfall will be normal (within 10% of long term mean), excess or deficient. The power regression model, however, is quantitative and takes into account the nonlinear nature of the interactions of the local and global forcings with the Indian monsoon. The 16 predictors used by these models include Pacific SST, surface temperature and pressure in India and other global locations, snow cover over the Himalayas and Eurasia, and the 500 hPa ridge position.

The temporal variations in the correlation between monsoon rainfall and some of the predictors, such as the interdecadal variability in the ENSO-monsoon relation, and the need to modify the list of predictors have been recognized by the modelers (Thapliyal 1990). Certain predictors have been discarded over time and new predictors have been added (Rajeevan 2001). Examples of temporal variations of correlation of monsoon rainfall with certain climate parameters related to, but not necessarily the same as, those used by the IMD, are provided in Fig. 10.20.

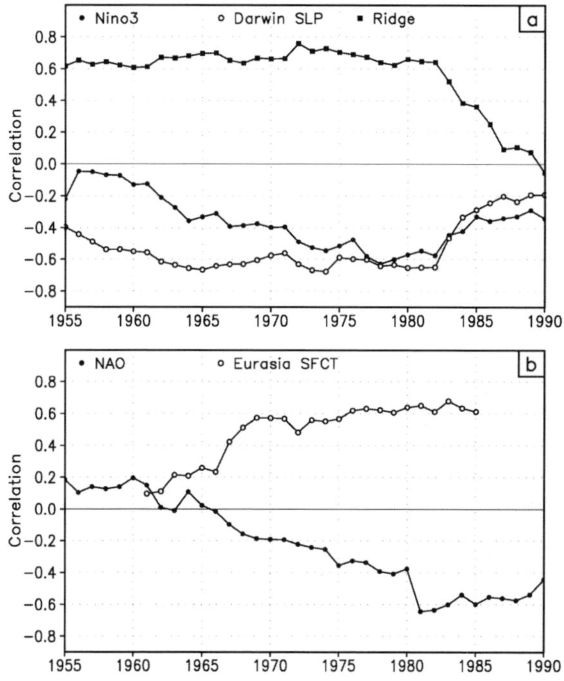

Fig. 10.20. Correlation of JJAS IMR seasonal anomalies with the seasonal anomalies of various indices in a 21-year moving window. The indices in panel (a) are Niño-3 index of MAM season (filled circles), the tendency of Darwin SLP from DJF to MAM seasons (open circles), and the April monthly mean 500hPa ridge at 75EE (filled squares). The indices in panel (b) are the NAO index of AM season (filled circles), and the Eurasian surface temperature anomaly of DJF season (open circles). The periods of all the predictors precede the JJAS monsoon season. The years in both panels are the center points of the window.

The three predictors shown in Fig. 10.20a are the MAM seasonal anomaly of Niño-3, (MAM-DJF) tendency of the Darwin SLP seasonal anomaly, and the April mean 500 hPa ridge at 75°E. The correlations of the 500 hPa ridge and the Darwin SLP tendency with the subsequent JJAS seasonal anomaly of IMR were somewhat strong and steady prior to 1982 (center of the 21-year window) (similar to the results of Shukla and Mooley 1987) but became weak rapidly after 1982. The correlation between Niño-3 index and IMR also declined rapidly after 1982 but was even weaker prior to 1965.

On the other hand, the correlations of the April-May (AM) seasonal anomaly of the NAO index (normalized pressure difference between a station in the Azores and one in Iceland) and the DJF Eurasian surface temperature seasonal anomaly (average over 30°E-50°E, 60°N-70°N) with the subsequent JJAS IMR have become substantially stronger since 1975 (Fig. 10.20b). The strengthening of the NAO-monsoon correlation and the weakening of the ENSO-monsoon correlation may be related because of the strengthening and poleward shift of the jet stream over the North Atlantic (Chang et al. 2001).

A complete list of the 16 predictors used by the IMD in 2000 and the performance of the IMD forecasts during 1988-2000 were provided by Rajeevan (2001). He showed that the root mean square error of the forecasts is 7.6% for this period during which the IMR has been normal (within 10% of long term mean) except for 1988.

10.4.2 General circulation models.

Prominent among the international projects to conduct GCM simulation of monsoon are the Tropical Ocean Global Atmosphere (TOGA) program's Monsoon Numerical Experimentation Group (MONEG) simulations, the Atmospheric Model Intercomparison Project (AMIP), the Climate Variability and Predictability (CLIVAR) program's Seasonal prediction Model Intercomparison Project (SMIP), and the CLIVAR Monsoon GCM Intercomparison Project, although not all of them were exclusively for monsoon studies. Based on the premise that the seasonal mean Indian monsoon has potential predictability because of forcing by slowly varying boundary conditions (Charney and Shukla 1981), the simulation experiments have been conducted with atmospheric GCMs (AGCMs) with observed SST specified as lower boundary conditions.

The 1987 El Niño and 1988 La Niña coincided with weak and strong Indian monsoon rainfall seasons, respectively, (see Fig. 10.13) and provided an opportunity for the MONEG modelers to simulate the monsoon with two starkly contrasting Pacific SST boundary forcings. Although there were systematic errors in the simulation of local features of the monsoon by most models, Palmer et al. (1992) provided an example of the ability of models to correctly simulate the contrasting interannual variability of the monsoon during 1987 and 1988. Their simulations, made with a T42 horizontal resolution European Center for Medium-Range Weather Forecasts (ECMWF) model, were 90 days long with initial conditions of

1 June 1987 and 1 June 1988. The ECMWF model simulated more rainfall over India during the summer months of 1988 than during 1987 although not close to observations. By specifying SST anomalies selectively over different oceanic regions, they concluded that the difference in the simulated rainfall over India between 1987 and 1988 was mostly due to the effects of the Pacific SST anomalies.

As part of the first AMIP experiment, 32 AGCMs were integrated over the period 1979-98 with the same observed SST specified as boundary conditions (Sperber and Palmer 1996). Some models were also integrated with different initial conditions but with same SST to estimate the potential predictability of the interannual variability. There was a wide spread in the interannual variability of the JJAS all-India monsoon rainfall among the model simulations for the AMIP period, even during 1987 and 1988. The spread in the simulations may be attributed to the differences in resolution and physical parameterizations of the models. The ensemble integrations of the ECMWF model showed that the all-India rainfall was higher in 1988 relative to 1987 while the rainfall variability during other years did not show any predictability. The rainfall climatology simulated by the ECMWF model showed less rainfall over the Indian continent and more over the Indian Ocean compared to observations. Sperber and Palmer (1996) noted that the dryness over India was perhaps due to dry air advection from northwest India and due to a weak simulated Somali Jet north of 12°N in the low level wind climatology of the ECMWF model. The link between the Indian rainfall and SST was found to be strong during ENSO periods. The Somali Jet was well simulated by many models resulting in better simulation of the rainfall over the west coast of India. However, dry advection from the north and northwest of India seems to prevent the moisture convergence necessary for monsoon rainfall over the Indian continent.

Recently, dynamical seasonal prediction (DSP), done by integrating GCMs with an ensemble of initial conditions, has been attempted to provide probabilistic seasonal forecasts. To assess the dynamical seasonal predictability of the Asian summer monsoon, the CLIVAR SMIP analyzed ensembles of summer monsoon hindcasts from seven GCMs for 1987, 1988 and 1993 corresponding to El Niño, La Niña and near normal SST conditions in the Pacific Ocean (Sperber et al. 2001). The models, ranging widely in resolution, were integrated using observed SST and observed initial conditions. The analysis investigated the link between subseasonal and interannual variability of the monsoon rainfall and 850 hPa winds for the JJAS season. The analysis found that the errors produced by the models in simulating the time mean rainfall and 850hPa flow were larger than the uncertainties in observations. Most models were able to simulate the EOFs associated with the dominant intraseasonal variations of 850hPa winds but not with the fidelity shown by the reanalysis data. The leading EOF, associated with the continental tropical convergence zone, was the best simulated among all the EOFs. The models generally failed to properly project the intraseasonal EOFs and PCs onto the interannual variability of the rainfall and 850 hPa winds. Some of the failures may indicate that the models were not setting up the observed teleconnections with the SST.

The CLIVAR Monsoon GCM Intercomparison Project assessed the performance of 10 AGCMs in simulating the Indian monsoon for the 20-year period of

1979-98 using observed SST and observed initial conditions. The climatological mean precipitation of the 20-year period has been analyzed by Kang et al. (2002). The JJAS climatological mean rainfall produced by all models is excessive over most of the Indian monsoon region including the Bay of Bengal and the Arabian Sea but deficient over certain parts of north India. The Center for Ocean-Land-Atmosphere Studies (COLA) AGCM was one of the models used in this project and its details and performance in simulating the interannual variability have been discussed by Krishnamurthy and Shukla (2001).

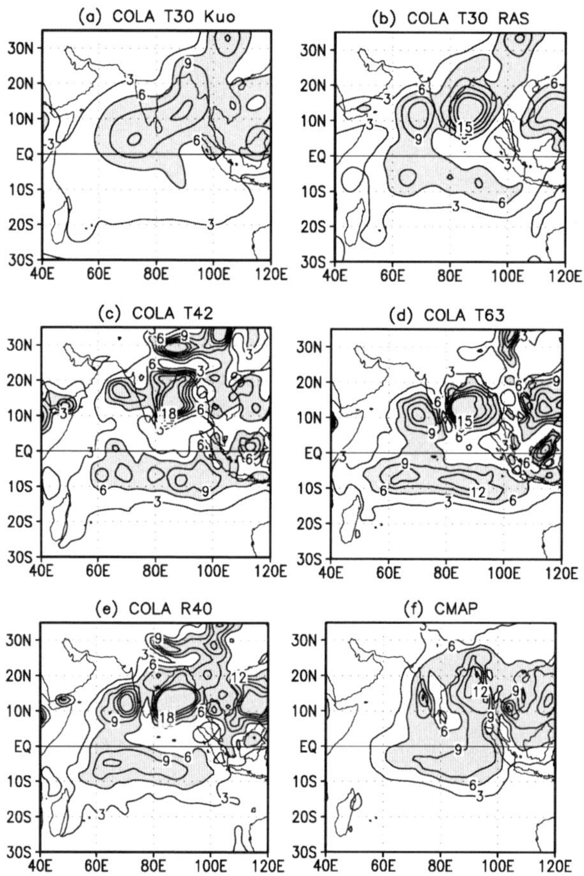

Fig. 10.21. JJAS seasonal climatological mean precipitation simulated by COLA AGCM with (a) T30 resolution with Kuo convection scheme, (b) T30 resolution with Relaxed Arakawa-Schubert convection scheme, (c) T42 resolution, (d) T63 resolution, and (e) R40 resolution. The observed mean from CMAP data is also shown (f). The contour interval is 3 mm day^{-1} and contours > 6 are shaded.

In Fig. 10.21, the climatological mean rainfall over the Indian monsoon region simulated by COLA AGCM used in the CLIVAR project is compared with observations and with the simulations by other versions of COLA model with different horizontal resolution. The models have horizontal resolution of T30, T42, T63 and R40. The T30 model, which uses a Kuo scheme for representing cumulus convection, produces very deficient rainfall all over the region (Fig. 10.21a) compared to observations (Fig. 10.21f). The rest of the models use a Relaxed Arakawa-Schubert (RAS) scheme instead of a Kuo scheme and simulate higher rainfall over the region. The rainfall over the Bay of Bengal and the Arabian Sea are quite excessive in all the model (with RAS) simulations, and the observed rainfall pattern over the equatorial Indian Ocean has not been properly captured. The R40 model (Fig. 10.21e) seems to have a somewhat better simulation over the Indian continent although with a maximum over the Tibetan region.

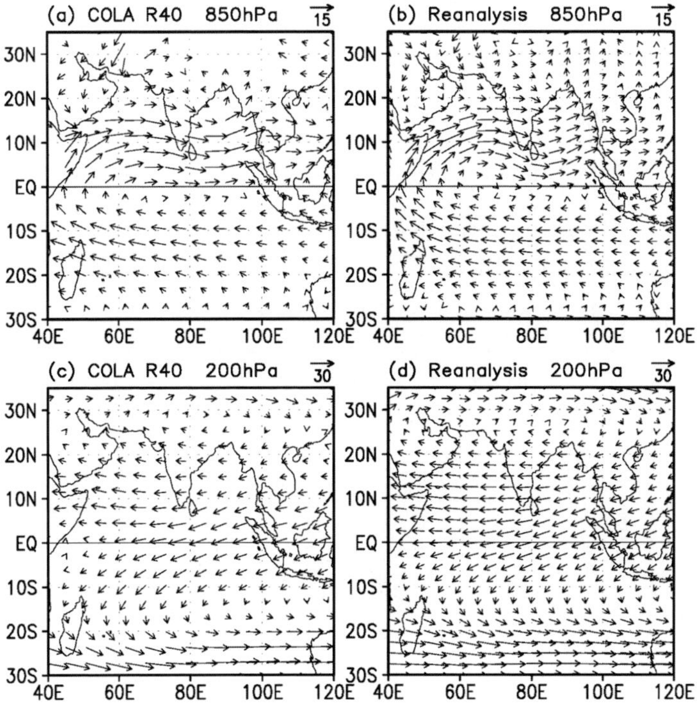

Fig. 10.22. JJAS seasonal climatological mean horizontal wind at (a) 850 hPa and (c) 200 hPa simulated by COLA AGCM with R40 resolution. The corresponding means from NCEP•NCAR reanalysis are shown in (b) and (d) respectively. Unit vectors are 15 m s^{-1} in (a) and (b) and 30 m s^{-1} in (c) and (d). (Reproduced from Krishnamurthy and Shukla 2001)

The JJAS climatological horizontal winds at 850 hPa and 200 hPa simulated by the R40 model are compared with the NCEP-NCAR reanalysis winds in Fig. 10.22. The model simulates the winds at the two levels fairly well, although, over the Arabian Sea and Bay of Bengal, the 850 hPa winds are more intense while the 200hPa winds are less intense.

The interannual variability of the R40 model simulation is compared with observations in Fig. 10.23, which shows the time series of JJAS seasonal anomalies of the IMR index for rainfall and the Monsoon Hadley (MH) index for circulation during 1979-98. The MH index is defined as the meridional shear anomaly between 850 hPa and 200hPa averaged over the same area as the EIMR region (Goswami et al. 1999) and is representative of the regional Hadley circulation. The correspondence between the time series of the model simulation and observation is very poor with correlation between the two time series being close to zero for both IMR and MH indices (Fig. 10.23). Krishnamurthy and Shukla (2001) have estimated that the SST-forced component of the variance of the rainfall is more dominant than the component due to internal dynamics. The poor skill displayed by the model in simulating the rainfall variability may be related to the systematic errors of the model in simulating the climatological mean monsoon circulation and rainfall, especially over the oceanic regions. It has been found that the simulation of the Indian monsoon rainfall and circulation is also difficult with coupled ocean-atmosphere models (e.g., Latif et al. 1994).

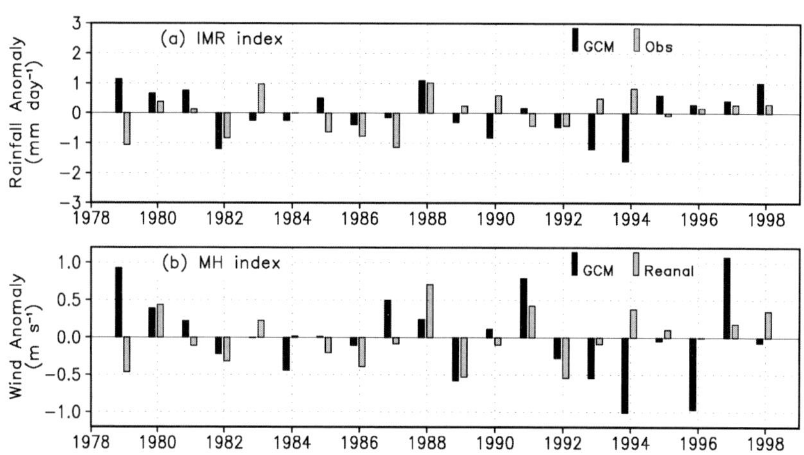

Fig. 10.23. Time series of JJAS seasonal anomaly (a) IMR index (mm day^{-1}) from the COLA AGCM integrations and from the observed data and (b) MH index (m s^{-1}) from the COLA AGCM integrations and from the NCEP-NCAR reanalysis (partly reproduced from Krishnamurthy and Shukla 2001).

Dynamical models do not have sufficiently high fidelity to accurately simulate the salient characteristics of the mean monsoon and monsoon variability. More observations and analyses are needed to understand the intraseasonal variability of the monsoon and its relation to the neighboring oceanic regions. Further studies are also required to understand the relationship of the monsoon with global climate features such as ENSO, NAO and Eurasian snow. The predictability of the monsoon depends on understanding the relative roles of internal dynamics and the influences of boundary conditions.

Acknowledgements

This work was supported by grants from the National Science Foundation (ATM-9814295), the National Oceanic and Atmospheric Administration (NA96-GP0056), and the National Aeronautics and Space Administration (NAG5-8202).

References

Angell, J. K. (1981) Comparison of variations in atmospheric quantities with sea surface temperature variations in the equatorial eastern Pacific. Mon. Wea. Rev., 109, 230-243.

Annamalai, H., and J. M. Slingo (2001) Active/break cycles: diagnosis of the intraseasonal variability of the Asian summer monsoon. Climate Dyn., 18, 85-102.

Bamzai, A., and L. Marx (2000) COLA AGCM simulation of the effect of anomalous spring snow over Eurasia on the Indian summer monsoon. Quart. J. Roy. Meteor. Soc., 65, 2575-2584.

Bamzai, A., and J. Shukla (1999) Relation between Eurasian snow cover, snow depth and the Indian summer monsoon: an observational study. J. Climate, 126, 3117-3132.

Bansod, S. D., S. V. Singh, and R. H. Kripalani (1991) The relationship of monsoon onset with subsequent rainfall over India. Int. J. Clim., 11, 809-817.

Barnett, T. P. (1983) Interaction of the monsoon and Pacific trade wind system at interannual time scales. Part I: The Equatorial zone. Mon. Wea. Rev., 111, 756-773.

Barnett, T. P. (1984) Interaction of the monsoon and Pacific trade wind system at interannual time scales. Part III: The tropical band. Mon. Wea. Rev., 112, 2388–2400.

Barnett, T. P. (1991) The interaction of multiple time scales in the tropical climate system. J. Climate, 4, 269–285.

Barnett, T. P., L. Dümenil, U. Schlese, E. Roeckner, and M. Latif (1989) The effect of Eurasian snow cover on regional and global climate variations. J. Climate, 46, 661-685.

Becker, B. D., J. M. Slingo, L. Ferranti, and F. Molteni (2001) Seasonal predictability of the Indian summer monsoon: what role do land surface conditions play? Mausam, 52, 175-190.

Bhatt, U. S. (1989) Circulation regimes of rainfall anomalies in the African•south Asian monsoon belt. J. Climate, 2, 1133–1145.

Bjerknes, J. (1969) Atmospheric teleconnections from the equatorial Pacific. Mon. Wea. Rev., 97, 163-172.

Blanford, H. F. (1884) On the connexion of Himalayan snowfall and seasons of drought in India. Proc. Roy. Soc. London, 37, 3-22.

Borgaonkar, H. P., and G. B. Pant (2001) Long-term climate variability over monsoon Asia as revealed by some proxy sources. Mausam, 52, 9-22.

Borgaonkar, H. P., G. B. Pant, and K. Rupa Kumar (1994). Dendroclimatic reconstruction of summer precipitation at Srinagar, Kashmir, India since the late 18th century. Holocene, 4, 299-306.
Brier, G. W. (1978). The quasi-biennial oscillation and feedback processes in the atmosphere-ocean-earth system. Mon. Wea. Rev., 106, 938–946.
Cadet, D., and G. Reverdin (1981) Water vapor transport over the Indian Ocean during summer 1975. Tellus, 33, 476-487.
Camberlin, P. (1997) Rainfall anomalies in the source region of the Nile and their connection with the Indian summer monsoon. J. Climate, 10, 1380–1392.
Chang, C. P., P. Harr, and J. Ju (2001) Possible roles of Atlantic circulations on the weakening Indian monsoon rainfall-ENSO relationship. J. Climate, 14, 2376-2380.
Charles, C. D., D. E. Hunter, and R. C. Fairbanks (1997) Interaction between the ENSO and the Asian monsoon in a coral record of tropical climate. Science, 277, 925-928.
Charney, J. G., and J. Shukla (1981) Predictability of monsoons. Monsoon Dynamics, J. Lighthill, Ed., Cambridge University Press, 99-109.
Clark, C. O., J. E. Cole, and P. J. Webster (2000) Indian Ocean SST and Indian summer rainfall: Predictive relationships and their decadal variability. J. Climate, 13, 2503-2519.
Clarke, A. J., X. Liu, and S. van Gorder (1998) Dynamics of the biennial oscillation in the equatorial Indian and far western Pacific Oceans. J. Climate, 11, 987-1001.
Clarke, A. J., and S. van Gorder (1999) The connection between the boreal spring Southern Oscillation persistence barrier and biennial variability, J. Climate, 12, 610–620.
Delworth, T. L., and S. Manabe (1988) The influence of potential evaporation on the variabilities of simulated soil wetness and climate. J. Climate, 1, 523-547.
Deser, C., M. A. Alexander, and M. S. Timlin (1999) Evidence for a wind-driven intensification of the Kuroshio current extension from the 1970s to the 1980s. J. Climate, 12, 1697-1706.
Dey, B., and O. S. R. U. Bhanu Kumar (1982) An apparent relationship between Eurasian spring snow cover and the advance period of the Indian summer monsoon. J. Appl. Meteor., 21, 1929–1932.
Dey, B., and O. S. R. U. Bhanu Kumar (1983) Himalayan winter summer snow cover area and summer monsoon rainfall over India. J. Geophys. Res., 88, 5471–5474.
Dey, B., S. N. Kathuria, and O. S. R. U. Bhanu Kumar (1985) Himalayan summer snow cover and withdrawal of the Indian summer monsoon. J. Appl. Meteor., 24, 865–868.
Dickson, R. R. (1984). Eurasian snow cover versus Indian monsoon rainfall – an extension of the Hahn-Shukla results. J. Appl. Meteor., 23, 171–173.
Dirmeyer, P. A. (1999) Assessing GCM sensitivity to soil wetness using GSWP data. J. Meteor. Soc. Japan, 77, 367-385.
Douville, H., F. Chauvin, and H. Broqua (2001) Influence of soil moisture on the Asian and African monsoons. Part I: Mean monsoon and daily precipitation. J. Climate, 14, 2381-2403.
Douville, H., and J. F. Royer (1996) Sensitivity of the Asian summer monsoon to an anomalous Eurasian snow cover within the Metéo-France GCM. Climate Dyn., 12, 449-466.
Fein, J. S., and P. L. Stephens (1987) Monsoons. John Wiley and Sons, 384 pp.
Fennessy, M. J., and J. Shukla (1999) Impact of initial soil wetness on seasonal atmospheric prediction. J. Climate, 12, 3167-3180.
Ferranti, L., J. M. Slingo, T. N. Palmer, and B. J. Hoskins (1999) The effects of land-surface feedbacks on the monsoon circulation. Quart. J. Roy. Meteor. Soc., 125, 1527-1550.
Findlater, J. (1969) A major low-level air current near the Indian Ocean during the northern summer. Quart. J. Roy. Meteor. Soc., 95, 362-380.
Gadgil, S., and G. Asha (1992) Intraseasonal variation of the summer monsoon I: Observational aspects. J. Meteor. Soc. Japan, 70, 517-527.
Gadgil, S., P. V. Joseph, and N. V. Joshi (1984) Ocean-atmosphere coupling over monsoon regions. Nature, 312, 141-143.
Gautier, C., P. Peterson, and C. Jones (1998) Variability of air-sea interactions over the Indian Ocean derived from satellite observations. J. Climate, 11, 1859-1873.

Ghosh, S. K., M. C. Pant, and B. N. Devan (1978) Influence of the Arabian Sea on the Indian summer monsoon. Tellus, 30, 117-125.

Goswami, B. N. (1995) A multiscale interaction model for the origin of the tropospheric QBO. J. Climate, 8, 524-534.

Goswami, B. N. (1998) Interannual variations of Indian summer monsoon in a GCM: External conditions versus internal feedbacks. J. Climate, 11, 501-522.

Goswami, B. N., and R. S. Ajaya Mohan (2001) Intraseasonal oscillations and interannual variability of the Indian summer monsoon. J. Climate, 14, 1180-1198.

Goswami, B. N., V. Krishnamurthy, and H. Annamalai (1999) A broad-scale circulation index for the interannual variability of the Indian summer monsoon. Quart. J. Roy. Meteor. Soc., 125, 611-633.

Goswami, B. N., D. Sengupta, and G. Suresh Kumar (1998) Intraseasonal oscillations and interannual variability of surface winds over the Indian monsoon region. Proc. Indian Acad. Sci. (Earth & Planet. Sci.), 107, 1-20.

Gowariker, V., V. Thapliyal, R. P. Sarker, G. S. Mandal, and D. R. Sikka (1989) Parametric and power regression models: New approach to long range forecasting of monsoon rainfall in India. Mausam, 40, 115-122.

Gowariker, V., V. Thapliyal, S. M. Kulshrestha, G. S. Mandal, N. Sen Roy, and D. R. Sikka (1991) A power regression model for long range forecast of southwest monsoon rainfall over India. Mausam, 42, 125-130.

Graham, N. E., and T. P. Barnett (1987) Sea surface temperature, surface wind divergence and convection over tropical oceans. Science, 238, 657-659.

Hahn, D., and S. Manabe (1975) The role of mountains in the south Asian monsoon circulation. J. Atmos. Sci., 32, 1515-1541.

Hahn, D., and J. Shukla (1976) An apparent relationship between Eurasian snow cover and Indian monsoon rainfall. J. Atmos. Sci., 33, 2461-2462.

Hartmann, D. L., and M. L. Michelsen (1989) Intraseasonal periodicities in Indian rainfall. J. Atmos. Sci., 46, 2838-2862.

Hastenrath, S. (1988) Climate and Circulation of the Tropics. Reidel, 455 pp.

He, H., J. W. McGinnis, Z. Song, and M. Yanai (1987) Onset of Asian summer monsoon in 1979 and the effect of the Tibetan plateau. Mon. Wea. Rev., 115, 1966-1995.

Hoskins, B. J., and M. J. Rodwell (1995) A model of the Asian summer monsoon. Part I: the global scale. J. Atmos. Sci., 52, 1329-1340.

Huang, B., and J. L. Kinter III (2001) The interannual variability in the tropical Indian Ocean and its relation to El Niño/Southern Oscillation. COLA Tech. Rep., 94, 48pp. [Available from Center for Ocean-Land-Atmosphere Studies, 4041 Powder Mill Road, Suite 302, Calverton, MD 20705, USA].

Ju, J., and J. Slingo (1995) The Asian summer monsoon and ENSO. Quart. J. Roy. Meteor. Soc., 121, 1133-1168.

Kang, I.-S., K. Jin, B. Wang, K.-M. Lau, J. Shukla, V. Krishnamurthy, S. D. Schubert, D. E. Waliser, W. F. Stern, A. Kitoh, G. A. Meehl, M. Kanamitsu, V. Ya. Galin, V. Satyan, C.-K. Park, and Y. Liu (2002) Intercomparison of the climatological variations of Asian summer monsoon precipitation simulated by 10 GCMs. Climate Dyn. (accepted).

Kiladis, G. N., and H. F. Diaz (1989). Global climatic anomalies associated with extremes in the Southern Oscillation. J. Climate, 2, 1069–1090.

Kinter III, J. L., K. Miyakoda, and S. Yang (2002) Recent change in the connection from the Asian monsoon to ENSO. J. Climate, 15 (in press).

Kirtman, B., and J. Shukla (2000) Influence of the Indian summer monsoon on ENSO. Quart. J. Roy. Meteor. Soc., 126, 213-239.

Kirtman, B., and J. Shukla (2001). Interactive coupled ensemble: A new coupling strategy for CGCMs. COLA Tech. Rep., 105, 40pp. [Available from Center for Ocean-Land-Atmosphere Studies, 4041 Powder Mill Road, Suite 302, Calverton, MD 20705, USA].

Kripalani, R. H., and A. Kulkarni (1997) Climatic impact of El Niño/La Niña on the Indian monsoon: A new perspective. Weather, 52, 39-46.

Kripalani, R. H., A. Kulkarni, and S. V. Singh (1997) Association of the Indian summer monsoon with the Northern Hemisphere mid-latitude circulation. Int. J. Climatol., 17, 1055-1067.

Krishna Kumar, K., B. Rajagopalan, and M. A. Cane (1999) On the weakening relationship between the Indian monsoon and ENSO. Science, 284, 2156-2159.

Krishnamurthy, V., and B. N. Goswami (2000) Indian monsoon-ENSO relationship on interdecadal timescale. J. Climate, 13, 579-595.

Krishnamurthy, V., and B. P. Kirtman (2001) Variability of the Indian Ocean: Relation to monsoon and ENSO. COLA Tech. Rep. 107, 40pp. [Available from Center for Ocean-Land-Atmosphere Studies, 4041 Powder Mill Road, Suite 302, Calverton, MD 20705, USA].

Krishnamurthy, V., and J. Shukla (2000) Intraseasonal and interannual variability of rainfall over India. J. Climate, 13, 4366-4377.

Krishnamurthy, V., and J. Shukla (2001). Observed and model simulated interannual variability of the Indian monsoon. Mausam, 52, 133-150.

Krishnamurti, T. N., and P. Ardunay (1980) The 10-20 day westward propagation mode and breaks in the monsoon. Tellus, 32, 15-26.

Krishnamurti, T. N., and H. N. Bhalme (1976) Oscillations of a monsoon system. Part 1. Observational aspects. J. Atmos. Sci., 33, 1937-1954.

Kutzbach, J. E., P. J. Geutter, W. F. Ruddiman, and W. L. Prell (1989) Sensitivity of climate to late Cenozoic uplift in southern Asia and the America west: Numerical experiments. J. Geophys. Res., 94, 18393-18407.

Lamb, P. J., and R. A. Peppler (1987) North Atlantic Oscillation: Concept and an application. Bull. Amer. Meteor. Soc., 68, 1218–1225.

Latif, M., A. Sterl, M. Assenbaum, M. M. Junge, and E. Maier-Reimer (1994) Climate variability in a coupled GCM. Part II: The Indian Ocean and monsoon. J. Climate, 7, 1449-1462.

Lau, K.-M., and P. H. Chan (1986) Aspects of the 40-50 day oscillation during the northern summer as inferred from outgoing longwave radiation, Mon. Wea. Rev., 114, 1354-1367.

Lau, K.-M., and H. T. Wu (2001) Principal modes of rainfall-SST variability of the Asian summer monsoon: A reassessment of the monsoon-ENSO relationship. J. Climate, 14, 2880-2895.

Lawrence, D. M., and P. J. Webster (2001) Interannual variations of the intraseasonal oscillation in the South Asian summer monsoon region. J. Climate, 14, 2910-2922.

Li, C., and M. Yanai (1996) The onset and interannual variability of the Asian summer monsoon in relation to land•sea thermal contrast. J. Climate, 9, 358-375.

Liu, X., and M. Yanai (2001) Relationship between the Indian monsoon rainfall and the tropospheric temperature over the Eurasian continent. Quart. J. Roy. Meteor. Soc., 127, 909-937.

Madden, R. A., and P. R. Julian (1972) Description of global-scale circulation cells in the tropics with a 40-50 day period. J. Atmos. Sci., 29, 1109-1123.

Mantua, J. N., S. R. Hare, Y. Zhang, J. M. Wallace, and R. C. Francis (1997) A Pacific interdecadal climate oscillation with impacts on salmon production. Bull. Amer. Meteor. Soc.,78, 1069-1080.

Matsuyama, H., K. Masuda (1998) Seasonal/interannual variations of soil moisture in the former USSR and its relationship to Indian summer monsoon rainfall. J. Climate, 11, 652-658.

Meehl, G. A. (1987) The annual cycle and its relationship to interannual variability in the tropical Pacific and Indian Ocean regions. Mon. Wea. Rev., 115, 27-50.

Meehl, G. A. (1992) Effect of tropical topography on global climate. Ann. Rev. Earth Planet. Sci., 20, 85-112.

Meehl, G. A. (1993) A coupled air-sea biennial mechanism in the tropical Indian and Pacific regions: Role of the ocean. J. Climate, 6, 31-41.

Meehl, G. A. (1994a) Coupled land-ocean-atmosphere processes and South Asian monsoon variability. Science, 266, 263–267.

Meehl, G. A. (1994b) Influence of the land surface in the Asian summer monsoon: External conditions versus internal feedbacks. J. Climate, 7, 1033-1049.

Meehl, G. A. (1997) The south Asian monsoon and the tropospheric biennial oscillation. J. Climate, 10, 1921–1943.

Mehta, V. M., and K. M. Lau (1997) Influence of solar irradiance on the Indian monsoon-ENSO relationship at decadal-multidecadal time scales. Geophys. Res. Lett., 24, 159-162.

Minobe, S. (1997) A 50-70 year climatic oscillation over the North Pacific and North America. Geophys. Res. Let., 24, 683-686.

Mooley, D. A., and B. Parthasarathy (1984) Fluctuations in all-India summer monsoon rainfall during 1871-1978. Climate Change, 6, 287-301.

Mooley, D. A., and J. Shukla (1987) Variability and forecasting of the summer monsoon rainfall over India. Monsoon Meteorology, C. P. Chang, and T. N. Krishnamurti, Eds., Oxford University Press, 26-59.

Nakamura, H., G. Lin, and T. Yamagata (1997) Decadal climate variability in the North Pacific during the recent decades. Bull. Amer. Meteor. Soc.,78, 2215-2225.

Nicholls, N. (1978) Air–sea interaction and the quasi-biennial oscillation. Mon. Wea. Rev., 106, 1505–1508.

Nicholls, N. (1979) A simple air–sea interaction model. Quart J. Roy. Meteor. Soc., 105, 93–105.

Nicholls, N. (1984) The Southern Oscillation and Indonesian sea surface temperature. Mon. Wea. Rev., 112, 424–432.

Nigam, S. (1994) On the dynamical basis for the Asian summer monsoon rainfall-El Niño relationship. J. Climate, 7, 1750-1771.

Palmer, T. N. (1994) Chaos and predictability in forecasting the monsoons. Proc. Indian Natl. Sci. Acad., 60A, N°. 1, 57-66.

Palmer, T. N., C. Brankovic, P. Viterbo, and M. J. Miller (1992) Modeling interannual variations of summer monsoons. J. Climate, 5, 399-417.

Pant, G. B. (1979) Role of tree ring analysis and related studies in paleoclimatology: preliminary survey and scope for Indian region. Mausam, 30, 439.

Pant, G. B. (1983) Climatological signals from the annual growth rings of selected tree species in India. Mausam, 34, 251.

Pant, G. B., and H. P. Borgaonkar (1984) Growth rate of Chir pines (Pinus roxburghii) trees in Kumaon area in relation to regional climatology. Himalayan Res. Dev., 3, 1-5.

Pant, G. B., H. P. Borgaonkar, and K. Rupa Kumar (1998) Climatic signals from tree rings: a dendroclimatic investigation of Himalayan spruce (Picea smithiana). Himalayan Geol., 19, 65-73.

Parthasarathy, B., A. A. Munot, and D. R. Kothawale (1994) All India monthly and seasonal rainfall series: 1871-1993. Theor. Appl. Climatol., 49, 217-224.

Parthasarathy, B., A. A. Munot, and D. R. Kothawale (1995) Monthly and seasonal rainfall series for all India, homogeneous regions and meteorological subdivisions: 1871-1994. Research Report No. RR-065, 113pp. [Available from Indian Institute of Tropical Meteorology, Homi Bhabha Road, Pune 411008, India.]

Philander, S. G. (1990) El Niño, La Niña, and the Southern Oscillation, Academic Press, San Diego.

Prell, W. L., and J. E. Kutzback (1987) Monsoon variability over the past 150,000 years. J. Geophys. Res., 92, 8411-8425.

Rajeevan, M. (2001) Prediction of Indian summer monsoon: Status, problems and prospects. Curr. Sci., 81, 1451-1457.

Ramamurthy, K. (1969) Some aspects of the break in the Indian southwest monsoon during July and August. Forecasting Manual, India Meteorological Department, Part IV-18.3.

Rao, K. G., and B. N. Goswami (1988) Interannual variations of sea surface temperature over the Arabian Sea and the Indian monsoon: A new perspective. Mon. Wea. Rev., 116, 558-568.

Rasmusson, E. M., and T. H. Carpenter (1983) The relationship between eastern equatorial Pacific sea surface temperatures and rainfall over India and Sri Lanka. Mon. Wea. Rev., 111, 517-528.

Rasmusson, E. M., X. Wang, and C. F. Ropelewski (1990) The biennial component of ENSO variability. J. Mar. Syst., 1, 71–96.

Rodwell, M. J., and B. J. Hoskins (1995) A model of the Asian summer monsoon. Part II: Cross-equatorial flow and PV behavior. J. Atmos. Sci., 52, 1341-1356.

Rogers, J. C. (1984) The association between the North Atlantic Oscillation and the Southern Oscillation in the Northern Hemisphere. Mon. Wea. Rev., 112, 1999-2015.

Ropelewski, C. F., M. S. Halpert, and X. Wang (1992) Observed tropospheric biennial variability and its relationship to the Southern Oscillation. J. Climate, 5, 594–614.

Ruddiman, W. F., and J. E. Kutzbach (1989) Forcing of late Cenozoic northern hemisphere climate by plateau uplift in southern Asia and the American west. J. Geophys. Res., 94, 18409-18427.

Ruddiman, W. F., W. L. Prell, and M. E. Raymo (1989) Late Cenozoic uplift in southern Asia and the American west: rationale for general circulation experiments. J. Geophys. Res., 94, 18379-18391.

Saji, N. H., B. N. Goswami, P. N. Vinayachandran, and T. Yamagata (1999) A dipole mode in the tropical Indian Ocean. Nature, 401, 360-363.

Shukla, J. (1975) Effect of Arabian sea-surface temperature anomaly on Indian summer monsoon: A numerical experiment with the GFDL model. J. Atmos. Sci., 32, 503-511.

Shukla, J. (1987) Interannual variability of monsoons. Monsoons, J. S. Fein, and P. L. Stephens, Eds., Wiley and Sons, 399-463.

Shukla, J., and M. J. Fennessy (1994) Simulation and predictability of monsoons. Proc. Int. Conf. on Monsoon Variability and Prediction, Tech. Rep. WCRP-84, World Climate Research Programmes, Geneva, Switzerland, 567-575.

Shukla, J., Y. Mintz (1982) Influence of land surface evapotranspiration on the Earth's climate. Science, 215, 1498-1501.

Shukla, J., and B. M. Misra (1977) Relationships between sea surface temperature and wind speed over the central Arabian Sea, and monsoon rainfall over India. Mon. Wea. Rev., 105, 998-1002.

Shukla, J., and D. A. Mooley (1987) Empirical prediction of the summer monsoon rainfall over India. Mon. Wea. Rev., 115, 695-703.

Shukla, J., and D. A. Paolino (1983) The Southern Oscillation and long-range forecasting of the summer monsoon rainfall over India. Mon. Wea. Rev., 111, 1830-1837.

Sikka, D. R. (1980) Some aspects of the large-scale fluctuations of summer monsoon rainfall over India in relation to fluctuations in the planetary and regional scale circulation parameters. Proc. Indian Acad. Sci. (Earth & Planet. Sci.), 89, 179-195.

Sikka, D. R., and S. Gadgil (1980) On the maximum cloud zone and the ITCZ over Indian longitudes during the southwest monsoon. Mon. Wea. Rev., 108, 1840-1853.

Singh, S. V., R. H. Kripalani, and D. R. Sikka (1992) Interannual variability of the Madden-Julian oscillations in Indian summer monsoon rainfall. J. Climate, 5, 973-978.

Slingo, J. M., and H. Annamalai (2000) 1997: The El Niño of the century and the response of the Indian summer monsoon. Mon. Wea. Rev., 128, 1778-1797.

Sperber, K. R., C. Brankovic, M. Déqué, C. S. Frederiksen, R. Graham, A. Kitoh, C. Kobayashi, T. Palmer, K. Puri, W. Tennant, and E. Volodin (2001) Dynamical seasonal prediction of the Asian summer monsoon. J. Climate, 129, 2226-2248.

Sperber, K. R., and T. N. Palmer (1996) Interannual tropical rainfall variability in general circulation model simulations associated with the Atmospheric Model Intercomparison Project. J. Climate, 9, 2727-750.

Sperber, K. R., J. M. Slingo, and H. Annamalai (2000) Predictability and the relationship between subseasonal and interannual variability during the Asian summer monsoon. Quart. J. Roy. Meteor. Soc., 126, 2545-2574.

Sud, Y. C., and W. E. Smith (1985) Influence of local land-surface processes on the Indian monsoon: A numerical study. J. Appl. Meteor., 24, 1015-1036.

Terray, P. (1995) Space-time structure of monsoon interannual variability. J. Climate, 8, 2595-2619.

Thapliyal, V. (1990) Long range prediction of summer monsoon rainfall over India: Evolution and development of new models. Mausam, 41, 339-346.

Thapliyal, V., and S. M. Kulshrestha (1992) Recent models for long range forecasting of southwest monsoon rainfall in India. Mausam, 43, 239-248.
Torrence, C., and P. J. Webster (1999) Interdecadal changes in the ENSO-monsoon system. J. Climate, 12, 2679-2690.
Trenberth, K. E. (1975) A quasi-biennial standing wave in the Southern Hemisphere and interrelations with sea surface temperature. Quart. J. Roy. Meteor. Soc., 101, 55–74.
Trenberth, K. E. (1976a) Fluctuations and trends in indices of the Southern Hemisphere circulation. Quart. J. Roy. Meteor. Soc., 102, 65–75.
Trenberth, K. E. (1976b) Spatial and temporal variations in the Southern Oscillation. Quart J. Roy. Meteor. Soc., 102, 639–653.
Trenberth, K. E. (1980) Atmospheric quasi-biennial oscillations. Mon. Wea. Rev., 108, 1370–1377.
Trenberth, K. E. (1990) Recent observed interdecadal climate changes in the Northern Hemisphere. Bull. Amer. Meteor. Soc., 71, 988-993.
Trenberth, K.E. and J.W. Hurrell (1994) Decadal atmosphere-ocean variations in the Pacific. Clim. Dyn., 9, 303-319.
Trenberth, K. E., D. P. Stepaniak, and J. M. Caron (2000) The global monsoon as seen through the divergent atmospheric circulation. J. Climate, 13, 3969–3993.
Van Loon, H. (1984) The Southern Oscillation. Part III: Associations with the trades and with the trough in the westerlies of the South Pacific Ocean. Mon. Wea. Rev., 112, 947–954.
Van Loon, H., and D. J. Shea (1985) The Southern Oscillation. Part IV: The precursors south of 15°S to the extremes of the oscillation. Mon. Wea. Rev., 113, 2063–2074.
Vernekar, A., D., J. Zhou, and J. Shukla (1995) The effect of Eurasian snow cover on the Indian monsoon. J. Climate, 8, 248–266.
Walker, G. T. (1923) Correlation in seasonal variations of weather, VIII: A preliminary study of world weather. Mem. India Meteor. Dep., 24, 75-131. [Also published in Sir Gilbert T. Walker - Selected Papers, Indian Meteorological Society, New Delhi, 1986, 120-178.]
Walker, G. T. (1924) Correlation in seasonal variations of weather, IX: A further study of world weather. Mem. India Meteor. Dep., 24, 275-332. [Also published in Sir Gilbert T. Walker - Selected Papers, Indian Meteorological Society, New Delhi, 1986, 179-240.]
Walker, J., and P. R. Rowntree (1977) The effect of soil moisture on circulation and rainfall in a tropical model. Quart. J. Roy. Meteor. Soc., 103, 29-46.
Wang, B. (1995) Interdecadal changes in El Niño onset in the last four decades. J. Climate, 8, 267-285.
Webster, P. J. (1987) The elementary monsoon. Monsoons, J. S. Fein and P. L. Stephens, eds., John Wiley and Sons, 3-32.
Webster, P. J., and L. Chou (1980) Low frequency transition of a simple monsoon system. J. Atmos. Sci., 37, 368-382.
Webster, P. J., V. O. Magaña, T. N., Palmer, J. Shukla, R. A. Tomas, T. M. Yanai, and T. Yasunari (1998) Monsoons: Processes, predictability, and the prospects for prediction. J. Geophys. Res., 103, 14451-14510.
Webster, P. J., A. W. Moore, J. P. Loschnigg, and R.R. Leben (1999) Coupled ocean-atmosphere dynamics in the Indian Ocean during 1997-98. Nature, 401, 356-360.
Webster, P. J., and S. Yang (1992) Monsoon and ENSO: Selectively interactive systems. Quart. J. Roy. Meteor. Soc., 118, 877-926.
Yanai, M., and C. Li (1994) Interannual variability of the Asian summer monsoon and its relationship with ENSO, Eurasian snow cover and heating. Proc. Int. Conf. on Monsoon Variability and Prediction, WMO/TD 619, Vol. I, World Meteorological Organization, Geneva, Switzerland, 27-34.
Yasuda, T., and K. Hanawa (1997) Decadal changes in the mode waters in the midlatitude North Pacific. J. Phys. Oceanogr., 27, 858-870.
Yasunari, T. (1979) Cloudiness fluctuations associated with the northern hemisphere summer monsoon. J. Meteor. Soc. Japan, 57, 227-242.

Yasunari, T. (1985) Zonally propagating modes of the global east–west circulation associated with the Southern Oscillation. J. Meteor. Soc. Japan, 63, 1013–1029.
Yasunari, T. (1989) A possible link of the QBO's between the stratosphere, troposphere and the surface temperature in the tropics. J. Meteor. Soc. Japan, 67, 483–493.
Yasunari, T. (1990) Impact of Indian monsoon on the coupled atmosphere/ocean system in the tropical Pacific. J. Meteor. Atmos. Phys., 44, 29-41.
Yasunari, T. (1991) The monsoon year - A new concept of the climatic year in the Tropics. Bull. Amer. Meteor. Soc., 72, 1331–1338.
Yeh, T.C., R. T. Wetherald, and S. Manabe (1984) The effect of soil moisture on the short-term climate and hydrology change - a numerical experiment. Mon. Wea. Rev., 112, 474-490.
Zhang, Y., J. M. Wallace, and D. S. Battisti (1997) ENSO-like interdecadal variability: 1900-93. J. Climate, 10, 1004-1020.
Zheng, Q., and K.N. Liou (1986) Dynamic and thermodynamic influences of the Tibetan plateau on the atmosphere in a general circulation model. J. Atmos. Sci., 43, 1340-1354.

11 Interactions between the Tropics and Extratropics.

A bit on theory, results and prospects for future predictability.

Xavier Rodó.

GRC – Grup de Recerca del Clima, Centre de Climatologia i Meteorologia, Parc Científic de Barcelona, Universitat de Barcelona. Baldiri i Reixach 4-6, Torre D, 08028 Barcelona, Catalonia
xrodo@porthos.bio.ub.es

Abstract. This chapter attempts to review current research on teleconnections between the Tropics and the Extratropics, for the most part at midlatitudes and focusing mainly, though not exclusively, on the Atlantic Ocean. El Niño-Southern Oscillation (ENSO) phenomenon, the largest interannual climate signal, is the first climate phenomenon shown to depend essentially upon coupled interactions of the dynamics of both ocean and atmosphere and the only one, at present, capable of yielding some (moderate) interannual predictability. In recent works, some regions in the Extratropics have appeared more sensitive to ENSO impacts in the present times. It is however, still unknown whether or not this situation is completely new, and if so, linked to the effects of global warming. Other important uncertainties not yet solved are the unknown impacts on the Extratropics, of large-scale climatic phenomena such as ENSO, in a future and more variable warmer world. The tropical atmospheric bridge hypothesis, though not the only hypothesis, provides a strong basis to address these sorts of interactions dynamically. Other mechanisms for the generation of interannual to interdecadal variability at midlatitudes that involve internal ocean dynamics in midlatitude oceans, might also provide useful insights to enhance climate predictability and will be briefly discussed in this chapter. As a last point, a brief overview is introduced on recent studies and techniques developed for the isolation and modeling of climate forcing on ecosystems. This extreme is achieved using newly-developed specific statistical tools and taking into account transient threshold-dependent relationships.

11.1 Introduction.

Interactions between the ocean and atmosphere contribute to the climate variability spectrum over a wide range of time and spatial scales. Among them, the ones that provide predictability at midlatitudes are very limited, in comparison with the amount of climate predictable fluctuations and the insight given by El Niño phenomenon in the Tropics. Gains in understanding and predicting ENSO provided the goal and the major successes of the Tropical Ocean-Global Atmosphere (TOGA) program (Neelin et al. 1998). Therefore, it has been recognized that

ENSO and consequently, the tropical Pacific (TP) Ocean, yields the largest potentially predictable signals in the Extratropics. Recently, some studies in extratropical oceans postulate the existence of decadal variability (thus predictable) in regions of the North Pacific and North Atlantic. These results, whether or not linked to ENSO, are still controversial. In the time range from seasons to a few years, potential for predictability is very small in the Extratropics, due to the low signal-to-noise ratios there (Hoerling and Kumar 1995).

In the northern Extratropics, some climatic phenomena, namely the 'so-called' 'oscillations' have been thoroughly studied in the search for predictability, mainly for Eurasia. In the US, ENSO already yields some zonal predictability in certain areas, at least both in the west and the southeast coasts. Among the phenomena studied in the Nothern Hemisphere (NH), there are a bunch of teleconnected patterns such as the East Atlantic (EA) pattern, the EA Jet pattern (EA-JET), the West Pacific pattern (WP), the East Pacific pattern (EP), the North Atlantic Oscillation (NAO), the North Pacific pattern (NP), the Pacific/North American pattern (PNA), the East Atlantic/Western Russia (EATL/WRU), and the Scandinavian patterns (SCAND) among some, and even some other minor teleconnection patterns might see the light in the years to come. In the Southern Hemisphere (SH), weaker patterns exist, such as the Pacific-South American (PSA) teleconnection. For a thorough revision, the reader should refer to the works by Wallace and Gutzler (1981) and Barnston and Livezey (1987). In the last few years, however, with the exception of the monsoon system and ENSO, the climatic phenomena that by far have deserved more attention have been the Pacific-North American (PNA) pattern for Northern North America and the North Atlantic Oscillation (NAO) in Europe and eastern USA.

The recent establishment of the Arctic Oscillation (AO), as a counterpart to its analogous in the SH, and the new hypothesis of the NAO (Fig. 11.1) being a regional manifestation of the AO (Kerr 1999), begin to pave the way towards the possibility of gaining some relative predictability for midlatitudes in the future in this direction (Thompson and Wallace, 2002, Thompson et al. 2002). However, some of these climatic phenomena mainly yield most of their predictability in the interdecadal scales and their utility for seasonal to interannual predictions remains low. Thompson and Wallace (2002) documented the remarkable similarity between the leading modes of month-to-month variability in the extratropical general circulation of the Northern and Southern Hemispheres. Both can be characterized as "annular modes" involving "seesaws" in atmospheric mass between the polar cap regions poleward of 60° latitude and the surrounding zonal rings centered near 45° latitude, as manifested in the respective leading empirical orthogonal functions (EOFs) of the sea-level pressure (SLP) field. Similarly, an out-of-phase variation was reported in westerly momentum in the ~35° and ~60° latitude belts. However, some of these AO signatures and other AO lays on trends in sea-level pressure (SLP) and surface air temperatures (SAT) over the NH, occur in conjunction with both ENSO effects and the "ENSO-like" interdecadal variability, as documented in Trenberth and Hurrell (1994) and Zhang et al. (1997). This fact considerably complicates the task of separating among all these contributing factors .

Fig. 11.1 shows the common pressure field distributions associated to positive (Fig. 11.1a) and negative (Fig. 11.1b) phases of the NAO, typically forming in the North Atlantic-European (NAE) sector. The former is characterized by strong westerly winds (arrow) driving winter storms well east into the European continent. This situation accompanies wet conditions in northern Europe and dry conditions in the Mediterranean area. Conversely, negative NAO situations are dominated by a weak North Atlantic pressure system, causing less rainfall in the north and an excess rainfall amount over the Iberian Peninsula and northern Africa, as winter storms are allowed to move freely into the Mediterranean basin.

Some time before, seminal works established, in the 80's, the link between TP SST changes (such as the ones associated to ENSO) and atmospheric patterns out of the Tropics (Horel and Wallace 1981; van Loon and Madden 1981). In the beginning, these early works described an association with a characteristic pattern in the North Pacific/North American region. Later studies (Lau and Nath 1990 and 1994, Graham et al. 1994) began to focus on the same subject, and with GCM's, simulated SST responses in other distant basins and laid the foundations of the 'tropical atmospheric bridge' hypothesis. From a general point of view, the key question to explore has to do with the onset of SST anomalies at distant geographic sites in association with peak phases in ENSO and the physical processes responsible therein. Specifically, this question deals with the relative role of the TP *vs* the extratropical oceans in the generation of these associated SST anomalies. The local –external or induced *vs* internal- atmosphere-ocean feedbacks, and the mediating processes linking these anomalies were for a long time and still are, the main hotspots of research on this subject.

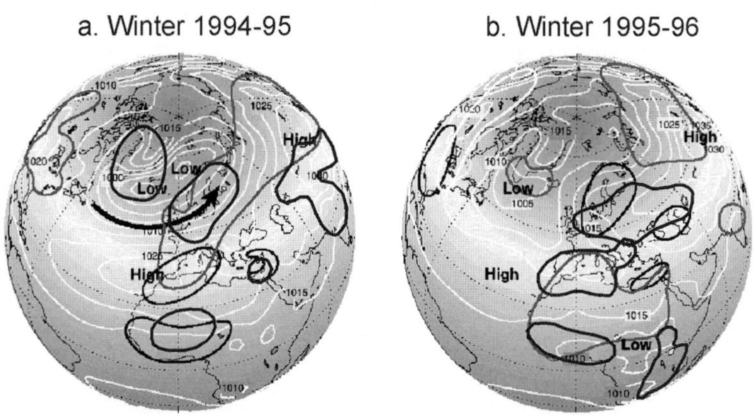

Fig. 11.1. Structure of the North Atlantic Oscillation phenomenon in its two typical phases. a) positive NAO, b) negative NAO. Arrow indicates strong westerly winds pushing winter storms eastward into the northern portions of the European continent (adapted from Kushnir 1999, with permission from Macmillan Magazines Ltd).

Fig. 11.2 shows the regressed patterns obtained for the surface winds, and the sensible and latent heat fluxes simulated in TOGA, a GCM experiment with monthly varying SST conditions in the TP (25°S and 25°N). In principle, the long timescales characteristic of the basin-scale ocean circulation are too long to account by themselves for this SST 'synchronicity' *sensu latto* (e.g. in some basins there is a characteristic delay of several months between the onset of SSTA in the TP and the appearance of similar SST changes in other oceanic regions). In addition to the patterns in the tropical oceans, there are other noticeable extratropical responses. Among these, the two cyclonic circulation anomalies in the North Pacific and the South Pacific convergence zone (Fig. 11.2a) are indeed remarkable, occurring in conjunction with the strong extratropical wind fields and positive fluxes in the midlatitude Atlantic (north and south), the Mediterranean Sea and the Southern Indian Ocean. In Fig. 11.2b, near the global ocean, the global heat and radiative fluxes appear linked to a tropical forcing. The main oceanic regions where associations to the Tropics have been described, are highlighted by boxes (Fig. 11.2). Arrows indicate geographical sources for the heatings or coolings in those regions. Among these are, for instance, the extratropical Pacific, the South China Sea, the midlatitude Atlantic, the Mediterranean Sea (Santoleri et al. 1995, Rodó 2001) and the Antarctic Circumpolar Wave (ACW) (Peterson and White 1998, Cai and Baines 2001). For the latter, studies suggested an ENSO influence on the ACW by advective oceanic teleconnections from the western subtropical South Pacific (Peterson and White 1998, Cai and Baines 2001). In particular, in the Southern Hemisphere, larger interannual teleconnections are found in the zonal mean flow, the Southern Hemisphere jet stream and storm tracks, rather than in the wavelike teleconnections. Thus, external influences such as those arising from the Tropics might be difficult to isolate. Despite this fact, some recent results seem to suggest a strong connection with the ACW, which for sure will be a matter of further research in the future.

However, in the search for ENSO-sensitive regions worldwide, signal identification procedures currently at use sometimes do not allow to appropriately capture the full range of responses occurring after a tropical forcing and therefore, some important impacts are not isolated. Some time ago, Jacobs et al. (1994), showed what looked like ENSO-associated responses in the North Pacific. In this situation Rossby waves were detected even a decade after they were generated. At least for the very-strong 1982-83 El Niño event, slow subsurface eastward propagation via equatorial Kelvin waves, that reflected on the American coastline, gave rise to a westward-propagating Rossby wave across the entire North Pacific. This Rossby wave propagation could be identified and tracked by monitoring sea-surface height (SSH) variations, though only a decade after its initiation. The picture however, is still more complicated, as some investigations showed how the Extratropics, and in particular the North Pacific might also play a crucial role in the generation of tropical decadal-scale variability. This process would involve the subsurface entrainment of subducted waters in the eastern North Pacific to the Tropics, altering the thermocline structure there and affecting the ulterior generation of warm anomalies conducting to stronger El Niños (Gu and Philander 1997,

Zhang et al. 1998) and perhaps also to ENSO regime shifts, such as the one in the late 1970s.

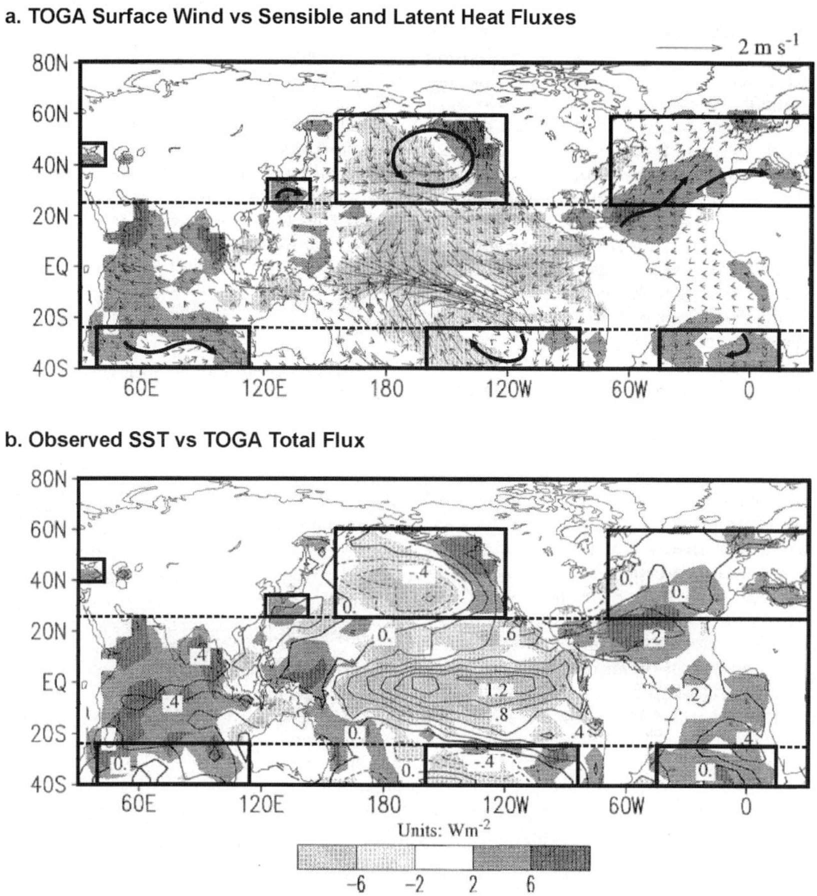

Fig. 11.2. Regression patterns of the scaled SST expansion coefficients associated with the leading singular value decomposition (SVD) mode for the TOGA experiment (results are averages over four different realizations for northern winters) versus a) zonal and meridional components of the wind field at the lowest layer of the model atmosphere and sum of latent and sensible heat fluxes from the atmosphere to the ocean (stippling), all simulated in the TOGA experiment; b) observed SST (contours, interval: 0.2°C) and TOGA-generated sum of downward latent heat, sensible heat, and radiative fluxes (stippling). Boxes approximately indicate main oceanic regions in the Extratropics where an association with a tropical forcing has been shown to exist. Adapted from Lau & Nath (1994).

Other limitations in methodological and modeling approaches concern nonlinear barotropic responses in the Extratropics, to forcing from SSTA in the Pacific. Similarly, the generation of transient dynamical responses between the Tropics and the Extratropics are not normally captured by standard statistical tools, nor properly reproduced by coupled models. Fortunately, in the last decades, and adding up to the signal-to-noise ratio controversy, an amplification of ENSO signals appears to have occurred that may help to clarify many of these ENSO forcings in the Extratropics (IPCC 2001). It is not known whether or not in conjunction or amplified by global warming. WETS (Workshop on Extra-Tropical SST anomalies), reviewed in Robinson (2000), stated that on timescales shorter than decades, in most places outside the Tropics, the atmosphere is the main generator of SST anomalies. On seasonal to interannual timescales, most variability in the extratropical atmosphere is intrinsic –generated by the internal nonlinear dynamics of the atmosphere-, and inherently unpredictable.

The development of new data-adaptative techniques for the analysis of local interactions in multiple series of spatial and temporal data, together with empirical and dynamical models capable of coping with these sorts of transient signals, are opening a new area of research in midlatitude climate studies. Otherwise, it is basically true that forcing from the tropical and equatorial regions manifest differently in different regions of midlatitudes, some being more sensitive than others. Though some of these locations have been recently identified, responses to these forcings are not always accurately simulated, even for these extratropical sensitive regions.

In the following sections and in addition to an insight to the dynamics underlying teleconnections between the Tropics and regions in the Extratropics, the main effort will center mainly in the area of improving the identification of nonstationary climatic signals, and to revise some recent approaches to modeling climate-ecosystems interactions.

In Sect. 11.2, a brief outline on the state of the art and the origins in teleconnection research is given. Sect. 11.3 focuses on some of the physical theory underlying tropical-extratropical interactions. Sect. 11.4. focuses on some of the current limitations on the signal identification techniques and some suggestions are given on how the approach to transient isolate signals should be constructed. Sect. 11.5 briefly refers to the current state of predictability research on tropical-extratropical forcing. In Sect. 11.6, some recent results on the feedbacks that may exist from midlatitudes to the Tropics at interdecadal timescales, are briefly discussed. Finally, Sect. 11.7 gives some ideas which might be the future areas of research on this topic in the years to come.

11.2 Brief background knowledge.

Observed relationships between theTropics and other latitudes which have in their origin an anomalous SST forcing from the Tropics were commonly termed teleconnections after the 1950s. It has not been however, since TOGA, that there is a widespread use of this term in climatological studies. As a routine, methods such as correlations and composite analysis are normally used to establish these sorts of relationships. This notwithstanding, in recent times, some concern has been shown regarding the ability of these techniques to detect weak signals or transients in nonlinear systems (Rodó 2001, Rodríguez-Arias and Rodó 2002).

After Bjerknes (1966) noted a teleconnection between ENSO warm events and the midlatitude circulation in the North Pacific and the North Atlantic, later discovered to affect North America and Europe, many studies were devoted to searching for ENSO signals throughout the world. Among those, Horel and Wallace (1981) and van Loon and Madden (1981) were probably the seminal studies, but many others followed seeking associations with surface temperature and precipitation (Rasmusson and Carpenter 1982, Kiladis and Díaz 1989, Ropelewski and Halpert 1989, Philander 1990, Dai et al. 1997). Recently, Dai and Wigley (2000) re-analyzed SOI-precipitation relationships, with precipitation data coming from merging recent satellite estimates of oceanic precipitation and old rain-gauge records, to drive a global climatology of ENSO-induced precipitation variations. They argued that associated changes in the monsoon systems (through the Hadley cells) over the Pacific, Indian and Atlantic Oceans and their interactions with midlatitude systems (particularly westerlies) generate coherent anomaly patterns over the Extratropics.

For Europe and the North Atlantic sector, several studies sought for possible associations with ENSO. Among those, for instance, Hamilton (1988) investigated the detailed regional sea-level pressure responses to warm ENSO events. However, despite the global impact of ENSO, there had been little hard evidence of ENSO impacts in Europe (Fraedrich and Muller 1992, von Storch and Kruse 1985, Zorita et al. 1992, von Storch et al. 1993). Ropelewski and Halpert (1987) and Kiladis and Díaz (1989), isolated two ENSO-sensitive regions in Northern Africa-Southern Europe (NAS) and the middle East and Israel, though due to sparse data, the strength of these relationships appeared to be weak. Later, other analyses (Rodó et al. 1997, Yakir et al. 1996), appeared to confirm the sensitiveness of these regions in the Mediterranean area to ENSO transient forcing.

Similar results arising in other regions, pointed again to the likely possibility that an intensification of these signals in the second half of the twentieth century in association with changes in ENSO itself, might be taking place, whether or not linked to global warming. For instance, May and Bengtsson (1998) described an impact of ENSO in the Atlantic/European sector pointing to the intensification (reduction) of the Aleutian low and the simultaneous reduction (intensification) of the Icelandic low during El Niño (La Niña) events, as the key determining factor. Conversely, other studies also began to address the appearance of lagged SST responses in separate basins (Lanzante 1996, Enfield and Mayer 1997). However, whether it is through the PNA, the Walker-Hadley system in the Atlantic or other alternative mechanisms, the transmission to midlatitudes is still a subject of ongoing debate.

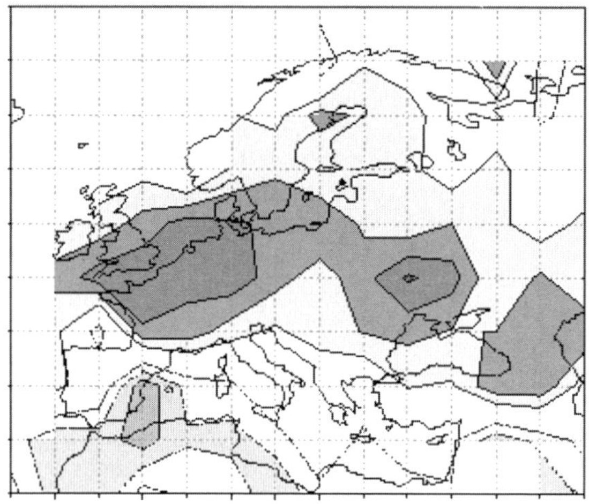

Fig. 11.3. Springtime (March through May) correlations between precipitation and spring averaged Niño-3 index in the European sector. A dipole structure in correlations with spring rainfall is evident between central and northern Europe and the Mediterranean area. (reprinted from Oldenborgh et al. 2000, with permission from Int. J. Clim.)

In Fig. 11.3, a dipole in correlations between rainfall on a grid in Europe and Niño-3 index clearly manifests coherence with former results obtained by other studies. These ongoing approaches are giving support to the seminal hypothesis of an ENSO impact on rainfall in selected areas in Europe, though there might be different mechanisms and timescales involved in these associations (Klein et al. 1999, Rodó 2001, Drevillon et al. 2001, Lau and Nath 2001). In recent studies, the tropical North Atlantic has been shown as a connecting bridge with the NAE sector, and perhaps there it lays the way by which forcing from the Pacific translates into higher latitudes in the Atlantic. A complete review of the tropical atmospheric bridge linking with the Extratropics, can be found in Lau and Nath (1990, 1994, 2001).

The role of the atmospheric circulation as a 'bridge' between SSTA in the TP and those appearing in midlatitude northern oceans, was first described by Weare et al. (1976) and Hsiung & Newell (1983) and later reviewed in Lau and Nath (1994). In episodes with peaking SSTA in the TP, extratropical perturbations in atmospheric temperature, humidity and wind fields induce changes in the latent and sensible heat fluxes across the air-sea interface of the midlatitude oceans. These anomalous fluxes, in turn, are described to originate extratropical SST changes (Lau and Nath, 1996). With this mechanism it has been possible to simulate the basic features of the observed SST variability in the North Pacific and western North Atlantic.

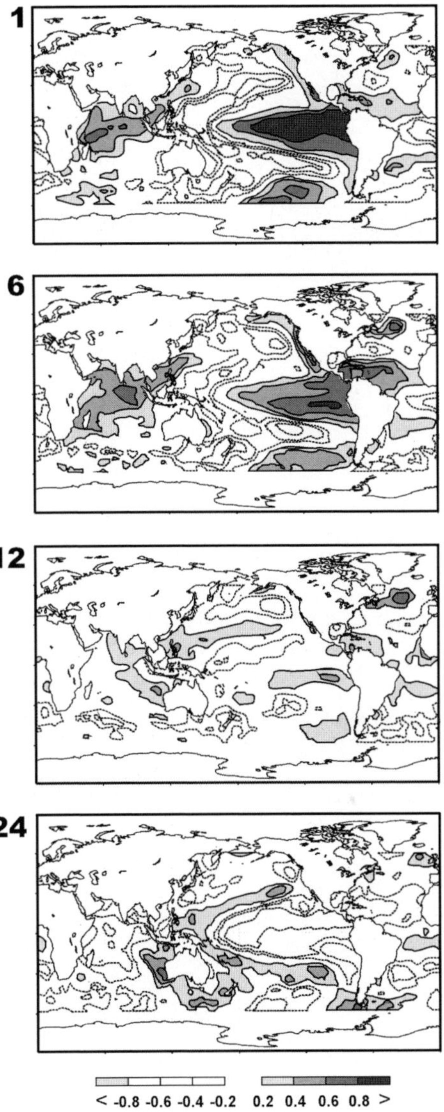

Fig. 11.4. Field correlations between an ENSO index consisting of ship-observed SSTA averaged for 5°N, 5°S, the South American coast and 180°W and SSTA, with the latter lagging ENSO 1, 6, 12 and 24 months (see Rodó 2001, for details). Contoured areas surpass $p<0.001$. (from Rodó 2001, reproduced with permission from Springer-verlag, Berlin-New York)

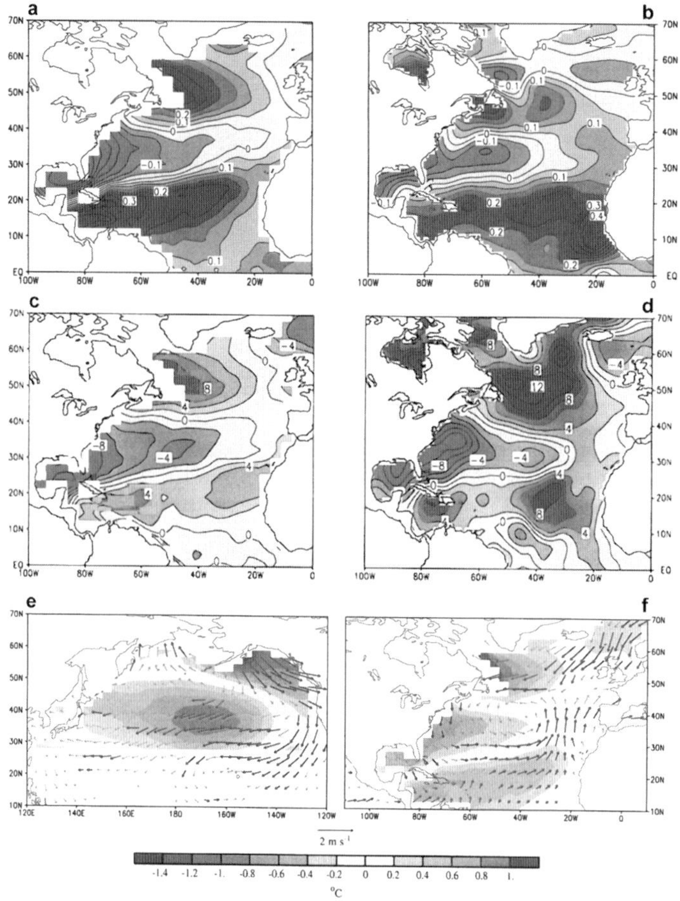

Fig. 11.5. Observed (right panels, a and c) and TOGA-ML simulated (left panels, b and d) fields of correlation between spring SSTA (March through May) and Niño-3 (January and February) index (upper panels, a and b), and sensible plus latent heat fluxes (middle panels, c and d). Lower panels (e and f) refer to the signatures in a TOGA-ML minus TOGA run for the cold composite (e) and warm composite (f) for SST/surface wind vectors. (Redrawn from Figs. 5,6 and 11 of Lau and Nath 2001, with permission of J. Climate).

In the 80's, Blackmon et al. (1984) reported that teleconnection phenomena could be grouped into three broad categories. Those in the NH, jet exit regions over the oceans and jet entrance areas over the continents seem to play a central role in the generation of some dipolar structures (e.g., those formed in the WP and

NP patterns). In these exit regions, interaction takes place with their own regional internal dynamics. This process can eventually either enhance or counteract the transmission of signals from the Tropics, thus originating an effect in some seasons and not in others. Anyhow, it is clear that these regions of midlatitudes offer a higher potential for predictability, even though on certain occasions, a manifest nonlinearity in responses has equally been documented (Hoerling and Kumar 1997).

The nonlinear approach to lagged responses to tropical forcing, as those reported in Lanzante (1996) and in Klein et al. (1999) for instance, may help to simulate sea surface temperatures arising from forcings originating in the TP. Continuing in this direction, Rodó (2001) traced the observed lagged atmospheric circulation changes occurring together with the delayed generation of anomalous fields in certain variables of interest. Structures that form in the temporal cascade, may help to explain barotropic transmission to midlatitudes. Fig. 11.4 shows the changes in SSTA fields linked to ENSO, investigated with delays of up to 2 years. A reversal of oceanic and atmospheric anomaly fields is evident in conjunction to changes of El Niño to La Niña conditions between 1986-87 and 1988-89. Concomitant reversals in atmospheric patterns during El Niño were also noted for satellite-measured total cloud cover and top-of-atmosphere absorbed solar radiation. These changes take place in conjunction with large-scale atmospheric circulation changes, whose motion can be traced with the aid of upper tropospheric relative humidity. Altered circulation patterns can anomalously affect several areas in mid-to-high latitudes. Among these, of particular relevance are parts of Europe (mainly the Mediterranean area and mid North Atlantic-European, NAE, sector) and the ACW in the SH. Attempts to simulate much of this behavior of the climate system in response to a tropical forcing during peak phases in ENSO recently yielded some success, as shown in Fig. 11.5. Using a TOGA-ML run in comparison with the spring season (March through April) in the north Atlantic, a noticeable degree of similarity with the observations is evident in the patterns reproduced.

In the NAE sector, the NAO drives seasonal variability in winter. However, only limited potential predictability at interdecadal scales has been recognized (Griffies and Bryan 1997). For the TP, also at longer timescales, a teleconnection with the North Pacific was noted in winter. The period of these fluctuations, though not accurately determined due to the lack of long SST records in the region, is however recognized to fluctuate around twenty years. Linkages between this region and the TP seem to have increased since the Pacific regime shift occurring in late seventies (Trenberth and Hurrell 1994). For instance the deeper Aleutian low emerging since those times, appeared to have profound ecosystem and societal impacts in the North Pacific ocean and nearby regions.

Concomitant with these predictability studies, the main question to address concerns the relative roles of the North Pacific ocean and atmosphere in forcing each other, and to assign a predominant role to one of the two. This fact would either imply or disregard the existence of an internal oceanic mode in the North Pacific, not necessarily linked to the TP and ENSO. This approach is illustrated in Fig. 11.6 on the basis of both simulated EOFs patterns from a mixed layer ocean model coupled to an AGCM and their similitude to the structure arising in the second EOF of SST measurements between 1945 and 1999.

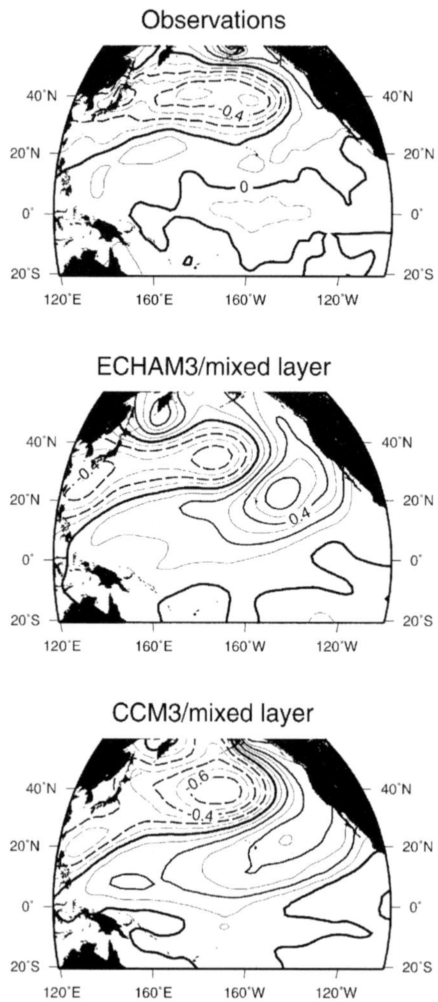

Fig. 11.6. Internal generation of the North Pacific Mode in the North Pacific, which appears not ultimately linked to a tropical forcing for the decadal timescales. Figure shows the second EOF from observations in the period 1945 to 1999 (top), and the leading EOFs simulated from two AGCM coupled to a mixed-layer ocean (middle and bottom). Reproduced from Pierce et al. 2001, with permission from Springer-Verlag, Berlin-New York.

11.3
Dynamics of tropical-extratropical connections: a summary of theoretical approaches

11.3.1
Forcing in the source

Bjerknes in the late sixties established a conceptual model on how the tropical ocean might affect the Extratropics (Bjerknes 1966). He described that a positive temperature anomaly in the TP strengthens the midlatitude zonal wind system there. The intensified Aleutian Low then anchors the phases of the prevailing waves in the upper westerlies. Consequently, a succession of positive and negative stationary wave anomalies appears downwind over North America/Southern Greenland and the north-western Europe and eastern north Atlantic. Later, significant responses with global pressure and temperature fields were found in regions of Europe and the North Atlantic (van Loon and Madden 1981, van Loon and Rogers 1981). In essence, Bjerknes initial hypothesis on the existence of coupled interactions of the dynamics of both ocean and atmosphere is still alive and the TOGA decade has, among other things, confirmed his initial findings (Neelin et al. 1998). At large scales for the tropical region, each medium is strongly controlled by the mutually imposed boundary conditions. While the large-scale upper ocean circulation is mainly determined by the past history of the wind stress, tropical atmopheric circulation from scales as short as seasonal onwards, is largely driven by SST (Fig. 11.7).

In the forcing region, the increased tropical heating during an El Niño event, originates changes in the zonal mean circulation in the Tropics and Subtropics, in addition to a local amplification of associated teleconnections further away. As a consequence, an increase in the zonal mean temperature takes place in the middle to upper troposphere during each event, in association with latent heat release in the region of enhanced convection and adiabatic heating in the descending branches of the anomalous Walker and Hadley circulations (Horel and Wallace, 1981). As a result, the zonal mean flow exhibits an enhanced Hadley circulation, together with increased equatorial easterlies in the tropical upper troposphere and enhanced subtropical westerlies.

Bjerknes hypothesis relies on the assumption that ENSO arises as a self-sustained cycle in which SST anomalies in the Pacific cause the trade winds to strengthen or slacken. This in turn, drives the ocean circulation changes that produce SSTA (Neelin et al. 1998). However, the mechanisms for the transition among phases were not clearly established by that time and have been the subject of considerable research. Now, it seems clear that the ocean, with its lower damping timescales, provides the memory that drives the transition to the different phases. When this occurs, SST boundary conditions in the TP eventually force the atmosphere, driving direct circulation cells thermodynamically. As a result, the Intertropical Convergence Zones (ITCZs) structures in the form of deep convective regions over the areas with warmest SST.

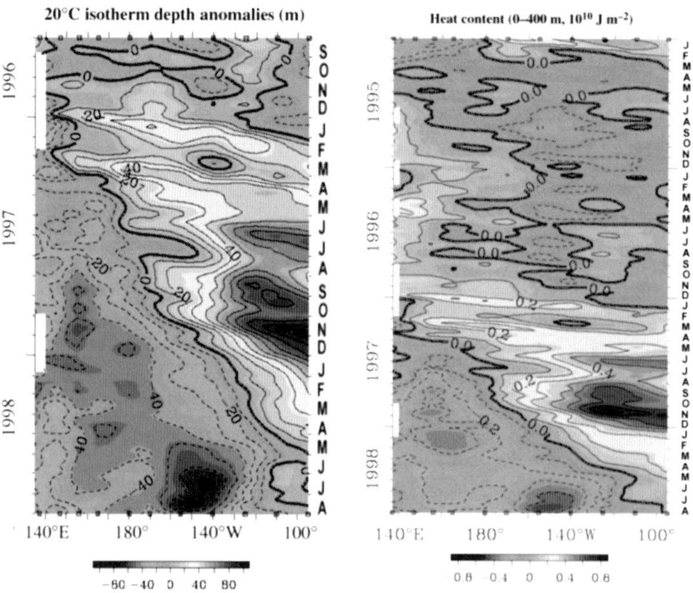

Fig. 11.7. Time vs longitude sections of anomalies in the 20°C isotherm depth (left) and observed heat content anomalies (from 0 to 400 m depth) averaged between 2°N and 2°S from the TAO array (right). Note the different temporal intervals included (redrawn from McPhaden 1999, with permission from Science).

For a thorough description of the essentials of ENSO, see Wallace et al. (1998). Summarizing the two key elements in the Bjerknes hypothesis, there is dependence on wind-driven ocean dynamics of the SST in the 'equatorial cold tongue region' (CT), and the Walker circulation response to anomalies in the SST pattern. The equatorial CT region in the eastern Pacific arises then from both the upwelling and the shallow thermocline occurring in that region, as opposed to a deep thermocline and warm SST west of the dateline. Conversely, an ENSO-like mode has been described –termed GR-, though its variance is mainly concentrated on the decadal time range. This mode (shown in Fig. 11.8 above and Fig. 11.9 below), with a similar spatial signature of that of CT has been sometimes erroneously assigned to ENSO.

Observations during pre-TOGA and early TOGA era initially regarded ENSO as an energy relaxation process arising in a deterministic ocean underlying a purely stochastic atmosphere. Later, the ideas of an atmosphere that couples to the ocean and the knowledge of the movements of heat through these media, furnished a more thorough interpretation of what is now understood as the dynamics of ENSO. First attempts to simulate this theory may be traced back to a linear stability study using a shallow water system coupled to a Gill atmospheric model

(Philander et al. 1984). In that experiment, SST was proportional to the thermocline depth and it was possible to simulate an eastward equatorial propagation. Concomitant with these studies, Cane and Zebiak began to obtain sustained oscillations, close to those observed in ENSO, in what was later known as the Cane-Zebiak (CZ) model. Another major accomplishment at that time, were the first ENSO forecasts made with a coupled ocean-atmosphere model (Cane et al. 1986), still an issue of interest in the restless search for accurate forecasts of seasonal-to-interannual variability.

When searching for ENSO signatures in the Extratropics, the first step always deals with the identification of typical ENSO frequency bands. Though ENSO has a spectrum with many frequencies, power tends to concentrate at the quasi-biennial (QB) (roughly more than two years) and quasiquadrennial (QQ) (though it may attain 3 and 6 years) bands. However, the existence of these two 'main' frequencies is still a subject of controversy due to the broad spectral peaks in which they are embedded. As far as we know from ENSO data and proxies, the contribution of both QB and QQ components has varied historically (both in timing and amplitude).

However, appropriate ENSO isolation has not been an easy task, mainly due to the high irregularity and its phase-locking to the seasonal cycle. The sources of this irregularity are currently assigned to emerge from either deterministic chaos on the one hand, or to atmospheric weather noise, or most likely, to a combination of both. For the former, the nonlinear interactions with the annual cycle may strongly affect the temporal evolution of ENSO, an issue that was for a long time overlooked in modeling attempts, and that is now recognized as crucial. For the latter, which may be of particular interest in the integration of ENSO signals within midlatitude 'noise', the key point is the short decorrelation timescales of synoptic systems, of the order of a month or so (Hasselmann 1976) and the high proportion of variance they represent. In this sense, it has been widely recognized that analyzing seasons other than winter in midlatitude studies, may help in rising the contribution of signals above a noise floor, that is variable in time.

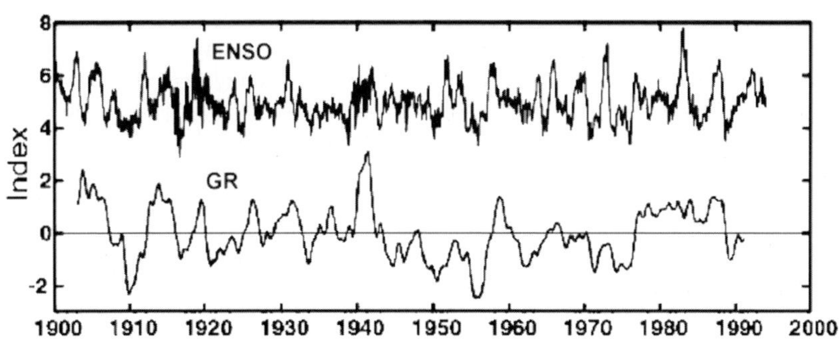

Fig. 11.8. Monthly series for ENSO and the ENSO-like mode (GR), from 1900. Note the slower frequency component in GR associated with its decadal signature (reproduced from Zhang et al. 1997, with permission from J. Climate)

As a consequence, a fundamental question arises regarding the relative roles of SSTA in distant places in the modulation of atmospheric variability at midlatitudes. The extratropical SSTA have since long been used as predictors in forecasts both in Europe and North America (Namias 1969). However, due to the limited length of historical data sets with few anomalous events, it is not possible to assess on the basis of instrumental data alone, whether tropical or extratropical SSTA are more effective contributors to atmospheric variability at midlatitudes. This fact urged the development of general circulation models (GCM), where the environment generating the anomalies was known and controlled. It was furthermore possible to simulate forcings at specific locations, with the use of arrays of experiments. Conversely to the refinement of models, until recently we have not seen a parallel enhancement of analytical tools, mainly of those allowing for constraints in time and space. This approach would definitely help to overcome the large temporal correlation existing between the tropical and extratropical SSTA. In GCM, such length limitations are solved by merely lengthening the integrations, or with the development of multiple integrations at a time.

Several attempts have been pursued with different success, to try to isolate and separate tropical from extratropical SSTA forcings in midlatitude flow patterns with AGCMs. The results primarily point to the TP being the strongest and most reliable driver, far beyond extratropical ocean. For instance, Lau and Nath (1994) showed responses to prescribed SSTA in various sites (in GOGA, TOGA and MOGA runs), with apparent and out-of-phase midlatitude ocean. Notwithstanding these results, atmospheric anomalies so reproduced at 500 mbar height, were much weaker than observed, as simulated responses at most, attained half of its real value, and in many situations, were located differently. The reason of these discrepancies may lie on either the lower amplitude resulting from the ensemble, with respect to individual runs, weaker responses to SST changes or an inaccurate representation of storminess. The coverage of transient forcing, acting only over an environmental threshold, is either removed or considerably diminished.

Trenberth and Hurrell (1994) stated, regarding these discrepancies, that altered synoptic systems might make it even more difficult to account for eddy transports. They also pointed to a differential atmospheric heating in the Extratropics, that instead of being local, may change as a function of atmospheric fluxes. In the Extratropics, the generation of synoptic systems is regional rather than local and this fact adds a large 'uncontrolled' component to the forcing. The issue of detecting any systematic effects in both models and the real atmosphere is, consequently, further complicated. As a matter of fact, in the Extratropics, it has been possible to accurately reproduce in atmospheric models certain decadal changes associated to ENSO, using specified SST (Kawamura et al. 1997). The outcome is completely different when synoptic systems –and thus shorter scales-, are taken into account. In such a situation, the sensible heat exchanged between the ocean and atmosphere, is realized locally, conversely to the oceanic latent heat lost through evaporation. The latter can only be assessed in the form of an increase in moisture, but the amount of atmospheric heating is not realized until it manifests as precipitation, often far from the original heating sources.

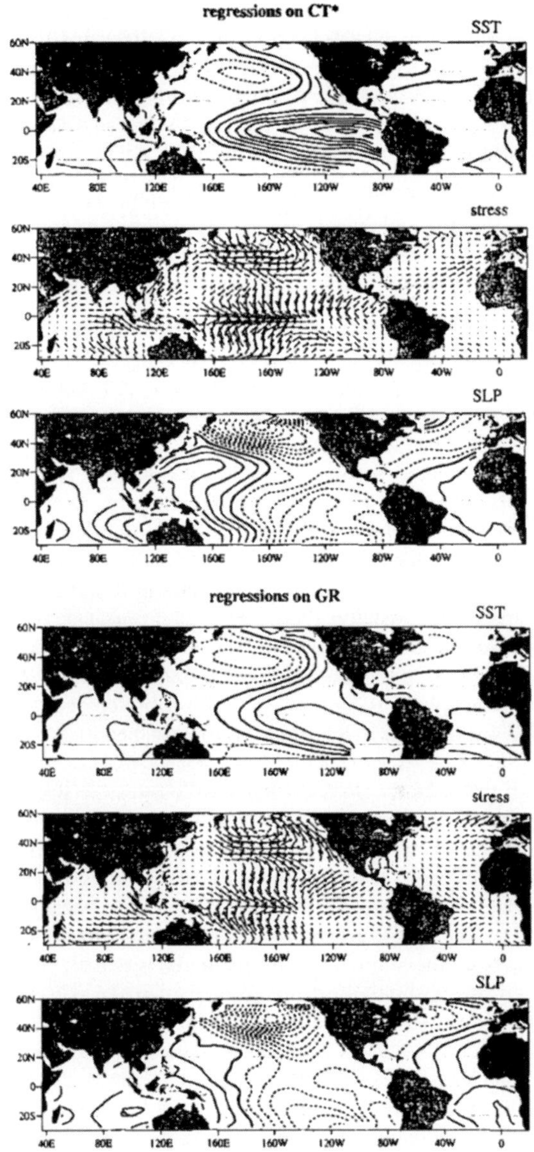

Fig. 11.9. Near-global fields regressed upon the monthly CT series and the monthly ENSO-like decadal GR series. Both CT and GR series are shown in Fig. 11.8. (reproduced from Zhang et al. 1997, with permission of J. Climate)

11.3.2
Transmission and interactions with midlatitude internal dynamics

Though still at present, Rossby wave propagation is believed to underlie the connections of the Tropics and Extratropics. Other important factors determining the extratropical circulation (Trenberth et al. 1998) have to do with the:

1. location and intensity of the tropical circulation anomaly.
2. effects of mean flow on planetary wave propagation and forcing.
3. interactions with midlatitude storm tracks.
4. interference from the internal chaotic variability of the midlatitude circulation.

In this respect, the mean winter tracks of some of the main teleconnection modes associated with the TP have been traced using storminess generation and previously discussed in this chapter (namely the CT, GR and PDO). These correlation structures, when positive, indicate an increase in storminess associated with warmer waters in the TP (Fig. 11.10).

ENSO has been related to variations of precipitation and temperature over much of the Tropics and Sub-tropics, and some mid-latitude areas. The search for ENSO signatures in the Extratropics has traditionally been a complicated task, mainly due to the difficulties in isolating forcing that have a discontinuous contribution which is commonly masked by the strong internal variability. Thus, whereas in the tropical atmosphere the decorrelation timescales of processes are so short as to allow for a proper separation from slower oceanic evolution, forced by SST, this is not the case in midlatitudes. There, by contrast, internal variability of both ocean and atmosphere is larger, and the task of separating fast components from the second from slower ones from the former, becomes a much more complicated issue.

In the case of the Atlantic Ocean, a dipole-like dynamic structure accounts for over a 40% of the total variance in the 1st SVD (Fig. 11.11). Observed and simulated climate impacts from teleconnections, which merge with the internal oceanic modes in each basin result, however, from the compositing between the 'forcing' component or 'signal' and the natural variability of the oceanic sub-system that accommodates it. The result would refer, *sensu latto,* to what we would call, statistically speaking, the 'noise'. In the particular situation of midlatitudes, the idea of noise might substantially differ from that in the equatorial and tropical regions, both in its structure and also, in the process pursued for its characterization.The isolation of signals embedded into noise has traditionally proved not to be an easy task. Conclusions drawn generally depend on the method of analysis and investigations do not always follow a similar standardized procedure. In fact, this question has traditionally received less attention than probably deserved and this is one of the reasons why signal extraction in the Extratropics has not paved its way as easily as in other regions of the world. The key question is, therefore, how to be able to distinguish between the two. Approaches in this respect have been many and historically diverse. Empirically, both white-noise and red-noise null models have been used to test the significance of fundamental signals despite their external or internal origin.

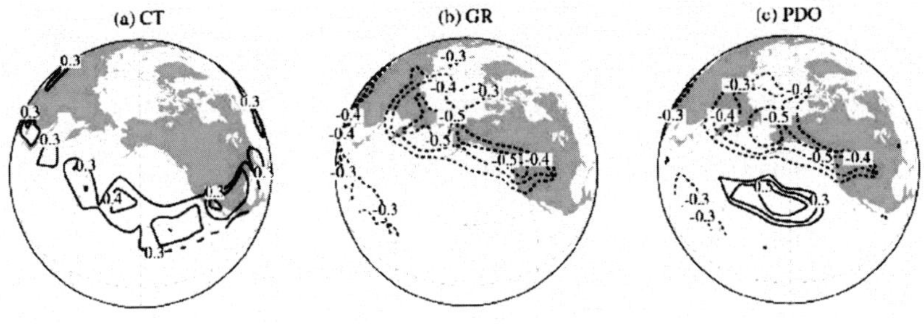

Fig. 11.10. Correlation maps showing winter mean storminess associated with: a) ENSO (the CT index), b) the decadal ENSO-like variability (GR-CT) and c) the PDO (reproduced from Bitz and Battisti 1999, with permission from J. Climate).

However, no attention has been given to the differential structures midlatitude 'noises' may reach, depending on the season and the year of analysis. Up to date, no time-evolving (seasonal, interannual, etc.) characterization of noise in a particular area of interest has been attempted and the same null-models –whether white or colored- are used no matter which latitudinal region we work. Fortunately, some attempts are currently being performed to approach the characterization of region-wide geophysical structures (Rodriguez-Arias and Rodó, in prep.)

In the case of GCM simulations, noise characterization is generally achieved by averaging and working with ensembles. It is then normally assumed this is an appropriate way of characterizing and eventually removing the 'obscuring' components. However, this procedure also introduces 'artificial' noise in the resulting series which is later submitted to signal extraction. In addition to this point, one must be cautious about the effect of increasing the intervals for time-averaging, because it results in a reduction in the contribution of white-noise, in favor of a more red-noise-like, low-frequency structure. Additional 'artificial' autocorrelation introduced and the resulting decrease in the number of degrees of freedom may affect any characterization of signals that are transient or have low amplitude in a climatic time series. In fact, a common way to work with extratropical climate fields, a view that emerges from tropical research, is to compute spatial ensembles or in the best case, work gathering seasonal to six-month ensemble data. In the Tropics, for instance, a single model realization is generally sufficient to establish the SST-forced interannual change. Conversely, for the Extratropics, a single realization is typically insufficient for capturing the boundary-forced signal since the externally-forced -and thus predictable- component of the shift in the mean state is within the envelope of the mean internal spread (Kumar and Hoerling 2000). A minimal ensemble with a sample size of at least 6 is needed to reliably detect the extratropical signals, according to GCM comparisons. As a result, model resolution is still inadequate for isolating signals or couplings that are transient in time or have a limited geographical spread. Modeled space-time estimates of teleconnected areas obtained from small ensembles of forced climate simulations, are normally "contaminated" by the model internal variability. Allen and

Tett (1999) point out that noise in the signal patterns will tend to make the standard detection algorithm (e.g., Hasselmann 1993, 1997) somewhat conservative. Methods for accommodating this source of noise have been available for more than a century (Adcock 1878). Some studies found, in addition, that while the question of which signals could be detected was generally unaffected, the estimated amplitude of individual signals was sensitive to this modification of the procedure. Another source of uncertainty currently encountered, refers to differences in signal patterns between different models. Fortunately, recent studies already begin to consider the sensitivity of detection and refer to their results, only with regard to these differences (Barnett et al. 2000, Hegerl et al. 2000).

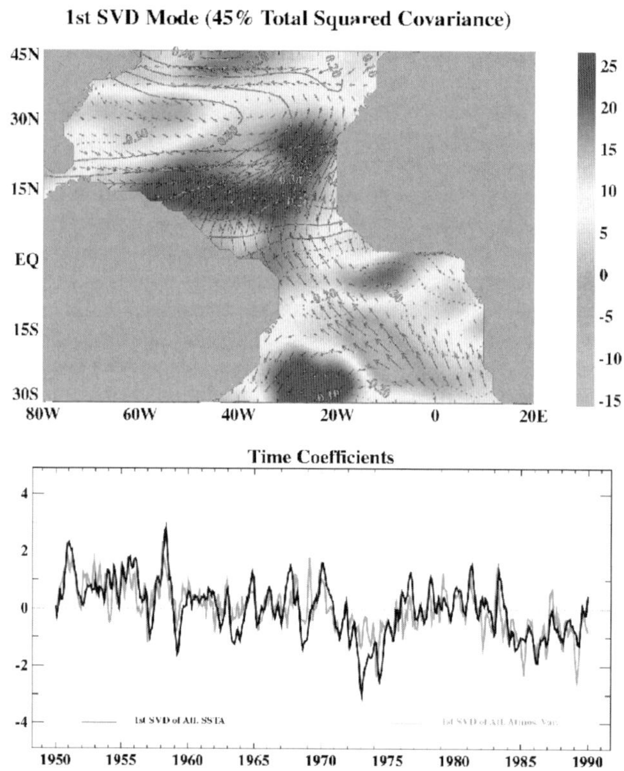

Fig. 11.11. The Atlantic dipole-like variability emerging from the analysis of the Comprehensive Ocean-Atmosphere Data Set (COADS). Upper panel: Spatial structure of the first SVD mode. Contours are SSTA (°C). Vectors depict wind stress anomalies. Shades indicate heat flux anomalies in Wm^{-2}. Bottom panel: Time series of the 1st SVD of Atlantic SSTA (black) and the 1st SVD of Atlantic atmospheric variation (gray). (reproduced from Chang et al. 1997, with permission of Nature).

Finally, another equally important source of uncertainty recently recognized, deals with the recognition that some seasons other than winter might be potentially important in producing SST anomalies in the Extratropics, also in responding to SST anomalies generated in distant sources (Robinson 2000). In this respect, in a previous section above, some of these points were addressed and the reader can refer to studies cited therein.

11.4 Analytical caveats: the transient approach

For the tropical regions, in interactions between the oceans and the atmosphere at large scales, each medium is strongly controlled by the boundary conditions imposed by the other. The large-scale upper ocean circulation is largely determined by the past history of the wind stress, with internal variability occurring primarily at space scales and timescales well separated from the seasonal-to-interannual timescales. Major features of the tropical atmospheric circulation, over timescales longer than a month or two, are largely determined by SST. Although the tropical atmosphere does have significant internal variability, the decorrelation timescales of these are short enough that a conceptual separation can often be made between atmospheric internal variability and that associated with slower ocean evolution communicated by SST. Conversely, for midlatitudes, the internal variability of both ocean and atmosphere is larger, and this distinction remains not so clear. This fact has traditionally posed considerable difficulties to the development of predictability studies for midlatitude regions, as signal-to-noise ratios are found to be low. However, the ENSO series is clearly 'non-stationary', which means, for instance, that El Niño is more prominent during some decades than in others. The same should apply for its effects in the Extratropics. Oceanographers do typically assume that this feature has to do with the depth of the thermocline, but if it only were a matter of merely increasing the depth of the thermocline and changing the phase speeds for waves, in some situations we would be in a completely different regime and, on some occasions, possibly one without El Niño (Fedorov and Philander 2000). Notwithstanding this, the lack of stationarity in ENSO produces, in specific situations, an activation of a cascade of changes in atmospheric circulation that, as a result, impacts the Extratropics in selected areas. These will tend to manifest as nonlinear discontinuous signals, for intervals of time during which the ENSO phenomenon will effectively drive atmospheric and oceanic conditions out of the Tropics. This forcing will ultimately affect ecosystem dynamics there. The signatures of such signals will be in the form of discontinous patterns that –when analysed by standard techniques- run the risk of being masked both by 'natural variability' and the analytical inclination to reduce dimensionality in dynamical systems research.

In this tedious search, some analytical techniques proved to be particularly useful in isolating transients. Among these, several wavelet and cross-wavelet decompositions for the frequency domain and the scale-dependent correlation (SDC) analysis (Rodó 2001, Rodríguez-Arias and Rodó *in press*) for both frequency and time domains, appear at present, to be among the most powerful tools currently at hand. Whereas many studies are devoted to wavelet applications to climatological studies, the prospects are just beginning for SDC. SDC analysis is an analytical and graphical tool recently developed to isolate and accurately characterize signals that are local in a time or spatial series and that vanish elsewhere. As the main difference it has with regard to spectral techniques such as wavelets, lies in the fact that it does not use the whole series to locate the transients, but iteractively searches for them with fragments of different size, its 'a priori' capability to identify such specific signatures appears to be higher. It has, in addition, an outstanding behavior when dealing both with very short time series and series with missing values: both often being the only data available in ecological and climatic studies. Conversely, in the frequency domain, such an approach would be problematic due to the strict length requirements on which the base algorithms rely.

In SDC, the main focus is to highlight hidden structures in one series or in the interaction between two time series (Fig. 11.12). This objective is accomplished first, by its data-adaptive character which makes it superior for instance, in the frequency domain to windowed Fourier analysis and most wavelet applications; and secondly, by means of its graphical representation: a square scatter plot (of size (n-s+1) x (n-s+1) that uses their [i,j] coordinates to locate and represent significant correlations at a particular window size.

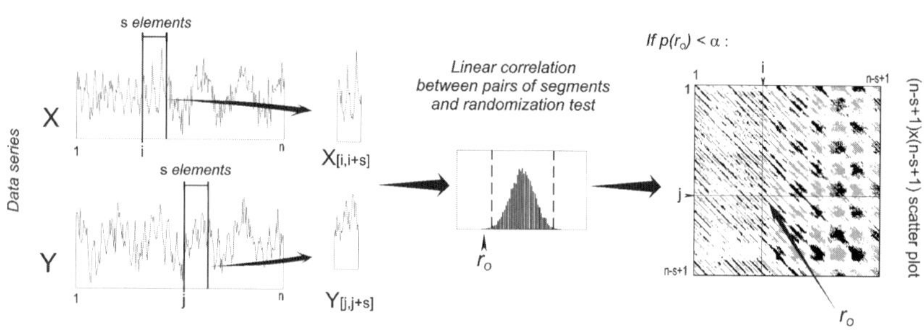

Fig. 11.12. Main description of the SDC methodology. X and Y refer to two data series with a characteristic length n and segment size s, which is the observation scale. SDC performs linear correlations (r_0) between fragments of size s ($X_{[i,i+s]}$ and $Y_{[j,j+s]}$). s-size is chosen to approach the scale of the process under investigation, when this parameter is known. Otherwise a random search for patterns is performed among a full battery of possible s. This is particularly useful when dealing with periodical and quasi-periodical processes (e.g. ENSO), for which the main periodicities are known. The randomization test applied in SDC uses m random permutations of the fragment belonging to the first series, to assess significance of each individual correlation. $\propto$ denotes the significance threshold. In the above example, X series combines two periodic signals (a high frequency oscillation in the first half and a low frequency signal in the second). The Y series combines two periods of different frequency that appear constantly throughout its length. Both series have a random error added (reproduced from Rodó 2001, with permission from Springer-Verlag, Berlin-New York).

SDC plots at the significance threshold selected by the user, appear useful at extracting and locating transient signals, be they periodic or not, that contribute for a limited duration to the total length of the series. These sorts of signals are normally embedded within the variance envelope of 'global' signals (those contributing to the whole extent of the series) (Rodríguez-Arias and Rodó, *in press*; Rodó *et al. submitted*).

The application of techniques such as SDC, that are capable of detecting these nonstationary patterns proves to be particularly useful in ecological and climatological studies. The main reason is that these techniques adapt fairly well to the underlying nature of the interactions that give rise to these structures. For instance, many ecological hypotheses deal with processes that interact during specific times and that have independent dynamics out of these intervals. In addition, some tropical processes are occasionally mediated by threshold-dependent relationships. This means, for instance, that for a certain controlling factor, only over a given value *a* and perhaps exclusively for a limited interval above this value, variable 1 will lay its signature on variable 2, but not in other situations. These sorts of interactions are also common when analyzing relationships between climatic series, or between these and affected ecosystems.

Fig. 11.13 presents an example of SDC analysis for real series. In this situation, SDC was applied to SSTA series of three regions extracted from a global grid, in the search for a dynamic explanation to ENSO connections between the Pacific basin, the tropical North Atlantic and the Western Mediterranean Sea (Rodó, 2001). As the ENSO forcing is discrete, which means it occurs in selected calendar months for the Northern Hemisphere and varies among episodes, it is specially appropriate for these kinds of searches. The results shown in the figure below are striking as they point out for the recent times, a detailed ENSO QQ structure in the Western Mediterranean, which was previously ignored. Conversely, SDC also denotes the complete absence of relationship at any of the scales analyzed, for the NAO and the same three oceanic regions.

Of particular use is the combination of different techniques that help to clarify the distinct signatures embedded in an oceanic or atmospheric field in a particular region. For instance, applying a decomposition to a series and reconstructing only for a particular "window" of interest –be it seasonal, quasi-biennial, quasi-quadrennial, interdecadal, etc., prior to the analysis with SDC-, helps to enhance the signal embedded into what can be considered as 'noise' or of no interest for a certain objective. This pretreatment can be optimally achieved through a wavelet decomposition or with the aid of Singular-Spectrum Analysis (SSA).

Since the climate system contains both nonlinear chaotic subsystems and sources of noise, even an impeccable model cannot reproduce a long natural time series (Cane et al. 1995). Since detailed predictions decades ahead are not possible, we must resort to statistical features to check models and theories of long-term natural variability. The instrumental record can suggest new hypothesis, but it is too short to allow for rigorous test outcomes. Conversely, time series long enough to check for interdecadal patterns are only possible from paleoclimatic proxies, but these often represent a nonlinear convolution of many climatic processes, both global or intrahemispheric, regional and local. As a result, new methods are required to extract this highly important information, in order to address a specific kind of record for each particular geographic location, its transfer function to climate. The reader can refer to Chap. 9, for a more thorough revision of paleoclimatic proxies.

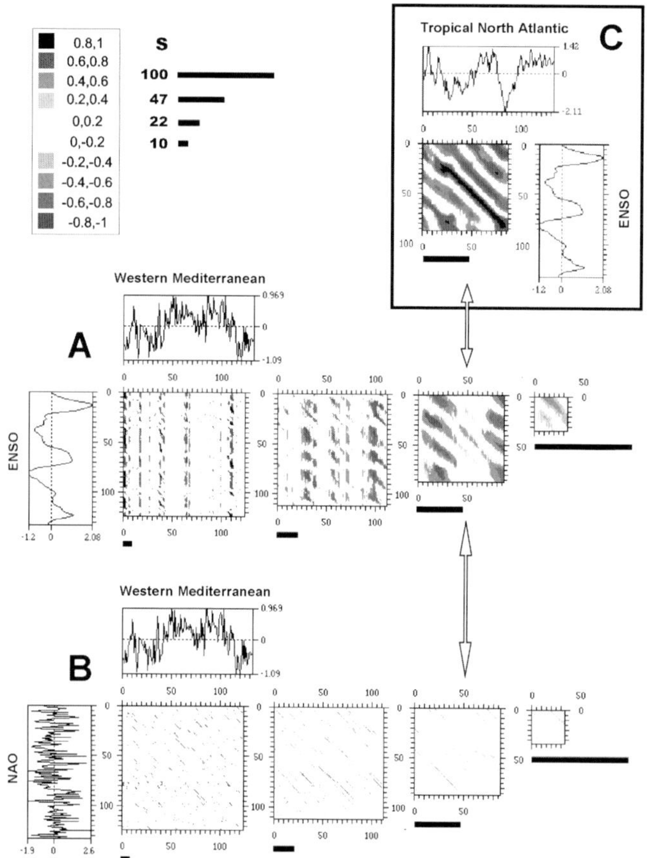

Fig. 11.13. Example of a Scale-dependent correlation (SDC) analysis applied to SSTA series in three oceanic regions: A) Western Mediterranean SSTA at 35°N and 7.5°E and ENSO index (see Rodó 2001, for details). Both series cover the period 1982-1992. The graphs in between the two series correspond to the different window sizes (from left to right, S = 10, 22, 47 and 100 months) used (which are equivalent to the horizontal bars plotted against each graph). Black corresponds to positive correlations and gray to negative. B)Idem as A) but between SSTA in the Western Mediterranean and NAO index. C) Idem but between SSTA in the Tropical North Atlantic at 12.5°N and 55°W and the Ei (right). Here, only the SDC plot for S = 47 was shown for comparison (reprinted from Rodó 2001, with permission from Springer-Verlag, New York-Berlin).

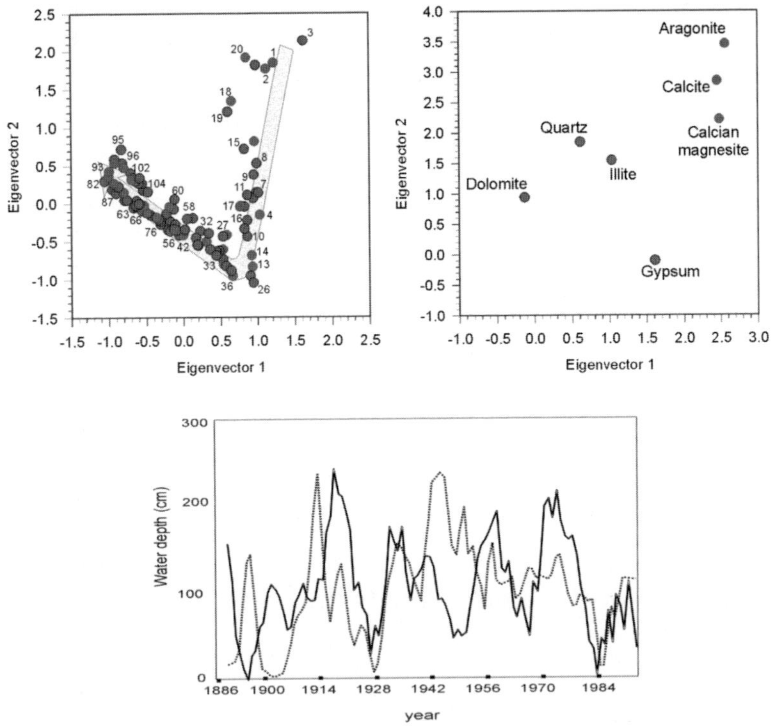

Fig. 11.14. An example of new statistical techniques applied to infer the sequential evolution of a hypersaline lake in a semi-arid region (Gallocanta lake, SE Spain), driven by climatic fluctuations. Upper panels: Relay Indices (RI) showing a mineral gradient, obtained by Correspondence Analysis. The arch disposition of the samples indicated by the underlying arrow, reproduces the sedimentary gradient. The first eigenvector is associated with the infilling process and the second displays the retraction phase. The position of the samples (numbers) on the second axis (left) gives the RI to be used in this particular reconstruction. Related mineral phases, which sequence is directly related to climate forcing, is represented in the upper right panel. Lower panel: mineral-inferred lake level from 1889 through 1992 obtained from RIs, and direct inference based on a multivariate time series model of climatic data (reproduced from Rodó et al. 2002, with permission from Elsevier Science B.V.)

In this tedious search, it is extremely important to make many more observational records and proxies, that overlap in time, to be able to test new hypotheses reliably. Such records are essential as no matter how well a model reproduces short-term variations, only a test against long records can assess their long-term reliability. Many observational ENSO-related studies are based on data that is irregularly distributed. This limitation pushed the use of alternative data sources, mainly satellite-based. This was particularly important over the oceans, were the coverage was very reduced. Among these studies, outgoing longwave radiation (OLR), has become extensively used (Chelliah and Arkin 1992). However, OLR is an analogous for convective precipitation and better estimates for nonconvective phenomena, as well as higher coverage, are still needed. Recently, the Argo project, a new survey in the Global Ocean Observing System (GOOS), is set to increase even more the global existing network of Argo floats throughout the ocean, up to about 3000 by 2005. Projects such as these will give us more clues in the way to increase our understanding of the ocean circulation, particularly at high latitudes in the Atlantic (www.argo.ucsd.edu and Nature 2002).

Similarly, in lacustrine sediments, other approaches to paleoclimatic information are still developing, with new use of long-known statistical tools, for instance, the use of the projections on the dominant axis of the correspondence space of both samples and mineral phases (e.g. Relay Indices in Fig. 11.14a). This approach proved particularly useful in arid locations, where instrumental records are limited (Fig. 11.14b).

11.5 Modeling and predictability for the Extratropics: a briefing on interactions between climate and ecosystems

Even though there has since long, existed a prevailing view of the atmospheric system varying linearly to extreme phases of ENSO, this paradigm has been questioned recently (Deser and Wallace 1990, Hoerling et al. 1997). Some time ago, Bradley et al. (1987) showed a linear response in temperature and precipitation over tropical land masses with respect to inverse states in the Southern Ocean. (SO). This result was to some extent also confirmed by later studies (Ropelewski and Halpert, 1989) and served to substantiate the search for climate associations to the SO, with the use of the SO reference of the differences between warm and cold anomalies. This and other common methodologies disregarded, however, other nonlinear responses in the climate system which are invisible to these techniques. Despite this fact, the inherent nonlinearity in relationships such as those between SST and deep atmospheric convection has since long, been recognized. This nonlinear response clearly shows up when dealing for example, with the thermodynamic control exerted by SST on deep convection in the Tropics, as these latter regions depend mainly on the underlying value of SST (Gadgil et al. 1984). This value must not be lower than 27°C to allow deep convection to take place.

As recently reported in ENSIP (the El Niño Simulation Intercomparison Project, Latif et al. 2001), ENSO modeling advanced considerably during the last decade, though several aspects of the simulated climatology and ENSO are not well captured by present-day coupled models (Anderson et al. 1998). In particular, a total of twenty-four coupled ocean-atmosphere models were intercompared and many of them overestimated western equatorial Pacific ENSO signals, while underestimated those in the eastern equatorial Pacific. Of course, it is easily acknowledged that being able to simulate ENSO signals in the Extratropics is far more difficult, and the modeling efforts still are at their beginning, as the question becomes more complicated because of the occurrence of both lagged responses and feedbacks that affect the coupled system. Table 11.1. highlights major discrepancies among several of the models tested.

Table 11.1. Simulation of the annual cycle of equatorial SST (2°N-2°S) (reproduced from Latif et al. 2001, with permission from Springer-Verlag, Berlin – New York).

Model	Range annual cycle in the east (°C)	Comments
GISST (obs.)	4.0	Annual cycle in the east, semi-annual in the west
BMRC	3.0	Phase shift in the east and the west
CCC	2.5	Too weak annual cycle in the east
CSSR	2.5	Too weak annual cycle in the east
CEA-DSM SACLAY	1.0	Too weak annual cycle in the east
CERFACS	4.5	Realistic
COLA	4.0	No westward phase propagation
DKRZ-OPYC	1.5	Too weak annual cycle in the east
DKRZ-LSG	1.0	Too weak annual cycle in the east
GFDL-R15	1.5	Too weak annual cycle in the east
GFDL-R30	3.5	Annual cycle extends far to the west
HAWAII	5.0	Too strong annual cycle
JMA	2.5	Too weak annual cycle in the east
LAMONT		Annual cycle prescribed
LMD/LODYC-2.5	1.5	Too weak annual cycle in the east
LMD/LODYC-TOGA	2.0	Too weak annual cycle in the east
LMD/LODYC-GLOBAL		Semi-annual cycle
MPI	2.5	Too weak annual cycle in the east
MRI	4.0	Annual cycle extends far to the west
NAVAL (NRL)	6.0	Too strong annual cycle
NCAR-CSM		Semi-annual cycle
NCAR-WM	2.0	Annual cycle extends far to the west
NCEP	4.0	Too weak semi-annual cycle in the west
UCLA	6.0	Too strong annual cycle near the east coast
UKMO	2.5	Too weak annual cycle in the east

Regarding signature climate lays on ecosystems, a major debate currently under way, concerns the relative importance of extrinsic factors in the modulation of ecosystem functioning. How to adequately isolate these forcings and the assessment of how important they are, still remains an elusive and controversial issue. The debate, when searching for processes generating cycles or fluctuations in populations, is served among 'externalists' or 'environmentalists' *vs* those more prone to look exclusively at internal dynamics as fundamental and conversely exclude any influence coming from external forcing. At present, there have not been many attempts seeking to integrate both approaches, assuming an interplay of both extrinsic and intrinsic factors. The main bulk of work with time series analysis of population data and climate, concentrated on animal and plant populations responding to weather and climate variability. For instance, a paradigmatic case long debated was the historical Canadian lynx data, for which the role played by climate variability is at present, still under discussion (Stenseth et al. 1999). Other current examples include the use of thresholds in environmental variables, as for instance, with SETAR and related models (Tong and Yeung 1988, Grenfell et al., 1998). Other attempts in the same direction, were also performed in the case of climate and disease dynamics, particularly for measles and for cholera in endemic regions (Pascual et al. 2000). Notwithstanding this, only a few examples of nonlinear statistical models combining the two sorts of forcing, are available at present in the literature. For instance, in the case of cholera in an endemic region of the Indian subcontinent such as Bangladesh, a nonlinear and nonparametric time series model was developed for predicting the spread of an epidemic which appears to be mainly environmentally driven in the present, and ultimately linked to regional climate variability, particularly ENSO, in the recent decades (Pascual et al. 2000, Rodó et al., submitted). In this situation, comparison of models that allow both for intrinsic and extrinsic dynamics, resulted in a combined influence from both drivers. When seasonality is incorporated but not the ENSO index, low-dimensional models are selected (that is, they have the smallest values of a cross-validation criterion, named *Vc*). The importance of seasonality is independently tested by comparing these models to their autonomous counterparts. Models with an equivalent or larger number of independent variables but no seasonality have larger values of *Vc*. The importance of ENSO is then examined by incorporating the ENSO index into the simplest seasonal model at different time lags between 0 and 12 months.

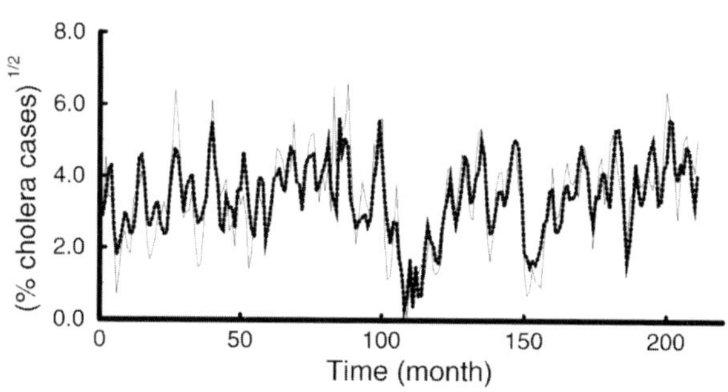

Fig. 11.15. The (square-root transformed) cholera data (black line) and the 2-months ahead prediction of the fitted model incorporating both seasonality and ENSO at a lag t-11 (gray line) (reprinted from Pascual et al. 2000, with permission from Science)

Final results yielded as the best fit, a low-dimensional model incorporating both ENSO and seasonality as extrinsic factors and previous disease levels as intrinsic ones. Fig. 11.15 shows the fitting of this nonlinear time series model to the original cholera series for Dhaka (Bangladesh). Black line shows the square-root-transformed cholera data and the red line is the result of the 2-months-ahead prediction of a model incorporating cholera at time *t*, a seasonal clock and ENSO at lag *t-11* months. Though the fitting is indeed remarkable for the fitted data, this fact does not necessarily ensure a similar accuracy for future disease incidence.

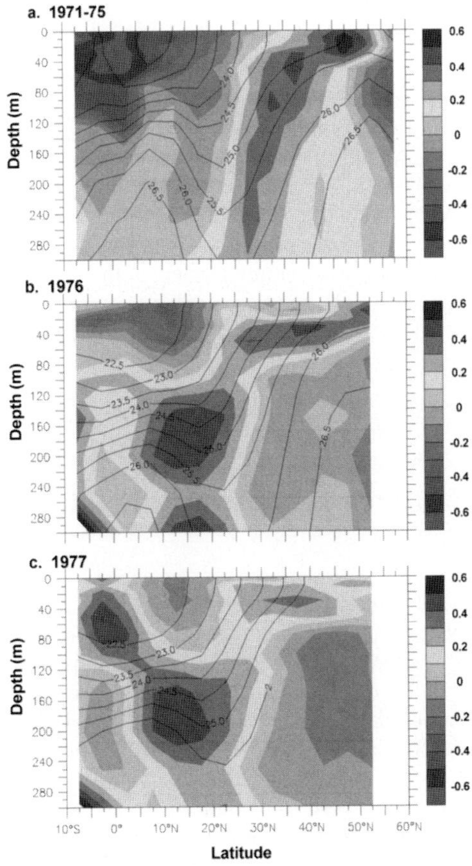

Fig. 11.16. Latitude-depth sections of oceanic temperatures in the upper 300 meters and isopycnal surfaces (contour lines are in Kg m^{-3}) for three different time intervals, showing the warm anomaly penetration from the North Pacific into the Tropics. (from Zhang et al. 1998, reprinted with permission from MacMillan Magazines Ltd.)

11.6 Feedbacks from the Extratropics

The Extratropics can also influence theTropics through the influx of water with anomalous temperature from higher latitudes, and give rise to continual interdecadal oscillations (Gu and Philander 1997). The links between both oceanic regions mainly involve the shallow, wind-driven meridional circulation and the subduction of water parcels in the eastern regions of the subtropical and midlatitude oceans. In parallel, model simulations show that warming in the Tropics leads to the intensification of extratropical westerlies that cool surface waters there and are windows to the equatorial thermocline (Gu and Philander, 1997). In addition to model simulations, Zhang et al. (1998) presented observational evidence that the upper-ocean warming in the Pacific and the 1976-77 regime shift in ENSO might have originated from decadal midlatitude variability. They noted how, at midlatitudes, a subsurface warm anomaly formed in the early 70s from subducted waters and penetrated through the Subtropics and into the Tropics (Fig. 11.16). This phenomenon seemed to have significantly contributed to an unusual warm surface-water anomaly, triggering a very strong El Niño in 1982-83.

The identification of this association, termed as the subsurface ocean 'bridge' may also help in the near future, to better predict decadal-scale changes (Zhang et al. 1998) and clarify the role of the North Pacific gyre region in the tropical variability.

11.7 Prospects for future research and the future evolution of ENSO

To advance more on the likely role of ENSO in a future climate is a key action of research in the years to come which will also have important implications for climate variability in the Extratropics. In this respect, instrumental records have been examined in the search for possible changes in ENSO, mainly over the past 120 years. This is the earliest date for which reasonably reliable reconstructions can be made. Three new reconstructions of SST in the eastern Equatorial Pacific that use optimum interpolation methods exhibit strong similarities (see Fig. 2.29 of IPCC 2001). The dominant 2 to 6 year timescale in ENSO is again apparent therein. Both the activity and periodicity of ENSO have varied considerably since 1871 with marked irregularity over time. There was an apparent "shift" in the temperature of the TP around 1976 to warmer conditions, already discussed in the Second Assessment Report (SAR), which appeared to continue until at least 1998. During this period, ENSO events were more frequent, intense or persistent. It is unclear whether this warm state still continues, with the prolonged La Niña from late 1998 until early 2001, and the prospects for a new El Niño in 2002. A number of recent studies have found changes in the interannual variability of ENSO over the last century, related in part to an observed reduction in ENSO variability between about 1920 and 1960 and to an increase in SSTs in the Niño-3 region as inferred from COADS, GOSTA and IGOSS databases. Various studies (Torrence and Compo 1998, Torrence and Webster 1998) show more robust signals in the quasi-biennial and 'classical' 3 to 4 year ENSO bands (3.4 and 7 years) during the first and last 40 to 50 years of the instrumental record. A period of very weak signal strength (with a near 5-year periodicity) occurs in much of the intervening epoch. The 1990s, which have received considerable attention, seem unusual relative to previous decades, as for the recent behavior of ENSO. Fig. 11.17 shows some of these changes for the last two decades, with an apparent inclination of the ENSO main axis in the Pacific, in association to a concomitant intensification of the quasi-quadrennial component in ENSO.

In conjunction with these interdecadal changes, a protracted period of low SOI occurred from 1990 to 1995, during which several weak to moderate El Niño events developed, with no intervening La Niña events (Goddard and Graham 1997). This sequential pattern of El Niño and La Niña was found by some studies to be statistically very rare (e.g., Trenberth and Hoar 1996). Whether global warming is, at present influencing El Niño, especially given the remarkable El Niño of 1997/98, is a key question to be solved in the decades to come (Trenberth, 1998).

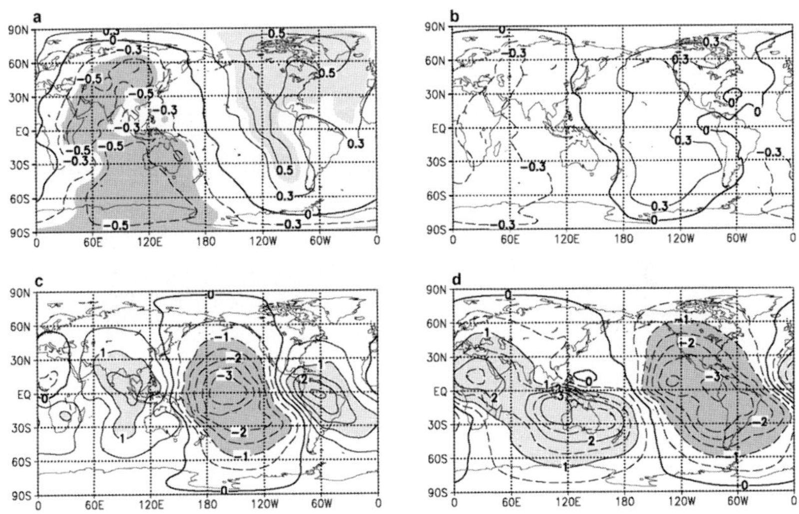

Fig. 11.17. Correlation between monsoon rainfall and velocity potential at 200 hPa (a) during 1958-80 and (b) 1982-1997. Significant correlations ($p<0.05$) are shaded. (c) composites of summer velocity potential ($x10^5$) at 200 hPa for the EN events occurring within (c) 1958-80 and (d) 1981-97. (Reprinted from Kumar et al. 2000 with permission from Science)

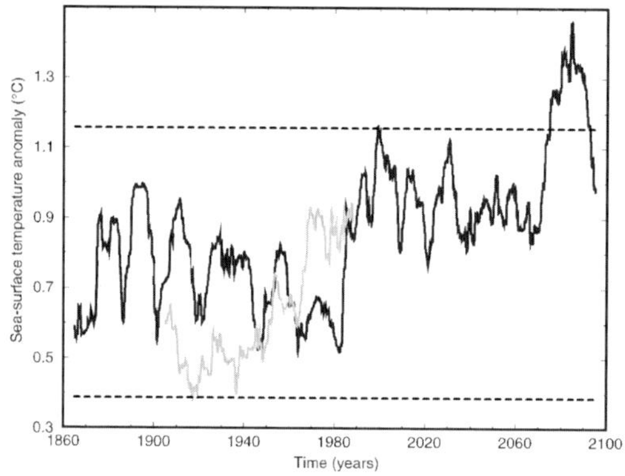

Fig. 11.18. Observed and simulated SSTA in the Niño 3 region during a transient greenhouse simulation. Both SST anomalies show an increase in interannual variability with the dashed line indicating present minimum and maximum standard deviations. Black line depicts standard deviations of Niño 3 SSTA and gray line refers to observed changes in standard deviations. (Redrawn from Timmermann et al. 1999, with permission from MacMillan Magazines Ltd.)

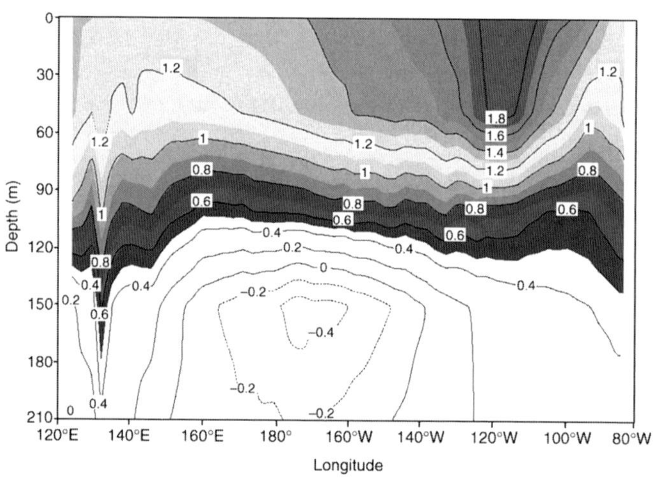

Fig. 11.19. Temperatures of the upper 210 m of the equatorial Pacific Ocean, as simulated from the same transient greenhouse simulation as in Fig. 11.18. Note both the warming trend in the surface and the cooling in the deeper waters, resulting in a stronger thermocline (reproduced from Timmermann et al. 1999, with permission from MacMillan Magazines Ltd.)

This fact is particularly important, as El Niño is equally said to affect global temperature itself and the relative importance of each of them on the other's dynamics is not yet completely understood. Projections of future behavior of ENSO, though still under controversy, tend to indicate an evolution towards more variability in El Niño conditions (Fig 11.18) and a stronger thermocline in the equatorial Pacific (Fig. 11.19). Among the factors that may have contributed most to the intensification of El Niño events in the last decades and that may equally trigger an amplification of its effects out of the Tropics, four of them are currently invoked (from McPhaden 1999):

1. Chaos: It is recalled by some theories to account for the irregularity of the ENSO cycle. Nonlinear resonances involving ENSO and the seasonal cycle have equally received special attention, but other chaotic interactions may affect as well (Tziperman et al. 1994, McPhaden, 1999).
2. Weather 'noise': Weather phenomena, inherently unpredictable more than about two weeks in advance, are a source of random forcing in the climate system. For instance, one source of weather variability in the Tropics is the Madden-Julian Oscillation (MJO), a wave-like disturbance in the atmosphere with a period of 30-60 days that originates over the Indian Ocean (Madden and Julian 1972). The TP was preconditioned to the onset of an El Niño by the build-up of excess heat in the western equatorial Pacific due to stronger than normal trade winds in 1995-96. However, beginning in late 1996, the MJO was particularly energetic, and several cycles of the wave amplified through nonlinear ocean-atmosphere interactions as they passed over the western Pacific. This process set in motion an array of positive feedbacks between the ocean and the atmosphere, which at turn, reinforced initial MJO-induced warming.
3. Pacific Decadal Oscillation (PDO): A naturally occurring oscillation of the coupled ocean-atmosphere system in the Pacific basin with a period of several decades (Mantua et al. 1997). In association with the PDO, SSTs have generally been higher in the TP from the mid-1970s. Since then, there have been more El Niños than La Niñas, two extremely strong El Niños have occurred and the early 1990s was a period of extended warmth in the TP. The PDO may then be one of the reasons for the observed decadal modulation of the ENSO cycle, because it affects the background conditions on which ENSO develops, though considerable research is still needed in this area.
4. Global warming trends: Possibly these are among the most real phenomena influencing the ENSO cycle, but at the same time, they are also very difficult to assert, due to the lack of an instrumental record of enough temporal and spatial extent. The last huge El Niño of 1997-98 coincided with record high temperatures. The 1997-98 ENSO event contributed to this increase, mainly because it is commonly observed that global mean temperatures generally rise a few tenths of a degree Celsius following the peak of El Niño, as the TP looses heat to the overlying atmosphere. Underlying these extreme temperatures, however, is a century-long warming trend that may well be due to anthropogenic greenhouse-gas warming (IPCC 2001). Some computer models suggest that global warming may be slowly heating up the eastern equatorial Pacific Ocean, as ob-

served over the past 25 years (Meehl and Washington 1996). Others propose that ENSO events might be stronger or more frequent in a warmer climate (Timmermann et al. 1999). The superposition of ENSO variations on increased warming due to CO_2 and a warm phase of the PDO could produce temperature fluctuations like those seen in the equatorial Pacific since the mid-1970s. Therefore, although we might expect global warming to affect the ENSO cycle, at the moment no firm conclusions can be drawn (McPhaden 1999).

In short, the strength of the latest very strong ENSO events may be traceable to interactions of the ENSO cycle with some combination of global warming trends, the PDO, the seasonal cycle and the MJO (McPhaden 1999). We will then be able to begin assessing the importance of any of these changes for the climate variability in the Extratropics.

In summary, I will recall Neelin et al. (1998) suggestions on the prospects of ENSO theory for the future. First, a simple consensus model must emerge that clarifies the dominant ENSO period in front of spurious non-oscillatory instabilities. This model should unavoidably cope with the influx of memory to the coupled system given by the subsurface oceanic domain. The inclusion of such a component will be of main interest for predictability studies related to ENSO signals, also in the Extratropics, where these slower components become particularly important and are often overlooked. More agreement is also needed among atmospheric models, refining the treatment given to tropical moist convection and radiation, so that eventually it might be possible to properly account for the atmospheric transport to the Extratropics, and for a good parameterization of ocean-atmosphere interactions between the discontinuous couplings that occasionally occur between the Tropics and midlatitudes. In conjunction to more efforts in these areas, signal-identification techniques will have to suffer a parallel evolution to be able to help assessing how much of all this has been achieved.

In this search, the basic knowledge of the role of the extratropical ocean in enhancing predictability remains a key question also to be further explored locally, for the different seasons, geographic areas and also for the wide range of possible temporal scales.

References

Adcock, RJ (1878) The Analyst (Des Moines, Iowa), 5, 53.

Allen MR, SFB Tett (1999) Checking for model consistency in optimal fingerprinting. Clim Dyn 15, 419-434.

Anderson DLT, ES Sarachik, PJ Webster, LM Rohstein (1998) The TOGA decade. J. Geophys. Res. (special issue) 14167-14510.

Bitz, C.M. and D.S. Battisti (1999) Interannual to decadal variability in climate and the glacier mass balance in Washington, western Canada, and Alaska. Journal of Climate, 12(11):3181-3196.

Blackmon ML, YH Lee, HH Hsu (1984) Time variation of 500 mb height fluctuations with long, intermediate, and short time scales as deduced from lag-correlation statistics. J. Atmos. Sci. 41, 981-991.

Barnett TP, GC Hegerl, T Knutson, SFB Tett (2000) Uncertainty levels in predicted patterns of anthropogenic climate change. J. Geophys.Res., 105, 15525-15542.
Barnston AG, RE Livezey (1987) Classification, seasonality and persistence of low-frequency atmospheric circulation patterns. Mon. Wea. Rev. 115, 1083-1126.
Bjerknes, J (1966) A possible response of the atmospheric Hadley circulation to equatorial anomalies of ocean temperature. Tellus 18, 820-829.
Cai W and PG Baines (2001) Forcing of the Antarctic Circumpolar Wave by ENSO teleconnections. J. Geophys. Res. 106(C5):9019-9038
Cai, W., P.G. Baines and A.B. Pittock (2001) Fluctuations of the relationship between ENSO and northeast Australian rainfall. Clym. Dyn. 17(5/6):421-432
Cane, M.A., S.E. Zebiak and S.C. Dolan (1986) Experimental forecasts of El Niño. Nature, 321,827-832.
Cane MA, Zebiak SE, Xue Y (1995) Model studies of the long-term behavior of ENSO. In: National Research Council (E.J. Barron, chairman) Natural climate variability on decade-to-century time scales. National Academy Press, Washington, pp. 442-455.
Chang, P., L. Ji and H. Li (1997) A decadal climate variation in the Tropical Atlantic Ocean from thermodynamic air-sea interactions. Nature, 385(6616):516-518.
Chelliah M, P Arkin (1992) Large-scale variability of monthly outgoing longwave radiation anomalies over the global tropics. J. Clim. 5, 371-389.
Dai A, Y Fung, A Del Genio (1997) Surface observed global land precipitation variations during 1900-1988. J. Clim. 10, 2943-2962.
Dai A, TML Wigley (2000) Global patterns of ENSO-induced precipitation. Geophys. Res. Lett. 27, 1283-1286.
Deser C, JM Wallace (1990) Large-scale atmospheric circulation features of warm and cold-episodes in the tropical Pacific. J. Clim. 3, 1254-1281.
Drévillon M, L Terray, P Rogel, C Cassou (2001) Mid latitude Atlantic SST influence on european winter climate variability in the NCEP reanalysis. Clim. Dyn. 18, 331-344.
Enfield DB, DA Mayer (1997) Tropical Atlantic sea surface temperature variability and its relation to El Niño-Southern Oscillation. J. Geophys. Res. 102, 929-945.
Fedorov, A.V. and S.G. Philander (2000) Is El Niño changing?, Science, 288(5473):1997-2002.
Fraedrich K, K Muller (1992) Climate anomalies in Europe associated with ENSO extremes. Int. J. Clim. 12, 25-31.
Gadgil SP, V Joseph, NV Joshi (1984) Ocean-atmosphere coupling over monsoon regions. Nature 312, 141-143.
Goddard, L. and N.E. Graham (1999). Importance of the Indian Ocean for simulating rainfall anomalies over eastern and southern Africa. J. Geophys. Res. 104 (D16):19099-19116.
Graham NE, TP Barnett, R Wilde, M Ponater, S Schubert (1994) On the roles of tropical and midlatitude SSTs in forcing interannual to interdecadal variability in winter Northern Hemisphere circulation. J Clim 7, 1416-1441.
Grenfell BT, et al., (1998) Noise and determinism in synchronized sheep dynamics. Nature 394, 674-677.
Griffies SM, K Bryan (1997) A predictability study of simulated North Atlantic multidecadal variability. Clim. Dyn. 13, 459-488.
Gu, D. and SGH. Philander (1997) Interdecadal climate fluctuations that depend on excjhanges between the tropics and extratropics. Science, 275:805-807.
Hamilton, K. (1988) A detailed examination of the extratropical response to tropical El Nino/Southern Oscillation events. J. Clim., 8(1):67-86.
Hasselmann, K (1993) Optimal fingerprints for the Detection of Time dependent Climate Change. J. Clim. 6: 1957-1971.
Hasselmann K (1997) Multi-pattern fingerprint method for detection and attribution of climate change. Clim. Dyn., 13: 601-612.
Hasselmann, K. (1976) Stochastic climate models. Tellus, 28:473-484

Hegerl GC, P Stott, M. Allen, JFB Mitchell, SFB Tett, U Cubasch (2000) Detection and attribution of climate change: Sensitivity of results to climate model differences. Clim. Dyn., 16, 737-754.
Hoerling MP, A Kumar, M Zhong (1997) El Niño, La Niña, and the nonlinearity of their teleconnections. J. Clim. 10, 1769-1786.
Horel JD, JM Wallace (1981) Planetary scale atmospheric phenomena associated with the Southern Oscillation. Mon Wea Rev 109, 813-829.
Hsuing, J. and R.E. Newell (1983) The principal nonseasonal modes of global sea surface temperature. J. Phys. Oceanogr., 13:1957-1967.
IPCC, 2001 Climate Change: The scientific Basis. Contribution of the Working Group I to the Third Assessment Report. Houghton, J.T., D.J. Ding, M. Griggs, M. Noguer, P.J. van der Linden, X. Dai, K. Maskell and C.A. Johnson (eds.). Cambridge University Press, Cambridge, 881 pp.
Jacobs GA, Hurlburt HE, Kindle JC, Metzger EJ, Mitchell JL, Teague WJ, Wallcraft AJ (1994) Decade-scale trans-Pacific propagation and warming effects of an El Niño anomaly. Nature, 370, 360-363.
Jacobs GA, JL Mitchell (1996) Ocean circulation variations associated with the Antarctic Circumpolar Wave. Geophys. Res. Lett., 23, 2947-2950.
Kawamura, R., M. Sugi, T. Kayahara and N. Sato (1997). Recent abnormal changes in wintertime atmospheric response to tropical SST forcing. Geophys. Res. Lett., 24(7):783-786.
Kerr, R. A., 1999: A new force in high-latitude climate. Science, 284, 241-242.
Kiladis GN, HF Díaz (1989) Global climatic anomalies associated with extremes in the Southern Oscillation. J. Clim. 2, 1069-1090.
Klein, S.A., B.J. Soden and N.C. Lau (1999) Remote Sea Surface Temperature during ENSO: Evidence for a Tropical Atmospheric Bridge. J. Clim., 12:917-932
Kumar A and MP Hoerling (1995) Prospects and limitations of seasonal atmospheric GCM predictions. Bull. Am. Met. Soc. 76, 335-345.
Kumar A. and M.P. Hoerling (2000). Analysis of a conceptual model of seasonal climate variability and implications for seasonal prediction. Bull. Am. Met. Soc., 81(2):255-264.
Kumar A., A.G. Barnston, P.T. Peng, M.P. Hoerling and L. Goddard (2000). Changes in the spread of the variability of the seasonal mean atmospheric states associated with ENSO. J. Clim., 13(17):3139-3151.
Kushmir, Y. (1999) Climatology – Europe winter prospects . Nature, 398:289-291.
Lanzante JL (1996) Lag relationships involving tropical sea surface temperatures. J. Climate 9, 2568-2578.
Latif, M., K. Sperber, J. Arblaster, P. Braconot, D. Chen, A. Colman, U. Cubasch, C. Cooper, P. Delecluse, D. DeWitt, L. Fairhead, G. Flato, T. Hogan, M. Ji, M. Kimoto, A. Kitoh, T. Knutson, H. Le Treut, T. Li, S. Manabe, O. Marti, C. Mechoso, G. Meehl, S. Power, E. Roeckner, J. Sirven, L. Terray, A. Vintzileos, R. Voss, B. Wang. W. Washington (2001). ENSIP: the El Nino simulation intercomparison project. Clim. Dyn., 18(3/4):255-276.
Lau NC and MJ Nath (1990) A general circulation model study of the atmospheric response to extratropical SST anomalies observed in 1950-79. J Clim 9, 965-989.
Lau NC and MJ Nath (1994) A modeling study of the elative roles of tropical and extratropical SST anomalies in the variability of the global atmosphere-ocean system. J Clim 7, 1184-1207.
Lau NC and MJ Nath (1996) The role of the "atmospheric bridge" in linking tropical Pacific ENSO events to extratropical SST anomalies. J. Clim., 9:2036-2057.
Lau NC and MJ Nath (2001) Impact of ENSO on SST variability in the North Pacific and North Atlantic: Seasonal dependence and role of extratropical sea-air coupling. J. Clim. 14(13), 2846-2866.
Madden, R. A., and P. R. Julian (1972) Description of global-scale circulation cells in the tropics with a 40-50 day period. J. Atmos. Sci., 29, 1109-1123.
Mantua NJ, SR Hare, Y Zhang, JM Wallace, RC Francis, Bull Am Met Soc 78, 1069-1079 (1997)

May W, L Bengtsson (1998) The signature of ENSO in the Northern Hemisphere midlatitude seasonal flow and high-frequency intraseasonal variability. Met. Atmos. Physics 69, 81-100.

McPhaden M, The child prodigy of 1997-98. Nature 398, 559-562 (1999)

Meehl GA and WM Washington (1996) El Nino-like climate change in a model with increased atmospheric CO sub(2) concentrations. Nature 382(6586): 56-60.

Namias, J. (1969) Seasonal interactions between the North Pacific Ocean and the atmosphere during the 1960's. Mon. Wea. Rev., 97:173-192

Nature (2002) Voyage of the Argonauts. R. Dalton (correspondent). Nature 415:954-955

Neelin JD, Battisti DS, Hirst AC, Jin F-F, Wakata Y, Yamagata T, Zebiak SE (1998) ENSO theory. J Gephys Res 103: 14261-14290

Oldenborgh, G.J. and G. Burgers (2000) On the El Niño teleconnection to spring precipitation in Europe. Int. Jour. Climatology: 20-565-574

Pascual M, Rodó X, Ellner S, Colwell R, Bouma M (2000) Cholera dynamics and El Niño-Southern Oscillation. Science, 289, 1766-1769.

Peterson RG, WB White (1998) Slow oceanic teleconnections linking the Antarctic Circumpolar Wave with the tropical El Niño-Southern Oscillation. J. Geophys. Res. 103, 24573-24583.

Philander, SGH, T. Yamagata and R.C. Pacanowski (1984) Unstable air-sea interactions in the tropics. J. Atmos. Sci., 41:604-613

Philander, SGH (1990) El Niño, La Niña and the Southern Oscillation. Academic Press.

Pierce DW, TP Barnett, N Schneider, R Saravanan, D Dommenget, M Latif (2001) The role of ocean dynamics in producing decadal climate variability in the North Pacific. Clim. Dyn., 18: 51-70.

Peterson RG, WB White (1998) Slow oceanic teleconnections linking the Antarctic Circumpolar Wave with the tropical El Niño-Southern Oscillation. J. Geophys. Res. 103, 24573-24583.

Rasmusson EM, TC Carpenter (1982) Variations in tropical sea surface temperature and surface wind fields associated with the Southern Oscillation/El Niño. Mon. Wea. Rev. 110, 354-384.

Robinson, W.A. (2000) Review of WETS – The Workshop on Extra-Tropical SST anomalies. Bull. Am. Met. Soc. , 81(3):567-577.

Rodó X, E Baert, FA Comín (1997) Variations in seasonal rainfall in Southern Europe during the present century: relationships with the North Atlantic Oscillation and the El Niño-Southern Oscillation. Clim. Dyn. 13, 275-284.

Rodó X, FA Comín (2000) Links between large-scale anomalies, rainfall and wine quality in the Iberian Peninsula during the last three decades. Glob. Change Biol. 6, 267-273.

Rodó X (2001) Reversal of three global atmospheric fields linking changes in SST anomalies in the Pacific, Atlantic and Indian oceans at tropical latitudes and midlatitudes. Clim. Dyn. 18: 203-217.

Rodó, X, M Pascual, AS Faruque (submitted) ENSO and cholera: a nonstationary link related to climate change?

Rodríguez-Arias, M.A. and X. Rodó (in prep). Geophysical fields modeling

Rodríguez-Arias MA, X Rodó (2002) On the characterization of transient dynamics in ecological series. Oecologia (in press).

Ropelewski CF, MS Halpert (1987) Global and regional-scale precipitation patterns associated with El Niño-Southern Oscillation. Mon. Wea. Rev. 115, 1606-1626.

Ropelewski CF, MS Halpert (1989) Precipitation patterns associated with the high index phase of the Southern Oscillation. J. Clim. 2, 268-284.

Santoleri R, E. Böhm, ME Schiano (1995) The sea surface temperature of the western Mediterranean sea: historical satellite thermal data. In La Violette PE (ed) Seasonal and interannual variability of the western Mediterranean sea. Am. Geophys. Union, 155-176.

Stenseth, NC, et al., 1999: Common dynamic structure of Canada lynx populations within three climatic regions. Science 285, 1071-1073.

Timmermann, A., J. Oberhuber, A. Bacher, M. Esch, M. Latiff and E. Roeckner (1999). Increased El Nino frequency in a climate model forced by future greenhouse warming. Nature, 398(6729):694-697.

Tong H, I Yeung (1988) On tests for SETAR-type non-linearity in partially observed time series. Tech. Rep., Institute of Mathematics, University of Kent.
Trenberth, K.E. (1998) Atmospheric moisture residence times and cycling: Implications for rainfall rates and climate change. Clim. Change, 39(4):667-694.
Trenberth, K.E. and T.J. Hoar (1996) The 1990-1995 El Nino-Southern Oscillation event: Longest on record. Geophys. Res. Lett., 23(1):57-60.
Trenberth KE, and J. W. Hurrell, 1994: Decadal atmospheric-ocean variations in the Pacific. Clim. Dyn., 9, 303-309.
Trenberth KE, GW Branstator, D Karoly, A Kumar, N-C lau, C Ropelewski (1998) Progress during TOGA in understanding and modeling global teleconnections associated with tropical sea surface temperatures. J. Geophys. Res. 103, 14291-14324.
Thompson DJ and JM Wallace (2002) Annular modes in the extratropical circulation. Part I: Month-to-month variability. J. Climate, in press.
Thompson DJ, JM Wallace, G Hegerl (2002) Annular Modes in the Extratropical Circulation. Part II: Trends. Journal of Climate, in press
Torrence, C. and G.P. Compo (1998). A practical guide to wavelet analysis. Bull. Am. Meteorol. Soc., 79(1):61-78.
Torrence, C. and P. Webster (1998). The annual cycle of persistence in the El Nino/Southern Oscillation. Quat. J. of the Roy. Meteorol. Soc., 124(550-part B):1985-2004.
Tziperman E, L Stone, H Jarosh, MA Cane, Science 264, 74-74 (1994)
van Loon H, RA Madden (1981) The Southern Oscillation. Part I: Global associations with pressure and temperature in northern winter. Mon. Wea. Rev. 109, 1150-1162.
van Loon H, JC Rogers (1981) The Southern Oscillation. Part II: associations with changes in the middle troposphere in the northern winter. Mon. Wea. Rev. 109, 1163-1168.
von Storch H, HA Kruse (1985) Extra-tropical atmospheric response to El Niño events. Tellus 37A, 361-377.
von Storch H, E Zorita, U Cubasch (1993) Downscaling of global climate change estimates to regional scales: an application to Iberian rainfall in wintertime. J. Clim. 6, 1161-1171.
Wallace JM, DS Gutzler (1981) Teleconnections in the geopotential height field during the northern hemisphere winter. Mon. Wea. Rev. 109, 784-812.
Wallace, J.M., E.M. Rasmusson, T.P. Mitchell, V.E. Kousky, E.S. Sarachik, H. von Storch (1998) The structure and evolution of ENSO-related climate variability in the tropical Pacific: Lessons from TOGA. J. Geophys. Res., 103(C7):14241-14259.
Weare, B.C., A. Navato and R.E. Newell (1976) Empirical orthogonal analysis of Pacific Ocean sea surface temperatures. J. Phys. Oceanogr., 6:671-678.
Yakir D, S Lev-Yadun, A Zangvil (1996) El Niño and tree ring growth near Jerusalem over the last 20 years. Glob. Change Biol. 2, 97-101.
Zhang, Y., J. M. Wallace and D. S. Battisti, 1997: ENSO-like interdecadal variability: 1900-93. J. Climate, 10, 1004-1020.
Zhang R-H, LM Rohstein, A Busalacchi (1998) Origin of upper-ocean warming and El Niño change on decadal scales in the tropical Pacific Ocean. Nature 391, 879-883.
Zorita E, V Kharin, H von Storch (1992) The atmospheric circulation and sea surface temperature in the North Atlantic area in winter: their interaction and relevance for Iberian precipitation. J. Clim. 5, 1097-1108.

Index

D

Druck- und Bindearbeiten: Legoprint, Italien